# Springer Series in Advanced Manufacturing

# Other titles in this series

Lihui Wang and Robert X. Gao (Eds.)

# Condition Monitoring and Control for Intelligent Manufacturing

With 255 Figures

 Springer

Lihui Wang, Ph.D., P.Eng.
Senior Research Officer
Integrated Manufacturing Technologies
  Institute
National Research Council of Canada
London, Ontario N6G 4X8
Canada

Robert X. Gao, Ph.D.
Professor
Department of Mechanical and
  Industrial Engineering
University of Massachusetts
Amherst, MA 01003
USA

*Series Editor:*
Professor D. T. Pham
Intelligent Systems Laboratory
WDA Centre of Enterprise in
  Manufacturing Engineering
University of Wales Cardiff
PO Box 688
Newport Road
Cardiff
CF2 3ET
UK

British Library Cataloguing in Publication Data
Condition monitoring and control for intelligent
    manufacturing. - (Springer series in advanced
    manufacturing)
    1.Machinery - Monitoring 2.Manufacturing processes -
    Automation
    I.Wang, Lihui II.Gao, Robert X.
    670.4'2
ISBN-10: 1846282683

Library of Congress Control Number: 2006924638

Springer Series in Advanced Manufacturing ISSN 1860-5168
ISBN-10:  1-84628-268-3          e-ISBN   1-84628-269-1          Printed on acid-free paper
ISBN-13:  978-1-84628-268-3

Printed in Germany

9 8 7 6 5 4 3 2 1

Springer Science+Business Media
springer.com

# Preface

With its history dating back to the 18th century, manufacturing has propelled civilization to its modern age around the world, and has influenced the development of the entire infrastructure upon which industrial societies are based. From power generation and machine tools to transportation and information distribution to consumer electronics – the influence of manufacturing is ubiquitously seen. While globalization continuous to shift and relocate the center of activities worldwide, manufacturing has maintained its central role in the economy of most industrial nations, accounting for nearly one-fifth of the GDP in the USA and Canada.

As a fundamental element in any manufacturing industry, manufacturing machines and related equipment provide the physical platform and technology needed to economically produce high-quality products in the required volume. Consequently, the working conditions under which these machines operate play a critical role in satisfying customers' demands. With the increasing level of complexity of manufacturing machines, there is an increased need for more effective and efficient techniques that monitor machine conditions in real time, detect the inception and progression of defects, and enable flexible maintenance scheduling, before a defect results in unexpected machine downtime. Such techniques are a prerequisite for realizing reliable, economical, environment-friendly, and ultimately, intelligent manufacturing and product quality control.

Over the past decades, the broad range of manufacturing processes and the rich physical variability associated with them have continued to drive the evolution of technologies for machine condition monitoring and control. While research and development efforts have been translated into a large volume of publications and impacted the present and future practices in the industry, there exists a gap in the literature for a focused collection of works that are dedicated to the field of condition monitoring and control for intelligent manufacturing.

To bridge such a gap and introduce the state-of-the-art to a broad readership, from academic researchers to practicing engineers, is the primary motivation for this book, and the fifteen chapters included herein provide an overview of some of the latest efforts in this vibrant field.

Taking a historical perspective, Chapter 1 presents a review of major research developments in the area of monitoring and control for several representative machining processes, such as turning, milling, drilling, and grinding. The discussion is extended in Chapter 2 to include monitoring for precision manufacturing of small feature sizes with high tolerances, using acoustic emission (AE) sensing as an enabling technique. With unmanned, automated machining in mind, Chapter 3 introduces a Multiple Principle Component-based approach to tool condition monitoring, with a focus on turning and drill chipping, and evaluates them through experimental case studies. In Chapter 4, the focus is on monitoring techniques for grinding, given the widespread applications of grinding in abrasive cutting and final surface finishing. The chapter also discusses future solutions to realize adaptive control for intelligent grinding.

Recognizing the importance of rolling bearings as one of the fundamental elements in rotary machines, Chapters 5–7 present a range of techniques for high-quality vibration sensing, dynamic structural modeling, signal feature extraction, and decision-making that are relevant to bearing and rotary machine condition monitoring, defect diagnosis, and remaining life prognosis. Some of the approaches employed, such as finite element modeling, are further discussed in Chapter 8, in the context of strain distribution and quality variation estimation for sheet metal stamping process monitoring. Chapter 9 continues the discussion on condition monitoring of rotary machines by taking a state-space representation approach, using a gearbox under varying loads as the object of study. Considering the commonalities shared by various types of signals encountered in manufacturing equipment and process monitoring, Chapter 10 presents a number of representative signal processing algorithms in various domains, and discusses their suitability for signal signature analysis.

Expanding the scope of sensor utilization from discrete to networked sensors, Chapter 11 introduces a medium-range active sensor network for position tracking and surveillance of moving objects, as commonly seen in applications involving robots. Given that the effectiveness and efficiency of sensor data analysis ultimately depend on the accessibility to such data from remote sites where decisions are made, the design of a flexible architecture that incorporates remote monitoring and control functions becomes critical to successful implementation of advanced techniques on the factory floor. Such a topic is addressed in Chapter 12, where a web-based, sensor-driven framework is introduced, in the context of *WISE-SHOPFLOOR*.

Combining the functions of cutting tool control and force sensing, Chapter 13 introduces a new type of nanofabrication probe based on integrated piezoelectric actuation and capacitance measurement, which can be used for diamond turning of complex surface profiles. In Chapter 14, a distributed architecture for sensor-to-network interface based on the IEEE 1451 standards is introduced. In view of the increasing demand for a new generation of networkable sensors and actuators for monitoring and control in manufacturing, effective application of such an interface enables as well as requires a decentralized intelligence scheme down to the sensor level. Therefore, research and development efforts related to sensor network standards will also impact the future development of the sensors industry. Finally, an Integrated System Health Management (ISHM) for rocket engine testing in mission-related aerospace applications is described in Chapter 15. Aerospace presents a specialty domain in manufacturing, with abundant unique challenges. Given that such a testing environment involves a spectrum of complex mechanical components, control networks, and multiple physical quantities that are also shared, to various degrees, by conventional manufacturing, techniques developed for ISHM may be directly applicable to advancing the state-of-the-art in the condition monitoring and control for intelligent manufacturing.

The editors would like to express their deep appreciation to all the authors for their significant contributions to the book. Their commitment, enthusiasm, and technical expertise are what made this book possible. We are also grateful to the publisher for supporting this project, and would especially like to thank Mr Anthony Doyle, Senior Editor for Engineering, and Ms Kate Brown, Editorial Assistant, for their constructive assistance and earnest cooperation, both with the publishing venture in general and the editorial details. We hope the readers find this book informative and useful.

London and Amherst
August 2005

Lihui Wang
Robert X. Gao

# Contents

**13    An Intelligent Nanofabrication Probe for Surface
Displacement/Profile Measurement** .......................................... 315

*Wei Gao*

**14    Smart Transducer Interface Standards for Condition Monitoring
and Control of Machines** ........................................................... 347

*Kang B. Lee*

**15    Rocket Testing and Integrated System Health Management**............ 373

*Fernando Figueroa and John Schmalzel*

# List of Contributors

**Ardevan Bakhtari**
Department of Mechanical &
Industrial Engineering
University of Toronto
Toronto, Ontario M5S 3G8
Canada

**Beno Benhabib**
Department of Mechanical &
Industrial Engineering
University of Toronto
Toronto, Ontario M5S 3G8
Canada

**Scott Billington**
Radatec, Inc.
Atlanta, GA 30308
USA

**David A. Dornfeld**
Laboratory for Manufacturing
Automation
University of California
Berkeley, CA 94720-1740
USA

**R. Du**
The Chinese University of Hong Kong
Hong Kong
China

**Mihaela Dumitrescu**
McMaster University
Hamilton, Ontario L8S 4L7
Canada

**Mo A. Elbestawi**
McMaster University
Hamilton, Ontario L8S 4L7
Canada

**Fernando Figueroa**
NASA Stennis Space Center
MS, 39529
USA

**Robert X. Gao**
Department of Mechanical &
Industrial Engineering
University of Massachusetts
Amherst, MA 01003
USA

**Wei Gao**
Tohoku University
Sendai, 980-8579
Japan

**Inkil Hwang**
Laboratory for Manufacturing
Automation
University of California
Berkeley, CA 94720-1740
USA

**Ichiro Inasaki**
Keio University
3-14-1 Hiyoshi, Kouhoku-ku,
Yokohama
Japan

**Bernhard Karpuschewski**
Otto-von-Guericke-University
Magdeburg
Post box 4120, D-39016 Magdeburg
Germany

**Thomas R. Kurfess**
George W. Woodruff School of
Mechanical Engineering
Georgia Institute of Technology
Atlanta, GA 30332-0405
USA

**Sherman Lang**
Integrated Manufacturing
Technologies Institute
National Research Council of Canada
London, Ontario N6G 4X8
Canada

**D.E. Lee**
Laboratory for Manufacturing
Automation
University of California
Berkeley, CA 94720-1740
USA

**Kang B. Lee**
National Institute of Standards and
Technology
Gaithersburg, MD 20899-8220
USA

**C. James Li**
Rensselaer Polytechnic Institute
Troy, NY 12180
USA

**Steven Y. Liang**
George W. Woodruff School of
Mechanical Engineering
Georgia Institute of Technology
Atlanta, GA 30332-0405
USA

**Viliam Makis**
Department of Mechanical &
Industrial Engineering
University of Toronto
Toronto, Ontario M5S 3G8
Canada

**Eu-Gene Ng**
McMaster University
Hamilton, Ontario L8S 4L7
Canada

**J.F.G. Oliveira**
University of Sao Paulo
Nucleus of Advanced Manufacturing
Sao Carlos
Brazil

**Peter Orban**
Integrated Manufacturing
Technologies Institute
National Research Council of Canada
London, Ontario N6G 4X8
Canada

**A. Parey**
Industrial Tribology, Machine
Dynamics & Maintenance
Engineering Center
Indian Institute of Technology
Hauz Khas, New Delhi 110016
India

**John Schmalzel**
Rowan University
Glassboro, NJ 08028
USA

**Weiming Shen**
Integrated Manufacturing
Technologies Institute
National Research Council of Canada
London, Ontario N6G 4X8
Canada

**Shuangwen Sheng**
Department of Mechanical &
Industrial Engineering
University of Massachusetts
Amherst, MA 01003
USA

**N. Tandon**
Industrial Tribology, Machine
Dynamics & Maintenance
Engineering Center
Indian Institute of Technology
Hauz Khas, New Delhi 110016
India

**A. Galip Ulsoy**
Department of Mechanical
Engineering
University of Michigan
Ann Arbor, MI 48109-2125
USA

**C.M.O. Valente**
University of Sao Paulo
Nucleus of Advanced Manufacturing
Sao Carlos
Brazil

**Lihui Wang**
Integrated Manufacturing
Technologies Institute
National Research Council of Canada
London, Ontario N6G 4X8
Canada

**Ruqiang Yan**
Department of Mechanical &
Industrial Engineering
University of Massachusetts
Amherst, MA 01003
USA

**Yimin Zhan**
Department of Mechanical &
Industrial Engineering
University of Toronto
Toronto, Ontario M5S 3G8
Canada

**Li Zhang**
Department of Mechanical &
Industrial Engineering
University of Massachusetts
Amherst, MA 01003
USA

# 1

# Monitoring and Control of Machining

A. Galip Ulsoy

University of Michigan
Ann Arbor, Michigan 48109-2125, USA
Email: ulsoy@umich.edu

## Abstract

This chapter reviews major research developments over the past few decades in the monitoring and control of machining processes (*e. g.* turning, milling, drilling, and grinding). The major research accomplishments are reviewed from the perspective of a hierarchical monitoring and control system structure, which considers servo, process, and supervisory control levels. The use and benefits of advanced signal processing and control methods (*e. g.* Kalman filtering, optimal control, adaptive control) are highlighted, and illustrated with examples from research work conducted by the author and his co-workers. Also included are observations on how significant the research to date has been in terms of industrial impact, and thoughts on how this research area might develop in the future.

## 1.1 Introduction

Machining, or metal removal, processes are widely used in manufacturing and include operations such as turning, milling, drilling, and grinding [1.1][1.2]. The trend toward automation in machining has been driven by the need to maintain high product quality while improving production rates and the potential economic benefits of automation in machining are significant [1.3]–[1.6].

As illustrated in Figure 1.1, a machining process (*e. g.* turning) is carried out on a machine tool (*e. g.* lathe) using a cutting tool for material removal on a workpiece to produce a part with a desired geometry. A process reference, set using productivity and quality considerations, and the process state are fed to the controller, which adjusts the desired process variables. These references are input to the servo controllers, which drive the servo systems (*e. g.* slides and spindles) and produce the actual process variables (*e. g.* cutting force). Sensor measurements of the process are then filtered and input to the monitoring and control algorithms. Typically, this complex operation must be carried out with high accuracy, high production rates, and low cost.

Monitoring and control of machining processes are becoming increasingly important [1.7][1.8]. In-process sensing and control techniques currently being developed can be viewed as a key component of the next generation of quality control (see Figure 1.2). In current industrial practice, quality is ensured in the product engineering cycle at two distinct stages. First, is the use of Taguchi type methods at the product design stage to ensure that quality is designed into the product [1.9]. Obviously this is done *before* the part is manufactured. The second is the use of statistical process control (SPC) methods at the inspection stage, *after* the part is manufactured, to check the quality of the manufactured part [1.10]. However, real-time sensing and control will introduce a third level of quality assurance, which can be implemented *during* machining (*i.e.* in-process). This will complement Taguchi and SPC methods, lead to the next generation of quality control, and eliminate the need for expensive post-process inspection. Such in-process quality assurance methods are currently being developed for many machining processes. However, widespread implementation in machining has not been achieved due to (1) the required trade-offs in machining between quality, productivity, and cost, (2) inadequate in-process sensing, and (3) lack of open-architecture control platforms.

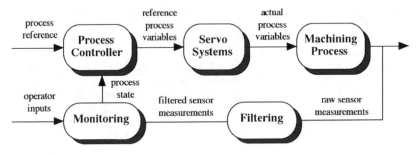

**Figure 1.1.** Schematic of monitoring and control of a machining processes

**Figure 1.2.** The next generation of quality control will involve quality assurance not only at the design and inspection stages, but also in-process quality control implemented at the machining stage

Machining operations date back to the 18th century, and represent a mature technology. However, there have been many improvements in machines, cutting tools, and materials that have led to significant benefits. One of the most important of these has been the introduction of automation during the past 3 to 4 decades. In the late 1950s, numerically controlled (NC) machine tools were developed under an Air Force contract to Parson's Machine Tool Company (Traverse City, Michigan) by the Servomechanisms Laboratory at MIT. These reprogrammable but hard-wired digital devices represented the state-of-the-art into the 1960s. During the 1960s and 1970s, both computer technology and NC machine tools continued to evolve. Computers became, not only more powerful, but also less expensive and more reliable. The servo control function (including interpolators for multi-axis coordination) became implemented using on-board computers rather that hard-wired digital circuits. These computer numerically controlled (CNC) systems, because of their powerful computing capabilities, led to advances in the interpolators and in the servo control loops [1.3].

The additional on-board computing power of CNC machines also led to research interest in implementing a higher level process control aimed at improving production rates and product quality. These process control systems are often referred to as "adaptive control" (AC) systems in the manufacturing community, but may or may not include adaptation in the sense in which the term is used in the control literature [1.11][1.12]. The first of these process control systems, which was designed under an Air Force contract by Bendix, was an ambitious system, which was only demonstrated in the laboratory [1.13][1.14]. It included advanced control features such as online optimization and adaptive control, and was termed an adaptive control with optimization (ACO) system. However, a major drawback of this system was the need for an accurate online measurement or estimate of tool wear. Such measurement methods were not available in the 1960s, and even today a reliable tool wear measurement method that can operate robustly in an *industrial* environment does not exist [1.15]–[1.17]. Subsequent systems were often less ambitious than the original Bendix concept, and were designed to operate on the constraints of the permissible workspace (*e. g.* maximum cutting force). These systems are often classified as adaptive control with constraints (ACC) and geometric adaptive control (GAC) [1.11][1.18]. Several commercial devices, of the ACC type, were also put on the market during the 1970s, but did not achieve widespread industrial acceptance [1.11]. Some of the problems, which led to the poor performance of these first-generation ACC systems, are discussed in a subsequent section. The need for measurement of wear in the Bendix system also generated considerable research interest in online tool wear measurement and estimation [1.19]. For a more detailed discussion on the state-of-the-art in machine tool automation through the 1970s, the reader is referred to [1.3][1.6][1.11] [1.19].

The 1980s saw increased research in the use of advanced control methods for control of manufacturing processes [1.6]. This can be related to several complementary factors: (i) the demand for better productivity, precision and quality in manufacturing continued to increase; (ii) the accomplishments of the 1970s had not solved the problems but clearly indicated the complexity of the machining system including the nonlinear, nonstationary and multivariable nature of the processes to be controlled; (iii) there were significant developments in computer technology on the one hand and in control and estimation theory on the other [1.5]. Methods such as optimal control, adaptive control, preview control, multivariable control, robust control, and state estimation have all been demonstrated on machining processes using the inexpensive and reliable microcomputer control technology of the 1980s [1.3]. Although, our focus here will be on machining processes such as turning, milling, drilling, and grinding, many of these developments are applicable to all manufacturing processes.

Figure 1.3 illustrates a convenient classification for the levels of control that one might encounter in a controller for machining. The lowest level is servo (or machine) control, where the motion of the cutting tool relative to the workpiece is controlled. Next is a process control level, where process variables such as cutting forces and tool wear are controlled to maintain high production rates and good part quality. Finally, the highest level is a supervisory one, which directly measures product-related variables such as part dimensions and surface roughness. The supervisory level also performs functions such as chatter detection, tool monitoring, machine monitoring, *etc.*

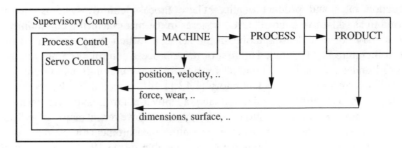

**Figure 1.3.** A controller for a manufacturing process, showing the servo, process and supervisory control levels

The hierarchical control structure illustrated in Figure 1.3 is actually a convenient simplification, in that machining processes have many (not only three) levels of control, each with its own characteristic iteration time. For example, servo control (axis motion control) loops may have sampling periods of 3 ms, process (*e. g.* force or torque) control loops may have sampling times of 30 ms, control of wear rate may require sampling at 3 s, tool changes may occur at intervals of 30 minutes, predictive machine maintenance at intervals of 30 days, *etc.* Typically, for predictable and efficient operation, all of these levels must be considered in designing a controller for a machining process. For example, performing predictive machine maintenance will improve the performance of the servo loops, and changing worn tools will improve the process controller performance. A multi-level hierarchical control scheme will make the process more predictable at the lower control levels, thus reducing the controller complexity and control effort required at those lower levels [1.20]. The design of such multi-level controllers will require a good understanding of various aspects of the machining process, which must be embodied in appropriate models. Future research on machining control can be expected to focus on these higher levels in the control hierarchy.

Such multi-level hierarchical monitoring and control systems for machine tools present serious industrial implementation challenges. The advent of open-architecture control has provided a natural framework for implementation of control systems in machine tools [1.21]–[1.27]. As illustrated in Figure 1.4, for a 5-axis CNC laboratory machine tool at the University of Michigan, there has been a gradual migration of the control technology. The original CNC machine tools had closed (*i. e.* proprietary "black box") commercial controllers, which were very difficult to modify for inclusion of additional monitoring and control functions. These systems have gradually evolved over the past 25 years into more open and more distributed control architectures, which facilitate the integration and upgrading of monitoring and control functions. However, despite major research investments (*e. g.* by the US Department of Defense in the 1980s and by the European Union in the 1990s), the commercial availability of open-architecture controllers for machine tools has been slow to develop, and has not led to widely accepted industry standards. These systems are now evolving into networked control systems, and will most likely capitalize on developments in other, higher-volume, real-time computing applications such as the automotive and home automation industries. Such systems are essential for the widespread implementation of monitoring and control for machining. The lack of such commercial open-architecture machine tool controllers has been, and continues to be, a major impediment to the widespread industrial use of monitoring and control beyond the servo control level.

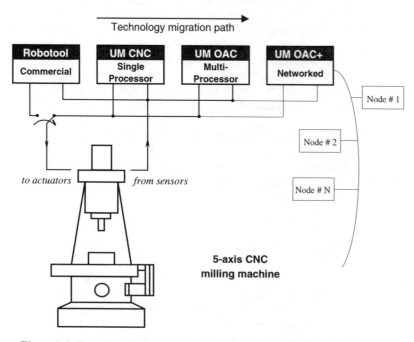

**Figure 1.4.** Evolution of open-architecture control technology for machine tools

The purpose of this chapter is to review major research developments in monitoring and control of machining processes conducted during the past thirty years, to assess the impact that this research has had on industrial practice, and to discuss the expected research directions for the next decade. In the next section of this chapter, machining processes are briefly described. Subsequent sections summarize the recent research accomplishments in monitoring of machining processes, and their control at the servo, process, and supervisory controller levels respectively. These developments are illustrated with examples from research work conducted by the author and his co-workers. Finally, thoughts on the impact of the research to date on industrial practice, and future research directions in automation of machining processes are given.

## 1.2 Machining Processes

Machining operations (*e. g.* turning, drilling, milling, grinding) are shape transformation processes where metal is removed from a stock of material (*i. e.*
workpiece) to produce a part. The objective is to produce parts with specified quality
at high production rates [1.28]. In this section we review the process phenomena
arising due to the interaction of the cutting tool and the workpiece (*e. g.* force
generation, chatter, tool wear and failure, burr and chip formation, heat generation)
and provide some background on those process phenomena. The focus is on the
development of an understanding of, and models for, process monitoring and control
[1.8].

The contact between the cutting tool and the workpiece generates forces, which
in turn create torques on the spindle and drive motors, and these torques lead to
power drawn from the motors. Excessive forces and torques can cause tool failure,
spindle stall (detected by monitoring the spindle speed), undesired structural deflections, *etc.* The cutting forces, torques, and power directly affect the other process
phenomena (*e. g.* tool wear and breakage). Therefore, these quantities are often
monitored as an indirect measurement of the process state and are regulated such
that productivity may be maximized while meeting machine tool and product quality
constraints.

*Cutting force* models tend to be quite complex and experimentation is required to
calibrate the parameters, as analytical models based entirely on first principles are
not available. However, the models used for controller design are typically simple.
The structure of the static cutting force is

$$F = K d^{\beta} V^{\gamma} f^{\alpha} \tag{1.1}$$

where $F$ is the cutting force, $K$ is the specific cutting force coefficient, $d$ is the depth-
of-cut, $V$ is the cutting speed, $f$ is the feed, and $\alpha$, $\beta$, and $\gamma$ are coefficients describing
the nonlinear relationships between the force and the process variables. The model
parameters in Equation (1.1) depend on the workpiece and cutting tool materials,
coolant, *etc.* and must be calibrated for each different operation. Static models are
used when considering a force *per spindle revolution* such as a maximum or average
force. Such models are suitable for interrupted operations (*e. g.* milling) where, in
general, the chip load changes throughout the spindle revolution and the number of
teeth engaged in the workpiece constantly changes during steady operation.

The first-order cutting force, assuming a zero-order hold equivalent, is

$$F = K d^{\beta} V^{\gamma} \frac{1+a}{z+a} f^{\alpha} \tag{1.2}$$

where $a$ is the discrete-time pole which depends upon the time constant and the
sample period, and $z$ is the discrete-time forward shift operator. The time constant,
in turn, is sensitive to the spindle speed since a full chip load is developed in
approximately one tool revolution [1.29]. In addition to the other model parameters,
$a$ must be calibrated for each different operation. First-order models are typically
employed when considering an instantaneous force, which is sampled several times

per spindle revolution. Such models are suitable for uninterrupted operations (e. g. turning) where, typically, a single tool is continuously engaged with the workpiece and the chip load remains constant during steady operation.

Load cells or dynamometers are often used in laboratory settings for cutting force measurements. However, they are impractical for industrial applications. This has been one of the major impediments to the industrial adoption of force-based monitoring and control techniques. Forces can also be predicted from the current of the feed axis drive, and torque can be monitored on the spindle unit with strain gauge devices, or laboratory dynamometers. Power from the spindle and axis motors is typically monitored using Hall effect sensors, which are easy to install and guard from the process. Due to the large masses these motors drive, the signal typically has a limited bandwidth [1.30]–[1.32].

Although the three major process variables (f, d, and V) affect the cutting forces, the feed is typically selected as the variable to adjust for force regulation. Typically, the depth-of-cut is fixed from the part geometry and the force–speed relationship is weak (i. e. $\gamma \approx 0$); therefore, these variables are not actively adjusted for force control. Reference force values are set in roughing passes to maximize productivity, while references are set in finishing passes to maximize quality (i. e. good surface finish, minimal tool deflection). In roughing passes, reference forces are due to tool failure and maximum spindle power constraints.

The forces generated when the tool and workpiece come into contact produce significant structural deflections. Regenerative *chatter* is the result of the unstable interaction between the cutting forces and the machine tool–workpiece structures, and may result in excessive forces and tool wear, tool failure, and scrap parts due to unacceptable surface finish [1.33]–[1.38]. The feed force for an orthogonal cutting process (e. g. turning thin-walled tubes) is typically described as

$$F(t) = Kd[f_n + x(t) - x(t - \tau)]$$
(1.3)

where $f_n$ is the nominal feed, $x$ is the displacement of the tool in the feed direction, and $\tau$ is the time for one tool revolution. It is typically assumed that the workpiece is much more rigid than the tool and that the force is proportional to the instantaneous feed and the depth-of-cut, and does not explicitly depend upon the cutting speed. The instantaneous chip load is a function of the nominal feed as well as the current tool displacement, $x(t)$, and the tool displacement at the previous tool revolution, $x(t - \tau)$. Assuming a single degree of freedom linear model, the vibration of the tool structure may be described by

$$m\ddot{x}(t) + c\dot{x}(t) + kx(t) = F(t)$$
(1.4)

where $m$, $c$, and $k$ are the effective mass, damping, and stiffness, respectively, of the tool structure. The stability of the closed-loop system formed by combining Equations (1.3) and (1.4) may be examined to generate the so-called stability lobe diagram and select appropriate process variables.

As an example, Figure 1.5 shows the stability lobes (i. e. regions of chatter (above the lobes) and chatter-free (below the lobes) milling operations) in the plane of spindle speed and axial depth-of-cut for a reconfigurable machine tool in two different configurations [1.38][1.39]. A cutting force model, as in Equation (1.3), and a structural dynamics model, as in Equation (1.4), are needed for each machine configuration to generate these stability lobes. The system experiences chatter for cutting conditions that fall above these lobes, as verified by the cutting experiment data shown in Figure 1.5.

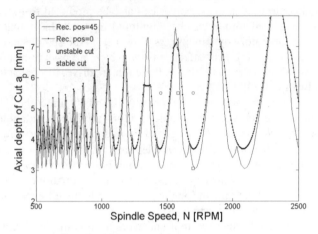

**Figure 1.5.** Analytical stability lobes showing the regions of chatter stability and instability for a reconfigurable machine tool in two configurations. Data points indicate experimental verification

Regenerative chatter can easily be detected by an operator due to the loud, high-pitched noise it produces and the distinctive "chatter marks" it leaves on the work-piece surface. However, automatic detection is difficult. The most common approach is to threshold the spectral density of a process signal such as sound [1.7]. These thresholding algorithms all suffer from the lack of an analytical method of selecting the threshold value, which is typically selected empirically and will not be valid over a wide range of cutting conditions. Several variables have been studied for use in the detection of chatter. These include the cutting force signal, displacement or acceleration of a point in the vicinity of the tool–workpiece interface, or the sound emitted from the machine [1.7][1.40]. While chatter has been investigated extensively, most of these efforts have been directed towards prediction of chatter rather than its detection. The approaches used for chatter detection mirror those employed for tool breakage detection, except that analysis is performed primarily in the frequency domain.

Chatter is typically suppressed by adjusting the spindle speed to lie in one of the stability lobe pockets, as shown in Figure 1.5 [1.36]–[1.38]. Feed has been shown to have a monotonic affect on the marginally stable depth-of-cut and is sometimes the variable of choice by machine tool operators [1.40]. The tool position may also be adjusted (e. g. depth-of-cut decreased) to suppress chatter, and while it is guaranteed to work, this approach is typically not employed since the part program must be rewritten and productivity is decreased. Spindle speed variation (SSV) is another technique for chatter suppression [1.41][1.42]. The spindle speed is varied about some nominal value, typically in a sinusoidal manner. Although SSV is a promising technique, the theory to guide the designer as to the optimal variation is not fully developed.

*Chip formation*, and chip entanglement, is another aspect of the machining process. The three major chip formation types are: discontinuous, continuous, and continuous with built up edge (BUE) [1.2]. Discontinuous chips arise when the operation continuously forms and fractures chips due to the inability of the work-piece to undergo large amounts of plastic deformation, while continuous chips do not fracture but rather form continuous ribbons. Continuous chips with BUE form when part of the chip welds to the tool due to high cutting temperatures and pressures. Continuous chips (with and without BUE) will interfere with the normal interaction between the tool and workpiece and cause poor surface finish as will discontinuous chips which do not clear the cutting zone. Therefore, chip control is the proper formation of chips, which clear the cutting zone and are directed towards the chip conveyor system for efficient removal. Research on chip formation process extends nearly a century, starting most notably with [1.43]. Theory has been developed to predict shear plane angle, chip velocity, chip curling and to provide chip breaking models [1.8].

Small, undesirable metal fragments left on the workpiece after the machining operation is complete are known as *burrs*. Burrs cause improper part mating, accelerated device wear, and decreased device performance. Since it is typically impossible to avoid the formation of burrs, the designer strives to reduce the complexity of the subsequent deburring operation by minimizing the burr strength and ensuring the burrs form at workpiece locations that are easy to access. Burrs form due to workpiece plastic deformation [1.44]. Burr measurement is typically performed offline by measuring the average height, base thickness, and toughness. Burr location, and its accessibility, are also important to note. Process variables are known to have a strong effect on the physical characteristics of burrs. If the depth-of-cut in a face milling operation is too small, the cutting tool will "push" the material over the side of the workpiece and form a large, strong burr on the workpiece edge. Without adequate models, one is left to empirical techniques to predict and control burr formation.

The friction between the cutting tool and workpiece generates significant heating in the cutting zone. The *cutting temperature* affects the tool wear rate and the workpiece surface integrity, and contributes to thermal deformation. The most basic temperature models estimate steady-state cutting temperatures and typically have the following nonlinear relationship with the process variables [1.45]

$$T = aV^b f^c \qquad\qquad (1.5)$$

where $T$ is the workpiece temperature and $a$, $b$, and $c$ are empirically determined constants. A comparison with experimental results shows most models to be qualitatively correct, but to quantitatively overestimate cutting temperatures and are unable to estimate cutting temperatures in operations with discontinuous chip formation [1.46]. The use of thermocouples and infrared data to measure cutting temperatures was investigated; however, cutting temperature measurements are rarely utilized in industrial settings [1.46]. Burr formation, chip formation and cutting temperature generation have received little attention from the monitoring and control community [1.47]–[1.49].

## 1.3 Monitoring

Process monitoring is the manipulation of sensor measurements (*e. g.* force, vision, temperature) to determine the state of the process. The machine tool operator routinely performs monitoring tasks; for example, visually detecting missing and broken tools and detecting chatter from the characteristic sound it generates. Unmanned monitoring algorithms utilize filtered sensor measurements which, along with operator inputs, determine the process state (Figure 1.1). The state of complex processes are monitored by sophisticated signal processing of sensor measurements which typically involve thresholding or artificial intelligence (AI) techniques [1.7] [1.50]–[1.51].

Monitoring is necessary for detection of a process anomaly so as to prevent machine damage by stopping the process, or to remove the anomaly by adjusting the process inputs (feeds and speeds). A process anomaly could be gradual such as tool/wheel wear, may be abrupt such as tool breakage, or may be preventable such as excessive vibration/chatter. Knowledge of tool wear is necessary for scheduling tool changes; detection of tool breakage is important for saving the workpiece and/or the machine; and identifying chatter is necessary for triggering corrective action. A major difficulty in machine tool monitoring is the lack of robust in-process sensors. Sensors can seldom be placed at the point of interest, and when located remotely they do not provide the clarity of measurement necessary. This limited sensing capability is often compensated by using multiple sensors to enhance reliability. Another difficulty in machine tool monitoring is the absence of accurate analytical models to account for changes in the measured variables by variations in the cutting conditions. Such changes are often attributed to process anomalies by the monitoring system, which result in false alarms.

Process monitoring is generally performed through the analysis of process measurements. For this purpose, a process variable or a set of variables (*e. g.* force, power, acoustic emission, feed motor current) is measured and processed online to be compared against its expected value. Any deviation from this expected value is attributed to a process anomaly. Expected values of measurements are either determined according to an analytical model of the process [1.52][1.53] or established empirically [1.51]. The advantage of analytical models is that they account for changes in the machine inputs such as feeds and speeds The disadvantage of analytical models is that they are often not accurate and need to be calibrated. Establishing the expected values of measurements empirically is simpler and more straightforward. However, the empirical values are only suitable for particular operations and cannot be extrapolated to others. This section focuses on monitoring techniques for tool failure and tool wear [1.54].

### 1.3.1 Tool Failure

Tool failure occurs when a significant portion of the tool breaks off, the tool shaft or cutting teeth severely fracture, or a significant portion of one or more teeth chip.

Broken tools drastically decrease productivity by creating unnecessary tool changes, wasting tools, and creating scrap parts, and may injure operators. Fracture is the dominant mode of failure for more than one quarter of all advanced tooling material. Therefore, online detection of tool breakage is crucial for fully automated machining. Ideally, a tool breakage detection system must detect failures rapidly to prevent damage to the workpiece, and must be reliable so as to eliminate downtime due to "false alarms".

The simplest way to detect a failed tool is to use a probe or vision system to inspect the cutting tool. While this inspection is typically performed offline, some techniques are being developed for online detection. However, chip and coolant interference is still a major obstacle to overcome. Consequently, many sensors have been used to detect tool failure indirectly, including acoustic emission, force, sound, vibration, *etc.* [1.15]. In these indirect methods, the signal magnitude, root mean square value, or magnitude of the power spectrum, among others, is inspected, typically via thresholding.

Several measurements have been reported as good indicators of tool breakage [1.7][1.15]. Among these, cutting force, acoustic emission, spindle motor current, feed motor current [1.32], and machine tool vibration have been investigated extensively for their sensitivity to tool breakage. In general, to utilize a measurement for tool breakage detection, two requirements need to be satisfied. First, the measurement must reflect tool breakage under diverse cutting conditions (*e. g.* variable speeds, feeds, coolant on/off, workpiece material). Second, the effect of tool breakage on the measurement (tool breakage signature) must be uniquely distinguishable, so that other process irregularities such as hard spots will not be confused with tool breakage. The tool breakage signature is commonly in the form of an abrupt change, in excess of a threshold value. Despite considerable effort [1.55], reliable signatures of tool breakage that are robust to diverse cutting conditions have not yet been found from individual measurements.

To improve the reliability of tool breakage signatures, pattern classification techniques have been utilized. One of the earliest efforts was by Sata and co-workers [1.56] who related features of the cutting force spectrum such as its total power, the power in the very low frequency range, and the power at the highest spectrum peak and its frequency to chip formation, chatter, and a built-up edge. It was shown that the cutting force measurement alone could provide sufficient information for unique identification of the above phenomena. Another important study was performed by Kannatey-Asibu and Emel, who applied statistical pattern classification to identify chip formation, tool breakage, and chip noise from acoustic emission measurements [1.57]. They reported a success rate of 90% for tool breakage detection. The only drawback to spectrum-based tool breakage detection is the computational burden associated with obtaining the spectrum, which often precludes its online application.

The alternative to single-sensor-based pattern classification is the multi-sensor approach using artificial neural networks for establishing the breakage patterns [1.58]. However, the utility of neural networks for tool breakage detection is limited by their demand for expensive training. A pattern classifier that requires less training than artificial neural networks is the Multi-valued Influence Matrix method, which has a fixed structure and has been shown to provide robust detection of tool breakages in turning with limited training [1.7]. Unsupervised neural networks have also been proposed for tool breakage detection in machining [1.7].

When a tool failure event has been detected, an emergency stop is typically initiated. A significant amount of time is spent not only changing the cutting tool and workpiece, but also restarting the machine tool or machining line. This loss of productivity can be avoided by an intelligent reaction to the tool failure event. For example, the cutting tool may be moved to the tool change position and vision may be utilized to examine the workpiece surface to verify whether or not the workpiece must be scraped. As another example, if a tooth chips in a milling cutter, optical techniques may be used to determine if the workpiece and tool are undamaged and, if so, the feed can be decreased and cutting may continue. There have been some studies to detect the onset of tool failure [1.7]. Typically, a process parameter such as the feed is adjusted; however, a reference force may also be adjusted if a force control scheme is being employed.

### 1.3.2 Tool Wear

The contact between the cutting tool and the workpiece and chips causes the shape of the tool to change (see Figure 1.6). This phenomenon, known as tool wear, has a major influence in machining economics, affects the final workpiece dimensions, and will lead to eventual tool failure. Wear on the face of the tool that contacts the workpiece is termed flank wear, whereas crater wear occurs on the tool face that contacts the chips. A typical tool wear (*i. e.* flank wear) curve is shown in Figure 1.7. The tool wears rapidly in the initial phase and then levels off to a constant rate during the steady phase. From an economic point of view, the designer would like to use the tool until just before it enters the accelerated wear phase where the tool will eventually fail.

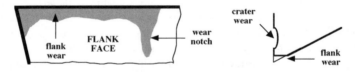

**Figure 1.6.** Wear on the face of a cutting tool

The three main tool wear mechanisms are *abrasion* between the cutting tool and workpiece, which is always present, *adhesion* of the chips or workpiece to the cutting tool, which removes cutting tool material and is more active as the cutting temperature increases, and *diffusion* of the cutting tool atoms to the chips or workpiece, which is typically active during the accelerated tool wear phase. The equation describing tool wear, developed by F. W. Taylor early in the twentieth century [1.43], and known as Taylor's tool wear equation, is

$$Vt_l^n = C \tag{1.6}$$

where $t_l$ is the tool lifetime and $C$ and $n$ are empirically determined constants. Modified Taylor equations include the effects of feed rate and depth-of-cut, as well as interaction effects between these variables. Increased testing is required to determine the extra model coefficients; however, these models are applicable over a wider range of cutting conditions. Models relating tool wear and cutting forces have also been developed [1.17][1.59]–[1.63].

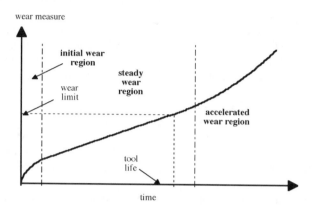

**Figure 1.7.** Typical flank wear versus time curve, showing regions of initial wear followed by a steady wear region and then an accelerated wear region leading to tool failure

Flank wear directly influences the dimensions and quality of the surface [1.7] [1.8]. Flank wear can affect the fatigue endurance limit by affecting surface finish, lubrication retention capability by changing the distribution of heights and slopes of the surface, and other tribological aspects by affecting the topography of the machined surface. Therefore, the information about the state of flank wear is needed to plan tool changes and avoid scrap, or manipulate the feed and speed in-process to control tool life [1.18].

Methods used for flank wear estimation can be classified as either *direct* or *indirect*. Direct methods measure flank wear either in terms of material loss from the tool, or by observing the worn surface using optical methods [1.62]–[1.67]. Direct methods are generally more reliable, although they are not convenient for in-process use in a harsh manufacturing environment. Indirect methods, on the other hand, estimate the flank wear by relating it to a measured variable such as the change in size of the workpiece, cutting force, temperature, or acoustic emissions [1.7]. The ideal measured variable in the indirect method is one that would be insensitive to process inputs. For example, non-contact methods have been recently developed for surface roughness measurement, which will undoubtedly have an impact on online estimation of tool wear [1.7].

Among the measurements for indirect flank wear estimation, acoustic emission (AE) and the cutting force have been the most popular due to their sensitivity to tool wear and reliability of measurement. The cutting force generally increases with flank wear due to an increase in the contact area of the wear land with the workpiece. Zorev [1.1] and De Filippi and Ippolito [1.68] were among the first who demonstrated the direct effect of flank wear on the cutting force, which motivated separation of the cutting force signal into two components, one associated with the unworn tool and the other associated with tool wear. The unworn tool component is estimated at the beginning of the cut with a new tool, and then subtracted from the measured force to estimate the wear affected component. This method can provide relatively accurate estimates of flank wear as long as the cutting variables (feed, speed, and depth of cut) remain unchanged. However, when the cutting variables change, due to such factors as geometric requirements of the part or manipulation of the operating parameters, the identification of the wear affected component is difficult. In such cases, either the effect of the manipulated cutting variable on the cutting force is estimated by a model [1.52] and separated to identify the wear-affected component [1.53][1.61]–[1.63], or the wear-affected component is estimated from small cutting segments where the cutting variables remain unchanged [1.17]. Recursive parameter estimation is used for identification, and requires "persistent excitation" of the cutting force to guarantee parameter convergence [1.12].

Like cutting force, acoustic emission has been studied extensively for flank wear estimation, and statistical properties of the AE signal have been shown to correlate with flank wear [1.64]. To define more clearly the effect of flank wear, statistical pattern classification of the AE signal in the frequency domain has been utilized [1.57] [1.69].

Despite the considerable effort towards estimation of flank wear from a single variable, single sensor measurements do not seem to be robust to varying cutting conditions. This has motivated integration of multiple measurements through artificial neural networks [1.7][1.58]. Generally, a neural network is trained to identify the tool wear pattern by supervised learning from samples of measurements taken at various levels of tool wear. Therefore, the ability of neural networks to form reliable wear patterns depends not only on their topology, but the extent of their training. In cases such as machining, where adequate data is not available to select the topology of the network, or to provide the tool wear patterns for a wide range of cutting conditions and material/tool combinations, these networks are not practical.

In [1.62][1.63], a hybrid, direct and indirect, tool wear monitoring technique was investigated. An adaptive observer was applied to estimate wear online and a vision system was used intermittently (e. g. between parts) to recalibrate the observer. A model of the machining process, based upon the physical mechanisms of tool wear, is utilized. The tool wear is a model state variable [1.60][1.70]. The measured output is the cutting force, and is used in conjunction with the model to estimate the state of wear of the tool. The difficulties in implementing this approach are the nonlinear nature of the process model and the variation of the model parameters [1.52][1.53]. Thus, an adaptive nonlinear state observer is required for effective estimation of the wear [1.61]–[1.63].

Although such a model-based wear estimation approach using force measure-ment works well enough to be used as the basis for tool replacement strategies, even under varying cutting conditions (e. g. feed or speed), it is not sufficiently accurate for compensation of geometric errors due to tool wear. A more accurate, and flexible, tool wear monitoring system can be developed by combining the force measurement based adaptive observer with direct optical measurement of the tool wear using a computer vision system [1.62][1.63]. The vision system is very accurate, but can only be used between parts when the tool is not in contact with the workpiece. The force-based adaptive observer can be used during cutting. By intermittently (i. e. between parts) recalibrating the adaptive observer using the vision system, a very accurate estimate of tool wear can be obtained even under large variations in the cutting conditions. Experimental results, shown in Figure 1.8, are from laboratory turning tests with large stepwise changes in the feed rate. The solid line is the tool wear estimate, the circles are the vision measurements used to recalibrate the observer, and the x's are the wear values measured using a tool maker's microscope. Errors in the flank wear estimate are less than 50 μm ($2 \times 10^{-3}$ inch) over the entire cutting period, and improve as the cutting proceeds towards the end of the tool life.

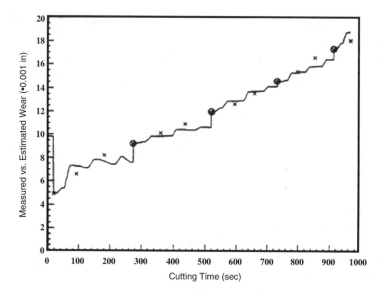

**Figure 1.8.** Measured (x and o) and estimated (——) flank wear curves [1.63]

The two main issues in tool wear regulation are compensation for tool wear and control of the tool wear rate. As the tool wears, the workpiece dimension may become out of tolerance. The tool position must be adjusted (typically through the part program) to compensate for the tool wear. From an economic point of view, it is desirable to regulate the tool wear rate such that the tool life corresponds to the scheduled tool change period in mass production, or to maximize the tool life in job shop situations.

## 1.4 Servo Control

The servo-level controllers move the machine axes, and can be classified into two cases: (i) point-to-point (PTP), and (ii) contouring. In a PTP system the controller moves the tool to a required position, where the process operation is performed. The trajectory between the points is not controlled. *Point-to-point control* is required in applications like drilling and spot welding, and is the simpler of the two cases. It has much in common with PTP servo control problems in other application areas, such as robotics. However, the PTP control problem in manufacturing does have some special considerations, which arise from the high accuracy requirements (*e. g.* 0.01 mm or better) and the interactions between the machine tool and the process as shown in Figures 1.1 and 1.3. Cutting forces and process heat generation must be considered to ensure accuracy in the PTP problem. The machine tool geometric errors affect PTP positioning accuracy, and are of several types: (i) volumetric errors due to slight irregularities in the components of the machine structure, (ii) volumetric errors due to the machine thermal effects, and (iii) volumetric errors due to the effects of cutting forces. The first type can be compensated for based upon offline measurements (and modeling) of the machine geometric errors by technologies such as laser interferometry. However, the second and third type require online compensation, typically based upon temperature and/or force measurements and models relating these measured variables to geometric errors in each axis of the machine tool. Typically, minimum variance, forecasting, or predictive control approaches are used in these applications [1.6][1.11][1.12]. Reductions in machine geometric and thermal errors can improve accuracies by an order of magnitude (*e. g.* from 200 to 20 µm) in medium to large machining centers [1.71].

The *contouring problem* is what really makes servo-level control difficult, and interesting, from a control point of view [1.6][1.72]–[1.74]. Contouring is required in processes such as milling, turning and arc welding. The task is one of tracking trajectories (*e. g.* lines, parabolas, and circles) very accurately in a spatial rather than temporal domain. The coordinated motion of multiple axes is required to generate accurate contours. Interpolators are used to generate the desired (reference) trajectories for each machine axis from the desired part geometry [1.3]. The goal is to reduce contour errors that depend on the axial errors. The interesting problem in machining is that reducing the contour error may or may not be the same as reducing individual axis errors. This is illustrated in Figure 1.9 for a two-axis system, where reducing the axis errors (*Ex* and *Ey*) does not necessarily reduce the contour error.

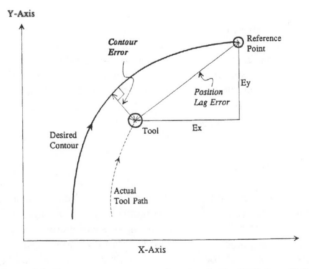

**Figure 1.9.** Contour errors versus axis errors in the X-Y plane [1.6]

Cross-coupling Control (CCC) methods have been developed to couple the machine axial controllers so that contour errors, rather than individual axis errors, can be reduced [1.72]–[1.76]. The cross-coupling concept, introduced by Koren [1.75], calls for the construction of the instantaneous contour error model in real time, and its utilization in a control law that generates a compensation signal to each of the control loops with the objective of reducing the contour error. The experimental results in Figure 1.10 for a three-dimensional conical cutting operation clearly indicate the advantages of the CCC approach; reductions in the maximum contour error of over 5:1 (from 250 µm to 40 µm) are achieved. The CCC method has been extended to 5-axis machines which include two rotational as well as three translation motion axes [1.72]–[1.76].

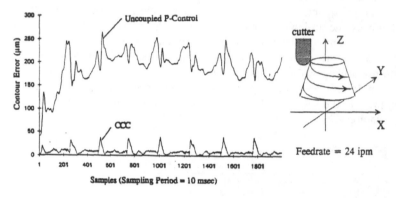

**Figure 1.10.** Contour errors in machining with and without cross-coupling control (CCC) [1.6]

To decrease tracking errors in contouring, feedforward and preview controllers can also be added to the control loops [1.74]. Preview controllers use the future reference trajectories that are available since the desired part geometry is known [1.77]. The concept of feedforward controllers is based on pole/zero cancellation. However, the closed-loop system may include unstable zeroes which preclude the use of pole/zero cancellation. Tomizuka has developed a zero phase error tracking controller that can be implemented even in the presence of unstable zeroes [1.78] [1.79]. Control of the depth-of-cut, in addition to feed and speed, has also been used to produce noncircular parts on a lathe [1.80]. A review of digital tracking controller design is given in [1.74].

## 1.5 Process Control

Process control is the manipulation of process variables (*e. g.* feed, speed, depth-of-cut) to regulate the processes. Machine tool operators perform online and offline process control by adjusting feeds and speeds to suppress chatter, initiating an emergency stop in response to a tool breakage event, rewriting a part program to increase the depth-of-cut to minimize burr formation, *etc*. Offline process control is performed at the process planning stage; typically by selecting process variables from a machining handbook or from the operator's experience. Computer-aided process planning is a more sophisticated technique which, in some cases, utilizes process models offline to select process variables. The drawbacks of offline planning are the dependence on model accuracy and the inability to reject disturbances. Adaptive control techniques [1.3], which include adaptive control with optimization, adaptive control with constraints, and geometric adaptive control, set process variables to meet productivity or quality requirements. There has also been significant research in AI techniques such as fuzzy logic, neural networks, knowledge base, *etc.* which require very little process information [1.7].

Figure 1.11 clearly illustrates the benefits of machining process monitoring and control [1.3]. The trend towards more frequent product changes has driven research in the area of reconfigurable machining systems [1.39]. Process monitoring technology will be critical to the cost-effective ramp-up of reconfigurable systems, while process control will provide options to the designer who reconfigures the machining system. While process control has not made significant headway in industry, there are currently companies, which specialize in developing process monitoring packages. Process monitoring and control technology will have a greater impact in future machining systems based on open-architecture systems [1.24], which provide the software platform necessary for the cost-effective integration of this technology.

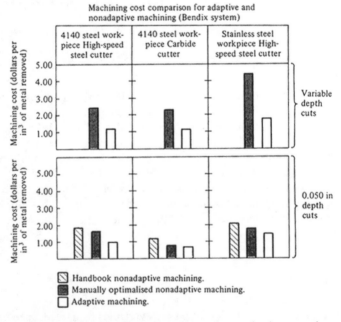

**Figure 1.11.** Comparison of adaptive and non-adaptive control

Successful implementation at the process control and supervisory levels requires realistic process modeling based upon a good physical understanding of the process to be controlled [1.11][1.17][1.61][1.81]–[1.85]. Some attempts at process control have utilized earlier ACO concepts [1.13][1.14]. For example, Watanabe [1.85] reported the development of an ACO system for milling. An ACO system for grinding was proposed and demonstrated [1.86]–[1.88] and subsequently generalized to the optimal locus control method by Koren [1.89]. As discussed in a previous section, the need for wear measurement or estimation in ACO has led to the development of model-based schemes for estimation of wear from indirect measurements such as cutting forces [1.17][1.59]–[1.63][1.70], temperature [1.90], and acoustic emissions [1.91][1.92].

Suboptimal process control systems were considered to be more practical than ACO systems, and systems of the ACC or GAC type were developed for various applications [1.3][1.11][1.81][1.88][1.93][1.94]. The emphasis has been on ACC systems, typically based upon force or torque measurements, rather than GAC systems based upon product dimensions and surface finish. This is primarily a result of the availability of less expensive and more reliable sensors for force and torque measurement. However, with recent advances in optical sensing techniques for part dimensions and surface finish, this trend can be expected to change. Use of optical sensing in GAC systems has already been demonstrated in turning and milling applications [1.3][1.94].

Despite developments in ACC systems, machining processes were still difficult to control due to process parameter variations, which could lead to problems such as tool breakage or even instability. Consequently, adaptive control methods were widely investigated for a variety of machining processes [1.11][1.21][1.83] [1.92][1.95]. The basic problem arises from the fact that the effective process gains and time constants depend upon process variables such as feed, speed, and depth-of-cut. Furthermore, process-related parameters such as tool geometry, work and tool material properties are difficult to characterize and can affect the process dynamics. This unpredictability of the process can lead to poor performance, tool breakage, or even instability with fixed gain process controllers [1.81]. These problems, due to process model uncertainty, were a primary reason for the poor industrial acceptance of the first generation of AC systems for machining. To address this problem, adaptive control schemes, combining online parameter estimation and control, were developed and applied to machining processes.

As an example of this class of controllers, a model reference adaptive force controller for a milling process is presented in [1.82][1.96]. A discrete-time model of a two-axis slot milling process can be written as [1.82]

$$\frac{F(z)}{V(z)} = \frac{b_0 z + b_1}{z^2 + a_1 z + a_2} \tag{1.7}$$

where $V(z)$ is a voltage signal proportional to the machine feed rate, and $F(z)$ is the measured resultant force. The parameters $a_1$, $a_2$, $b_0$, and $b_1$ depend upon the spindle speed, feed, depth-of-cut, workpiece material properties, *etc.* Thus, they must be estimated online for effective control of the resultant force based upon manipulation of the feed rate (*i. e.* $V(z)$). The goal is to maintain the resultant force at the reference level, $R(k)$, which is selected to maintain high metal removal rates without problems of tool breakage. The process zero at $-(b_1/b_0)$, although the values of $b_0$ and $b_1$ can vary, was found to be inside the unit circle for all cutting situations at the 50 ms sampling period that was used [1.82]. Therefore, direct adaptive control methods can be employed. A model reference adaptive controller was designed and evaluated in laboratory tests for two different workpiece materials and for changing depths of cut [1.21]. The controller design was straightforward, and gave satisfactory performance despite the process variations. The main difficulty is eliminating bias in the parameter estimates due to the "runout" noise in the force measurements [1.97].

Experimental results are shown in Figure 1.12 for machining of a 1020 CR steel workpiece with step changes in depth-of-cut from 2 to 3 to 4 mm. The desired force value of 400 N is maintained despite these large changes in depth-of-cut at a spindle speed of 550 rev/min. Initially the feed rate saturates, and the reference value of 400 N cannot be achieved at the 2 mm depth-of-cut. Also initial transients occur due to the parameter adaptation. This controller also gave good results when machining other materials (*e. g.* aluminum) without any additional controller tuning [1.21].

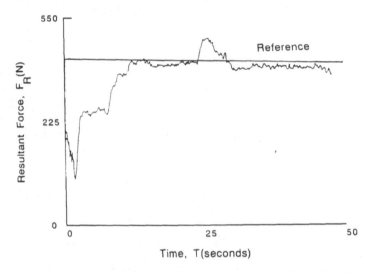

**Figure 1.12.** Model reference adaptive control of cutting force in milling [1.96]

Many force control studies, like the example above, are based on adaptive control [1.11]. However, model-based and robust control techniques have recently been gaining attention [1.98]–[1.101]. Adaptive techniques consider a linear relationship between the force and the feed and view changes in process variables and other process phenomena as changes in the cutting force parameters. Model-based techniques directly incorporate the nonlinear model. Robust control techniques [1.100][1.102] have also gained attention. These techniques incorporate the cutting force model and require bounds on the model parameters. Regardless of the control approach, saturation limits must be set on the commanded feed. A lower saturation of zero is typical since a negative feed will disengage the cutting tool from the workpiece; a non-zero lower bound may be set due to process constraints. An upper bound is set due to process or servo constraints.

Two machining force controllers are designed and implemented for the following static cutting force

$$F = 0.76d^{0.65} f^{0.63} \tag{1.8}$$

where $\gamma = 0$ in Equation (1.1) and $F$ is a maximum force per spindle revolution in a face milling operation. For control design, the model is augmented with an integral state to ensure constant reference tracking and constant disturbance rejection. A model-based design is applied [1.98]. The control variable is $u = f^{0.63}$ and the design model (with an integral state) is

$$F(z) = \theta \frac{1}{z-1} u(z) \tag{1.9}$$

where, $\theta = 0.76d^{0.65}$ is the gain. The nonlinear model-based controller utilizes process information (in this case, depth-of-cut) to directly account for known process changes. The Model Reference Control (MRC) approach is applied and the control law is

$$u(z) = \frac{1}{z-1} \frac{1+b_0}{\theta} [F_r(z) - F(z)] \tag{1.10}$$

where $F_r$ is the reference force and $b_0$ is calculated, given a desired closed-loop time constant and sample period. The commanded feed is calculated from the control variable

$$f = \exp\left[\frac{\ln(u)}{0.63}\right] \tag{1.11}$$

Therefore, the lower saturation on the control variable is chosen to have a small non-negative value. The experimental results for the nonlinear model-based controller are shown in Figure 1.13.

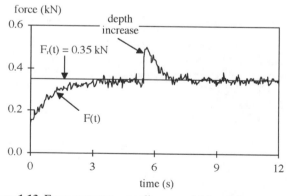

**Figure 1.13.** Force response – nonlinear model-based force controller

Next, an adaptive force controller is designed. The control design model, including an integral state, is

$$F(z) = \theta \frac{1}{z-1} f(z) \tag{1.12}$$

where $\theta$ is the gain and is assumed to be unknown. The MRC approach is applied and the control law is

$$f(z) = \frac{1}{z-1} \frac{1+b_0}{\hat{\theta}} [F_r(z) - F(z)] \tag{1.13}$$

The term $\hat{\theta}$ is an estimate of the gain, obtained using the recursive least squares method [1.12]. At the $i$th time iteration, the estimate is calculated as

$$\hat{\theta}(i) = \hat{\theta}(i-1) + K(i)\varepsilon(i) \tag{1.14}$$

where

$$K(i) = \frac{P(i-1)f(i)}{\left[1 + f(i)P(i-1)f(i)\right]} \tag{1.15}$$

$$P(i) = \left[1 - K(i)f(i)\right]P(i-1) \tag{1.16}$$

$$\varepsilon(i) = F(i) - f(i)\hat{\theta}(i-1) \tag{1.17}$$

The parameter $P$ is the covariance and the parameter $\varepsilon$ is the residual. Estimating the model parameters online is a method for accounting for model inaccuracies; however, the overall system becomes complex, and chaotic behavior can result.

The experimental results for the adaptive controller are shown in Figures 1.14 and 1.15. Both adaptive and model-based approaches successfully regulate the cutting force while accounting for process changes in very different ways. The adaptive technique is useful when an accurate model is not available, but is more complex compared to the model-based approach.

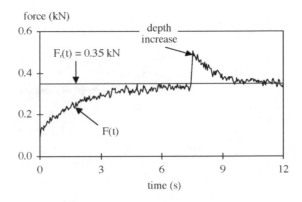

**Figure 1.14.** Force response – adaptive force controller

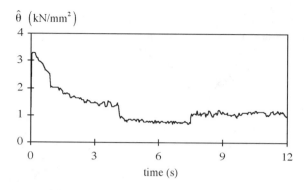

**Figure 1.15.** Force model gain estimate – adaptive force controller

## 1.6 Supervisory Control

The supervisory level is the area that has received the least research attention to date, and can be expected to be a major focus of future research [1.48][1.103]–[1.106]. There are in fact many levels of control in a machining process control hierarchy, and what we refer to here as supervisory control includes functions such as selection of control strategies [1.46], sensor fusion [1.58], generation of reference commands for the process control level [1.107]–[1.109], tool breakage monitoring [1.110], chatter detection [1.33][1.111], and machine monitoring [1.30][1.31][1.112]. Supervisory-level strategies are aimed at compensating for factors not explicitly considered in the design of the servo- and process-level controllers. For example, knowledge of the part geometry as contained in the computer-aided design system can be used to determine the reference values of process variables. Furthermore, the supervisory level can monitor for tool and/or machine failures or degradation. Process phenomena, such as chatter and chip entanglement, can also be monitored. Information from various process-related sensors can be integrated to improve the reliability and quality of sensor information. Finally, all of this information can be used to achieve online optimization of the machining process.

As an example of supervisory control in machining, consider a through-hole drilling operation (see Figure 1.16). A successful control strategy must ensure good hole quality, prevent drill breakage, minimize burr formation at the exit, and also maintain high metal removal rates. A strategy which achieves these goals, based upon process optimization methods, is described in [1.48]. The control system includes a servo control level where the spindle speed and the drill feed rate are accurately controlled. A process control level implements a torque controller to prevent drill breakage.

The supervisory controller, based upon the drill position (depth-of-hole), selects an appropriate strategy for use at the process control level as well as the appropriate reference signals for that strategy. Specifically, the supervisory level selects a feed and speed control strategy at the entrance phase based upon hole quality (hole location error) considerations. The feed and speed references are selected based upon the required hole location error constraint. After the hole is initiated, the supervisory controller switches to a speed and torque control strategy with the reference values determined from tool breakage and tool wear considerations. Finally, as the operation nears drill breakthrough, the supervisory controller switches back to a feed and speed control strategy to minimize burr formation. Thus, the supervisory controller selects control strategies and their associated reference levels for optimization of the drilling process.

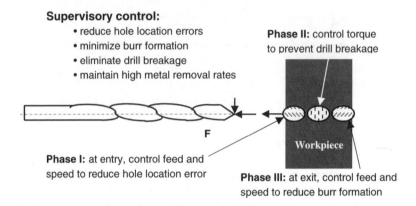

**Figure 1.16.** Supervisory control of a through-hole drilling operation

The results of using such a supervisory control strategy are illustrated in Table 1.1. This table compares experimental results obtained using four types of controllers: (1) no control (conventional approach where nominal feed and speed values are selected but not controlled during drilling), (2) feedback control of feed and speed, (3) feedback control of torque and speed, and (4) the supervisory controller (which combines (2) and (3) as described above). The comparison is made in terms of: (1) machining time for one hole, (2) burr rating, (3) hole location error (in terms of the pooled standard deviation of the hole location error), and (4) percentage of holes drilled with stoppage events. The machining time per hole is given in seconds, the burr rating ranges from 1 = very little burr formation to 5 = large burrs, the hole quality is given in terms of a pooled standard deviation with smaller values indicating better hole location accuracy. The percentage of stoppage events is for holes for which the torque exceeded a maximum allowable value for drill breakage and the process was stopped. In this application a hole location error pooled standard deviation of $< 4.5 \times 10^{-3}$ in, and a burr rating of $< 1.75$ are required. It is also desired that there be no stoppage events, and that the machining time be minimized subject to these hole quality, burr, and breakage constraints.

The supervisory control strategy is the only one which meets the required hole quality and burr constraints, while eliminating stoppage events. It also yields a machining time very comparable to the uncontrolled or feed/speed control cases, and only slightly longer than the torque/speed controller. The results are average values based upon a statistical study involving the drilling of 20 holes with each strategy in a randomized order [1.48][1.49][1.112]. These results clearly illustrate the potential advantages of a supervisory control strategy over each of the individual process control strategies (*i.e.* feed, speed, or torque control), and additional experimental results are presented in [1.48][1.49][1.112].

**Table 1.1.** Comparison of drilling control strategies

|  | No Controller | Feed/Speed Controller | Torque/Speed Controller | Supervisory Controller |
|---|---|---|---|---|
| **Machining Time (s)** | 11.11 | 11.28 | 9.79 | 11.71 |
| **Burr Rating** | 2.93 | 2.94 | 2.26 | 1.58 |
| **Hole Location Quality (in)** | 4.43E$^{-3}$ | 4.53 E$^{-3}$ | 6.28 E$^{-3}$ | 4.25 E$^{-3}$ |
| **Event Stoppages (%)** | 25 | 15 | 0 | 0 |

## 1.7 Concluding Remarks

This chapter has reviewed research developments of the past several decades in the monitoring and control of machining processes. The research in this area world-wide has been extensive, and this chapter has highlighted major developments, and included examples from the research work of the author and his co-workers. An overview of machining, as relevant to monitoring and control, was presented and introduced important aspects of metal removal processes, including force, torque, power, tool wear and breakage, vibrations and chatter, chip and burr formation, temperature generation, *etc*. Important issues in cutting process monitoring, including tool wear, tool breakage and chatter, were described. Examples included the chatter stability lobes for a reconfigurable machine tool, and a hybrid tool wear estimation method that combines direct (*i. e.* vision) and indirect (*i. e.* force) sensing techniques.

Sensing and control technologies have played an increasingly important role in machining since the introduction of CNC systems in the 1960s. In reviewing these developments, the servo, process and supervisory control levels have been considered. At the servo control level, advances in both point-to-point and contouring control cases have been described: introduction of feedforward and preview techniques, as well as the example of the cross-coupling controller which minimizes contour errors rather than individual axis errors. At the process control level, where process variables such as force or torque are controlled, adaptive and model-based control techniques have been employed to address the problem of process parameter variations and uncertainty. Examples of adaptive and nonlinear force controllers for milling were presented. At the supervisory control level, additional control and monitoring capabilities (*e. g.* sensor fusion, chatter detection, tool and machine condition monitoring, online process optimization) have been incorporated. A supervisory controller for through-hole drilling (which simultaneously ensures good hole quality, eliminates breakage, reduces burr formation, and maximizes throughput) was presented.

To date, the impact of these research developments on industrial practice, most of them demonstrated in the laboratory, has been modest. This is in part due to the unfavorable experience with the first generation of AC systems and their components introduced in the 1970s, and in part because these systems are still difficult to apply without considerable knowledge and experience on the part of the operator. The keys to enabling widespread industrial application of sensing and control technologies in machining in the future are the development of robust in-process sensing methods and open-architecture systems which facilitate the integration of new technologies into commercial products in a cost-effective manner. Such systems hold the promise of achieving in-process quality assurance, system flexibility, and high production rates in a cost-effective manner. Global competition will force the machine tool industry, which in the USA has seen a serious economic downturn since the 1980s, to develop new state-of-the-art products. These new products will, in general, be machines whose accuracy and productivity is enhanced through the use of sensing and control technology.

Future research on the monitoring and control of machining processes will continue to benefit from developments in control theory and signal processing, such as robust control, networked control systems, discrete and hybrid controls, wavelets, artificial intelligence, expert systems, fuzzy control, neural networks and many others. Fast optical sensing methods will allow us to close the loop on part dimensions, surface finish and wear; not only process variables like force or tool holder vibrations. These developments in sensing will be the necessary ingredient for in-process quality control. There will also be developments in computer software and hardware towards modularity and an open system architecture. Software modules for rapid modeling and compensation for various error sources (*e. g.* friction, backlash, thermal deformations, geometric errors) will need to be developed. The trend in machining is toward higher speeds, and this trend will require appropriate developments in the control strategies associated with machining processes (*e. g.* cutting will occur during transients in contouring operations). One can expect to see the fruits of current efforts towards standardization in hardware, software and interfacing. Such technological developments will be extremely important for the future of the machine tool industry.

Advances in the modeling of cutting mechanics will be required; in particular, analytical models based on first principles, which may be applied to a wide variety of cutting conditions, are needed. Currently, models are determined empirically and typically contain nonlinear terms, which account for unmodeled effects. Further, the cost-effective design of process monitoring and control technology will require simulation tools, which simulate not only the cutting mechanics and the monitoring and control modules, but also the machine tool structure and servo mechanisms [1.113]. Such a comprehensive simulator will allow the designer to investigate process monitoring and control technology in a realistic environment. Other issues in process monitoring include the use of increasingly sophisticated sensors and the placement of these sensors in harsh machining environments. Advances in sensor technology to integrate the sensors with the machine tool or cutting tool and research into using CNC-integral sensors (*e. g.* drive current) will address these issues. As process monitoring becomes more reliable, sensors become available, and system integration in an open environment feasible, process control will become more prevalent.

Much of the research in monitoring and control of machining processes to date has been concerned with regulating a single process via a single process variable. Future research will be concerned with utilizing multiple process variables to control single and multiple processes. During the ramp-up phase of a machining system, process controllers will provide an effective means of determining near-optimal process variables for complex operations. The part program can be modified to incorporate the new process variable time histories and then process controllers may be utilized in the production phase to reject disturbances.

## Acknowledgment

The author gratefully acknowledges the numerous co-workers who contributed to the research work summarized here, and thanks the University of Michigan's Engineering Research Center for Reconfigurable Manufacturing Systems (National Science Foundation Grant EEC95-92125), and its industrial partners, for their financial support.

## References

[1.1]   Zorev, N. N., 1966, *Metal Cutting Mechanics*, Pergamon Press, Oxford, England.
[1.2]   DeVries, W. R., 1992, *Analysis of Material Removal Processes*, Springer-Verlag, NY.
[1.3]   Koren, Y., 1983, *Computer Control of Manufacturing Systems*, McGraw-Hill, NY.
[1.4]   Koren, Y., 1997, "Control of Machine Tools," *ASME J. of Manufacturing Science and Engineering*, Vol. 119, pp. 749–755.
[1.5]   Ulsoy, A. G. and W. R. DeVries, 1989, *Microcomputer Applications in Manufacturing*, John Wiley and Sons, NY.
[1.6]   Ulsoy, A. G. and Y. Koren, 1993, "Control of Machining Processes," *ASME J. of Dynamic Systems, Measurement and Control*, Vol. 115, No. 2(B), June, pp. 301–308.
[1.7]   Danai, K., 2002, "Machine Tool Monitoring and Control," in the *Mechanical Systems Design Handbook*, O. Nwokah and Y. Hurmuzlu (eds.), CRC Press, Boca Rotan, FL, pp. 75–84.
[1.8]   Landers, R. G., Ulsoy, A. G. and R. Furness, 2002, "Monitoring and Control of Machining Operations," in the *Mechanical Systems Design Handbook*, O. Nwokah and Y. Hurmuzlu (eds.), CRC Press, Boca Raton, FL, pp. 85–104.
[1.9]   Taguchi, G., 1989, *Quality Engineering in Production Systems*, McGraw-Hill, NY.
[1.10]  Shewhart, W. A., 1986, *Statistical Method from the Viewpoint of Quality Control*, Dover, New York.
[1.11]  Ulsoy, A. G., Koren, Y. and F. Rasmussen, 1983, "Principal developments in the adaptive control of machine tools," *ASME J. of Dynamic Systems, Measurement, and Control*, Vol. 105, No. 2, pp. 107–112.
[1.12]  Åström, K. J. and B. Wittenmark, 1995, *Adaptive Control*, Addison-Wesley, NY.
[1.13]  Centner, R., 1964, Final report on the development of an adaptive control technique for a numerically controlled milling machine, *USAF Technical Documentary Report ML-TDR-64-279*, August.
[1.14]  Huber, J. and R. Centner, 1968, "Test results with an adaptively controlled milling machine," *ASTME Paper No. MS68–638*.
[1.15]  Tlusty, J. and G. C. Andrews, 1983, "A Critical Review of Sensors for Unmanned Machining," *CIRP Annals*, Vol. 32, No. 2, pp. 563–572.
[1.16]  Yen, D. W., Wright, P. K., 1983, "Adaptive control in machining – a new approach based on the physical constraints of tool wear mechanisms," *ASME J. of Engineering for Industry*, Vol. 105, February, pp. 31–8.

[1.17]    Koren, Y., Ko, T. R., Ulsoy, A. G., Danai, K., 1991, "Flank wear estimation under varying cutting conditions," *ASME J. of Dynamic Systems, Measurement, and Control*, Vol. 113, No. 2, pp. 300–307.

[1.18]    Koren, Y. and A. G. Ulsoy, 1989, "Adaptive Control in Machining," in the *Metals Handbook: Machining*, J. R. Davis (ed.), Volume 16, 9th Edition, ASM Int., Metals Park, OH, pp. 618–626.

[1.19]    Ulsoy, A. G. and Y. Koren, 1989, "Applications of Adaptive Control Theory to Machine Tool Process Control," *IEEE Control Systems Magazine*, Vol. 9, No. 4, pp. 33–37.

[1.20]    Altintas, Y. and J. Peng, 1990, "Design and analysis of a modular CNC system," *Computers in Industry*, Vol. 13, pp. 305–16.

[1.21]    Anonymous, 1980, *Proceedings of the Machine Tool Task Force Conference*, Chicago.

[1.22]    Ashley, S, 1990, "A Mosaic for Machine Tools," *Mechanical Engineering*, September, pp. 38–43.

[1.23]    National Center for Manufacturing Sciences (NCMS), 1990, *Next Generation Workstation/Machine Controller (NGC): Requirements Definition Document (RDD)*, Prepared for Air Force Systems Command, August.

[1.24]    Pritschow, G., Daniei, C. H., Jurghans, G. and W. Sperling, 1993, "Open systems controllers - a challenge for the future of the machine tool industry," *CIRP Annals*, 42, 449.

[1.25]    Koren, Z. J. Pasek, A. G. Ulsoy, and U. Benchetrit, 1996, "Real-Time Open Control Architectures and System Performance," *CIRP Annals*, Vol. 45, No. 1, pp. 377–380.

[1.26]    Park, J., Z. J. Pasek, Y. Shan, Y. Koren, K. G. Shin, and A. G. Ulsoy, 1996, "An Open-Architecture Real-Time Controller for Machining Processes," *Manufacturing Systems*, Vol. 25, No. 1, pp. 23–27.

[1.27]    Schofield, S. and P. Wright, 1998, "Open Architecture Controllers for Machine Tools, Part 1: Design Principles," *ASME J. Manufacturing Science and Engineering*, Vol. 120, No. 2, May, pp. 417–424.

[1.28]    Tlusty, J., 1986, "Dynamics of High Speed Machining," *ASME J. of Engineering for Industry*, Vol. 108, pp. 59–67.

[1.29]    Koren, Y. and O. Masory, 1981, Adaptive control with process estimation, *CIRP Annals*, 30, 373.

[1.30]    Stein, J. L. and K. C. Shin, 1986, "Current monitoring of field controlled DC spindle drives," *ASME J. of Dynamic Systems, Measurement and Control*, Vol. 108, pp. 289–95, December.

[1.31]    Stein, J. L. and C.-H. Wang, 1990 "Analysis of power monitoring on AC induction drive systems," *ASME J. of Dynamic Systems, Measurement and Control*, Vol. 112, pp. 239–48, June.

[1.32]    Altintas, Y., 1992, "Prediction of cutting forces and tool breakage in milling from feed drive current measurements," *ASME J. of Engineering for Industry*, Vol. 114, pp. 386–392.

[1.33]    Hanna, N. H. and S. A. Tobias, 1974, "A Theory of Nonlinear Regenerative Chatter," *ASME J. of Engineering for Industry*, pp. 247–254.

[1.34]    Tlusty, J., 1985, "Machine Dynamics." *Handbook of High-speed Machining Technology*, R. I. King, ed., Chapman and Hall, New York.

[1.35]    Chen, S. G., Ulsoy, A. G. and Y. Koren, 1997, "Computational Stability Analysis of Chatter in Turning," *ASME J. of Manufacturing Science and Engineering*, Vol. 119, No. 4(A), Nov., pp. 457–460.

[1.36]    Ulsoy, A. G., 1998, "Dynamic Modeling and Control of Machining Processes," Chapter 2 in *Nonlinear Dynamics of Material Processing and Manufacturing*, F. C. Moon (ed.), Wiley, NY, pp. 33–55.

[1.37]   Li, C.-J., A. G. Ulsoy, and W. J. Endres, 2003, "Effect of Flexible-Tool Rotation on Regenerative Instability in Machining" *ASME J. of Manufacturing Science and Engineering,* Vol. 125, No. 1, Feb., pp. 39–47.

[1.38]   Dhupia, J., Powalka, B., Katz, R. and A. G. Ulsoy, 2005, Dynamics of the Arch-Type Reconfigurable Machine Tool, *Proc. CIRP Reconfigurable Manufacturing Conference,* May, Ann Arbor, MI.

[1.39]   Koren, Y., Heisel, U., Jovane, F., Moriwaki, T., Pritchow, G., Ulsoy, A. G. and H. VanBrussel, 1999, "Reconfigurable Manufacturing Systems," *CIRP Annals,* Vol. 48, No. 2, pp. 527–540.

[1.40]   Landers, R. G. and A. G. Ulsoy, 1996, "Chatter analysis of machining systems with nonlinear force processes," *ASME International Mechanical Engineering Congress and Exposition,* DSC 58, ASME, NY, 183.

[1.41]   Radulescu, R., Kapoor, S. G. and R. E. DeVor, 1997, An investigation of variable spindle speed face milling for tool-work structures with complex dynamics, part 2: physical explanation, *ASME J. of Manufacturing Science and Engineering,* Vol. 119, pp. 273.

[1.42]   Pakdemirli, M. and A. G. Ulsoy, 1997, "Perturbation Analysis of Spindle Speed Variation in Machine Tool Chatter," *J. of Vibration and Control,* Vol. 3, No. 3, Aug., pp. 261–278.

[1.43]   Taylor, F. W., 1907, On the art of cutting tools, *Trans. ASME,* 28.

[1.44]   Gillespie, L. K., 1976, *Deburring Capabilities and Limitations,* Society of Manufacturing Engineers, Dearborn, MI.

[1.45]   Chu, T. H. and Wallbank, J., 1998, Determination of the temperature of a machined surface, *ASME J. of Manuf. Science and Engineering,* Vol. 120, pp. 259.

[1.46]   Stephenson, D. A., 1991, Assessment of steady-state metal cutting temperature models based on simultaneous infrared and thermocouple data, *ASME J. of Engineering for Industry,* Vol. 113, pp. 121.

[1.47]   D'Errico, G. E., Calzavarini, R. and L. Settineri, 1994, "Experiments on self tuning regulation of cutting temperature in turning process," *Proc. of the IEEE Conference on Control Applications,* 2, IEEE, Piscataway, 1165.

[1.48]   Furness, R., Ulsoy, A. G. and C. L. Wu, 1996, "Supervisory Drilling Control," *ASME J. of Engineering for Industry,* Vol. 118, No. 1, Feb., pp. 10–19.

[1.49]   Furness, R., Wu, C. L. and A. G. Ulsoy, 1996, "Statistical Analysis of the Effects of Feed, Speed, and Wear on Hole Quality in Drilling," *ASME J. of Manufacturing Science and Engineering,* Vol. 118, No. 3, Aug., pp. 367–375.

[1.50]   Byrne, G., Dornfeld, D., Inasaki, I., Ketteler, G., Konig, W. and R. Teti, 1995, Tool condition monitoring (TCM) – the status of research and industrial application, *CIRP Annals,* Vol. 44, pp. 541.

[1.51]   Du, R., Elbestawi, M. A. and S. M. Wu, 1995, Automated monitoring of manufacturing processes, part 1: monitoring methods, *ASME J. of Engineering for Industry,* Vol. 117, pp. 121.

[1.52]   Danai, K. and A. G. Ulsoy, 1987, "An Adaptive Observer for On-Line Tool Wear Estimation in Turning – Part I: Theory," *Mechanical Systems and Signal Processing,* Vol. 1, No. 2, Apr., pp. 211–225.

[1.53]   Danai, K. and A. G. Ulsoy, 1987, "An Adaptive Observer for On-Line Tool Wear Estimation in Turning - Part II: Results," *Mechanical Systems and Signal Processing,* Vol. 1, No. 2, Apr., pp. 227–240.

[1.54]   Koren, Y., Ulsoy, A. G. and K. Danai, 1986, "Tool Wear and Breakage Detection Using a Process Model," *CIRP Annals,* Vol. 35, No. 1, Aug., pp. 283–288.

[1.55]   Altintas, Y., Yellowley, I. and J. Tlusty, 1988, "The Detection of Tool Breakage in Milling Operations," *ASME J. of Eng. for Industry,* Vol. 110, No. 3, pp. 271–277.

[1.56]   Sata, T., Matsushima, K., Nagakura, T. and E. Kono, 1973, "Learning and

Recognition of the Cutting States by the Spectrum Analysis," *CIRP Annals*, Vol. 22, pp. 41–42.

[1.57]   Kannatey-Asibu, E. and E. Emel, 1987, "Linear Discriminant Function Analysis of Acoustic Emission Signals for Cutting Tool Monitoring," *Mechanical Systems and Signal Processing*, Vol. 4, pp. 333–347.

[1.58]   Rangwala, S. and D. A. Dornfeld, 1990, "Sensor integration using neural networks for intelligent tool condition monitoring," *ASME J. of Engineering for Industry*, Vol. 112, pp. 219–28, August.

[1.59]   Danai, K. and A. G. Ulsoy, 1986, "A Model Based Approach for Tool Wear Estimation in Turning," *SME Manufacturing Technology Review*, Vol. 1, pp. 49–54.

[1.60]   Danai, K. and A. G. Ulsoy, 1987, "A Dynamic State Model for On-Line Tool Wear Estimation in Turning," *ASME J. of Engineering for Industry*, Vol. 109, No. 4, Nov., pp. 396–399.

[1.61]   Park, J. J. and A. G. Ulsoy, 1992, "On-Line Tool Wear Estimation Using Force Measurement and a Nonlinear Observer," *ASME J. of Dynamic Systems, Measurement, and Control*, Vol. 114, No. 4, Dec., pp. 666–672.

[1.62]   Park, J. J. and A. G. Ulsoy, 1993, "On-Line Flank Wear Estimation Using an Adaptive Observer and Computer Vision, Part I: Theory," *ASME J. of Engineering for Industry*, Vol. 115, No. 1, Feb., pp. 30–36.

[1.63]   Park, J. J. and A. G. Ulsoy, 1993, "On-Line Flank Wear Estimation Using an Adaptive Observer and Computer Vision, Part II: Experiment," *ASME J. of Engineering for Industry*, Vol. 115, No. 1, Feb., pp. 37–43.

[1.64]   Cook, N. H. and Subramanian, K., 1978, "Micro-Isotope Tool Wear Sensor," *CIRP Annals*, Vol. 27, No. 1, pp. 73–78.

[1.65]   Kannatey-Asibu Jr., E. and D. A. Dornfeld, 1982, "A Study of Tool Wear in Metal Cutting using Statistical Analysis of Acoustic Emission," *Wear*, Vol. 76, No. 2, pp. 247–261.

[1.66]   Cook, N. H., 1980, "Tool Wear Sensors," *Wear*, Vol. 62, pp. 49–57.

[1.67]   Dan, L. and J. Mathew, 1990, "Tool wear and failure monitoring techniques for turning – a review," *Int. J. of Machine Tools and Manufacture*, Vol. 30, pp. 579.

[1.68]   De Filippi, A. and R. Ippolito, 1969, "Adaptive Control in Turning: Cutting Forces and Tool Wear Relationships for P10, P20, P30 Carbides," *CIRP Annals*, Vol. 17, pp. 377–379.

[1.69]   Houshmand, A. A. and E. Kannatey-Asibu, 1989, "Statistical Process Control of Acoustic Emission for Cutting Tool Monitoring," *Mechanical Systems and Signal Processing*, Vol. 3, No. 4, pp. 405–424.

[1.70]   Koren, Y., 1978, "Flank wear model of cutting tools using control theory," *ASME J. of Engineering for Industry*, Vol. 100, No. 1, February.

[1.71]   Chen, J. S., Yuan, J. X., Ni, J. and S. M. Wu, 1991, "Real-time compensation for time-variant volumetric error on a machining center," *ASME Winter Annual Meeting*, Atlanta, Georgia, December.

[1.72]   Lo, C. C. and Y. Koren, 1992, "Evaluation of servo-controllers for machine tools," *Proceedings of the American Control Conference*, Chicago, IL, pp. 370–374, June.

[1.73]   Koren, Y. and C. C. Lo, 1991, "Variable-gain cross-coupling controller for contouring," *CIRP Annals*, Vol. 40, No. 1, pp. 371–374.

[1.74]   Tomizuka, M., 1993, "On the design of digital tracking controllers," *ASME J. of Dynamic Systems, Measurement and Control*, June.

[1.75]   Koren, Y., 1980, "Cross-Coupled Computer Control for Manufacturing Systems," *ASME J. of Dynamic Systems, Measurement, and Control*, Vol. 102, No. 4, pp. 265–272.

[1.76]   Koren Y., Lo, C. C., 1992, "Advanced controllers for feed drives," *CIRP Annals*, Vol. 2.

[1.77]   Tomizuka, M., Dornfeld, D. A. and X. Q. Bian, 1984, "Experimental evaluation of the preview servo scheme for a two-axis positioning system," *ASME J. of Dynamic Systems, Measurement and Control*, Vol. 106, March, pp. 1–5.

[1.78]   Tomizuka, M., 1987, "Zero phase error tracking algorithm for digital control," *ASME J. of Dynamic Systems, Measurement and Control*, Vol. 109, pp. 65–68.

[1.79]   Tsao, T. C. and M. Tomizuka, M., 1987, "Adaptive Zero Phase Error Tracking Algorithm for Digital Control," *ASME J. Dynamic Systems, Measurement, and Control*, Vol. 109, pp. 349–354.

[1.80]   Tsao, T. C., Chen, Y. Y. and M. Tomizuka, 1990, "Noncircular turning of workpieces with sharp corners," *ASME J. of Engineering for Industry*, Vol. 112, May, pp. 181–3.

[1.81]   Masory, O. and Y. Koren, 1980, "Adaptive control system for turning," *CIRP Annals*, Vol. 1.

[1.82]   Lauderbaugh, L. K. and A. G. Ulsoy, 1988, Dynamic Modeling for Control of the Milling Process," *ASME J. of Engineering for Industry*, Vol. 110, No. 4, Nov., pp. 367–375.

[1.83]   Tomizuka, M., Zhang, S., 1988, "Modeling and conventional/adaptive PI control of a lathe cutting process," *ASME J. of Dynamic Systems, Measurement and Control*, Vol. 110, December, pp. 350–354.

[1.84]   Daneshmend, L. K. and H. A. Pak, 1986, "Model reference adaptive control of feed force in turning," *ASME J. of Dynamic Systems, Measurement and Control*, Vol. 108, pp. 215–22, September.

[1.85]   Watanabe, T., 1986, "A model-based approach to adaptive control optimization in milling," *ASME J. of Dynamic Systems, Measurement and Control*, Vol. 108, pp. 56–64.

[1.86]   Amitay, G., Malkin, S. and Y. Koren, 1981, "Adaptive Control Optimization of Grinding," *ASME J. of Engineering for Industry*, Vol. 103, No. 1, pp. 102–111.

[1.87]   Malkin, S. and Y. Koren, 1984, "Optimal infeed control for accelerated spark-out in plunge grinding," *ASME J. of Engineering for Industry*, Vol. 106, pp. 70–4, February.

[1.88]   Hahn, R. S., 1986, "Improving performance with force-adaptive grinding," *Manufacturing Engineering*, Vol. 97, October, pp. 73.

[1.89]   Koren, Y., 1989, "The Optimal Locus Approach with Machining Applications," *ASME J. Dynamic Systems, Measurement and Control*, Vol. 111, pp. 260–267.

[1.90]   Chow, J G., Wright, P. K., 1988, "On-line estimation of tool/chip interface temperatures for a turning operation," *ASME J. of Engineering for Industry*, Vol. 110, pp. 56–64, February.

[1.91]   Kannatey-Asibu, Jr. E. and D. A. Dornfeld, 1981, "Quantitative relationships for acoustic emission from orthogonal metal cutting," *ASME J. of Engineering for Industry*, Vol. 103, No. 3, pp. 330–340.

[1.92]   Liang, S. Y. and D. A. Dornfeld, 1989, "Tool wear detection using time series analysis of acoustic emission," *ASME J. of Engineering for Industry*, Vol. 111, pp. 199–205, August.

[1.93]   Watanabe, T. and S. Iwai, 1983, "A control system to improve the accuracy of finished surfaces in milling," *ASME J. of Dynamic Systems, Measurement and Control*, Vol. 105, pp. 192–199.

[1.94]   Wu, C. L., Haboush, R. K., Lymburner, D. R. and G. H. Smith, 1986, "Closed-loop machining control for cylindrical turning," *Modeling, Sensing and Control of Manufacturing Systems*, ASME, New York, November, pp. 189–204.

[1.95]   Masory, O. and Y. Koren, 1983, "Variable-gain adaptive control systems for machine tools," *Journal of Manufacturing Systems*, Vol. 2, No. 2, pp. 165–174.

[1.96]   Lauderbaugh, L. K. and A. G. Ulsoy, 1989, "Model Reference Adaptive Force Control in Milling," *ASME J. of Engineering for Industry*, Vol. 111, No. 1, pp. 13–21.

[1.97]   Turkay, O. and A. G. Ulsoy, 1988, "Frequency versus time domain parameter estimation: application to a slot milling operation," *Mechanical Systems and Signal Processing*, Vol. 2, No. 3, pp. 265–277.

[1.98]   Landers, R. G. and Ulsoy A. G., 1996, "Machining force control including static, nonlinear effects," *Japan-USA Symp. on Flexible Automation*, 2, ASME, NY, 983.

[1.99]   Landers R. G. and A. G. Ulsoy, 2000, "Model-Based Machining Force Control," *ASME J. Dynamic Systems, Measurement and Control*, Vol. 122, No. 3, Sept., pp. 521–527.

[1.100]  Kim, S. I., R. A. Landers, and A. G. Ulsoy, 2003, "Robust Machining Force Control with Process Compensation," *ASME J. of Manufacturing Science and Engineering*, Vol. 125, No. 3, Aug., pp. 423–430.

[1.101]  Landers, R. G., Ulsoy, A. G. and Y. H. Ma, 2004, "A Comparison of Model-Based Machining Force Control Approaches," *J. of Engineering Manufacture*, Vol. 44, No. 7-8, June, pp. 733–748.

[1.102]  Rober, S. J., Shin, Y. C. and O. D. I. Nwokah, 1997, A digital robust controller for cutting force control in the end milling process, *ASME J. of Dynamic Systems, Measurement, and Control*, Vol. 119, pp. 146.

[1.103]  Ivester, R. W. and K. Danai, 1996, "Intelligent Control of Machining under Modeling Uncertainty," *CIRP Manufacturing Systems*, Vol. 25, No. 1, pp. 73–79.

[1.104]  Ivester, R., Danai, K. and S. Malkin, 1997, "Cycle Time Reduction in Machining by Recursive Constraint Bounding," *ASME J. of Manufacturing Science and Eng.*, Vol. 119, No. 2, pp. 201–207.

[1.105]  Landers, R. G. and A. G. Ulsoy, 1998, "Supervisory Machining Control: Design Approach and Experiments," *CIRP Annals*, Vol. 47, No. 1, pp. 301–306.

[1.106]  Landers, R. G. and A. G. Ulsoy, 2001, "Supervisory Control of a Face Milling Operation in Different Manufacturing Environments," *Trans. on Control, Automation and Systems Engineering*, Vol. 3, No. 1. March, pp. 1–9.

[1.107]  Spence, A., Altintas, Y. and D. Kirkpatrick, 1990, "Direct calculation of machining parameters from a solid model," *Computers in Industry*, Vol. 14, pp. 271–80, July.

[1.108]  Chryssolouris, G., Guillot, M., 1990, "A comparison of statistical and AI approaches to the selection of process parameters in intelligent machining," *ASME J. of Engineering for Industry*, Vol. 112, May, pp. 122–131.

[1.109]  Thangaraj, A., Wright, P K., 1988, "Computer-assisted prediction of drill-failure using in-process measurements of thrust force," *ASME J. of Engineering for Industry*, Vol. 110, May, pp. 192–200.

[1.110]  Tlusty, J., Smith, S. and Zamudio, C., 1990, "New NC Routines for Quality in Milling," *CIRP Annals*, Vol. 39, pp. 517–521.

[1.111]  Isermann, R., 1984, "Process Fault Detection Based on Modeling and Estimation Methods - A Survey," *Automatica*, Vol. 20, No. 4, pp. 387–404.

[1.112]  Furness, R., Ulsoy, A. G. and C. L. Wu, 1996, "Feed, Speed, and Torque Controllers for Drilling," *ASME J. of Engineering for Industry*, Vol. 118, No. 1, Feb., pp. 2–9.

[1.113]  Chen, S. G., Ulsoy, A. G. and Y. Koren, 1998, "Machining Error Source Diagnostics Using a Turning Process Simulator," *ASME J. of Manufacturing Science and Engineering*, Vol. 120, No. 2, May, pp. 409–416.

# Precision Manufacturing Process Monitoring with Acoustic Emission

D.E. Lee[1], Inkil Hwang[1], C.M.O. Valente[2], J.F.G. Oliveira[2] and David A. Dornfeld[1]

[1] Laboratory for Manufacturing Automation, University of California
Berkeley, CA 94720-1740, USA
Email: dornfeld@me.berkeley.edu

[2] University of Sao Paulo, Nucleus of Advanced Manufacturing
Sao Carlos, Brazil
Email: cmov@sc.usp.br

**Abstract**
Demands in high-technology industries such as semiconductor, optics, MEMS, etc., have predicated the need for manufacturing processes that can fabricate increasingly smaller features reliably at very high tolerances. In-situ monitoring systems that can be used to characterize, control, and improve the fabrication of these smaller features are therefore needed to meet increasing demands in precision and quality. This paper discusses the unique requirements of monitoring of precision manufacturing processes, and the suitability of acoustic emission (AE) as a monitoring technique at the precision scale. Details are then given on the use of AE sensor technology in the monitoring of precision manufacturing processes; grinding, chemical mechanical planarization (CMP) and ultraprecision diamond turning in particular.

## 2.1 Introduction

Current demands in high-technology industries such as semiconductor, optics, MEMS, *etc.*, have predicated the need for manufacturing processes that can fabricate increasingly smaller features reliably at very high tolerances. This increasing demand for the ability to fabricate features at smaller lengthscales and at greater precision can be represented in the Taniguchi curve (see Figure 2.1), which demonstrates that the smallest achievable accuracy (and, as a consequence, smallest reproducible feature) decreases as a function of time [2.1].

In-situ monitoring systems that can be used to characterize, control, and improve the fabrication of these smaller features are therefore needed to meet increasing demands in precision and quality. Sensor-based monitoring yields valuable information about the manufacturing process that can serve the dual purpose of process control and quality monitoring, and will ultimately be the part of any fully-automated manufacturing environment. However, a high degree of confidence and reliability in characterizing the manufacturing process is required for any sensor to be utilized as a monitoring tool.

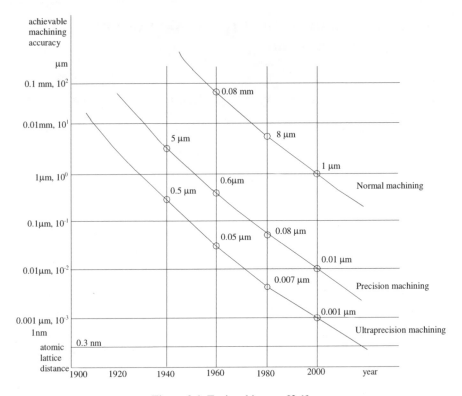

**Figure 2.1.** Taniguchi curve [2.1]

An example of the need for process monitoring in an automated manufacturing environment is described below. Figure 2.2 shows a polycrystalline oxygen-free high conductivity (OFHC) copper workpiece machined to a surface finish of about 50 nm $R_a$ with a single point diamond ultraprecision turning process. While the workpiece appears to have a good optical-quality finish to the naked eye, upon closer inspection of the surface with a contact-mode atomic-force microscopy (AFM) scan, variations in the surface topography can be observed. Step heights of roughly 10 nm in height can be seen, as well as variation in the $R_a$ value on the order of several tenths of a nanometer from region to region. It was later revealed through chemical etching of the workpiece surface that the surface finish and topography was directly influenced by each crystalline grain and their individual orientation. Hence, for ultraprecision scale machining of polycrystalline materials, the workpiece material can no longer be treated as a homogenous continuum as in conventional machining theory. The polycrystalline material at microscale is a "composite" of sorts, with each grain having a varying degree of machinability and resulting surface and edge topography after machining. The variations in the surface shown in Figure 2.2 may not be acceptable for meeting the requirements of ultraprecision optics, hence predicating the need for robust monitoring systems that can detect such process-induced defects in an in-situ fashion.

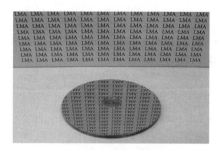

~50 nm $R_a$ surface finish

(a) Actual workpiece

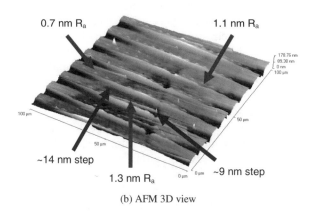

(b) AFM 3D view

**Figure 2.2.** (a) Ultraprecision machined polycrystalline copper workpiece, (b) Contact AFM scan of machined workpiece

As demonstrated in a previous review paper [2.2], acoustic emission (AE) has demonstrated a high degree of confidence in characterizing various phenomena related to material removal, particularly at the microscale, hence lending credence to its suitability for precision manufacturing process monitoring. This chapter serves to demonstrate sensitivity of AE at the three different manufacturing regimes outlined in the Taniguchi curve; the normal/conventional, precision and ultraprecision scales.

## 2.2 Requirements for Sensor Technology at the Precision Scale

In material removal processes at the precision scale, the undeformed chip thickness can be on the order of a few microns or less, and can approach the nanoscale in some cases. At these length scales, the surface, subsurface, and edge condition of machined features and the fundamental mechanism for chip formation are much more intimately affected by the material properties and microstructure of the workpiece material, such as ductile/brittle behavior, crystallographic orientation of the material at the tool/chip interface, and microtopographical features such as voids, secondary phases, and interstitial particulates [2.3][2.4]. Characterizing the surface, subsurface, and edge condition of machined features at the precision scale, as well as tracking relevant process parameters such as material removal rate (MRR), tool contact/cycle time, and state of tool condition (level of pad wear in chemical mechanical planarization, wheel loading in grinding, tool wear in machining, for instance) are of increasing importance for monitoring, evaluating, and controlling the manufacturing process.

A great deal of work has been devoted to evaluating the applicability and need of sensors for a variety of manufacturing processes, with previous reviews by Dornfeld *et al.* [2.2][2.5] focusing on recent sensing techniques with respect to future manufacturing requirements. Likewise, an early review by Tlusty *et al.* examined the role of cutting force and AE as a means for condition monitoring and adaptive control [2.6], and Tonshoff *et al.* proposed techniques for condition monitoring strategies such as intelligent machining and multi-sensor systems [2.7]. One of the most comprehensive reviews of overall sensor technology and data processing techniques for automated machining was conducted by Byrne *et al.* [2.8]. Likewise, Iwata [2.9] surveyed the requirements of the Japanese machine tool industry for sensors and process monitoring. However, a key note in the above papers was their focus on monitoring at the conventional scale, and not necessarily at the precision or ultraprecision scales.

A variety of sensors (each having a degree of applicability depending on the level of precision required or type of phenomena or control parameter that needs to be measured) are used to capture necessary information about the manufacturing process. Figure 2.3 diagrams several different classes of sensors and their applicability to both level of precision (starting at the conventional scale and moving down the ordinate to ultraprecision material removal levels) and type of control parameter. It is important to note however that the boundaries for each sensor type are not necessarily "hard" boundaries, and that continually improving sensor technology will allow sensors to cover a wider range of control parameters with increasing sensitivity (as of this writing, the recent introduction of microcutting load cells is a good example). Conventional measuring techniques such as load cells are suitable for measuring conventional-scale control parameters such as cutting and thrust forces at ranges from several to hundreds of newtons. However, conventional sensors may not have the necessary signal-to-noise (S/N) ratio and sensitivity required to adequately and reliably characterize surface finish, subsurface damage, and cutting forces at the ultraprecision scale due to the extremely low cutting forces (~0.1 N and lower) and power consumption present in ultraprecision machining.

As illustrated in Figure 2.3, AE is highly desirable due to its relatively superior S/N ratio and sensitivity at the ultraprecision scale when compared to conventional load cell technology, with different levels of AE detectable even at extremely low depths of cut [2.4]. This is true even when the chip formation process diminishes and the only remaining tool–material interaction involves rubbing and burnishing of the surface along different grain orientations. Another particular advantage of AE is that it tends to propagate at frequencies (typically the kHz/MHz range) well above the characteristic frequencies attributed to machining (such as spindle RPM) or natural structural modes in the machine tool structural loop, minimizing the introduction of noise into the AE signal from the above phenomena. Hence, AE is particularly well-suited because of its ability to detect microscale deformation mechanisms within a relatively "noisy" machining environment.

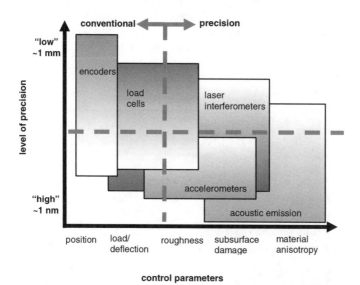

**Figure 2.3.** Sensor application versus level of precision and control parameters

## 2.3 Sources of AE in Precision Manufacturing

Machinists have long used audible feedback during machining as a means of monitoring the cutting process, with skilled machinists able to judge even minute variation in tool wear and surface finish simply by listening to the machining process. The unique sound given by metals during plastic deformation is nothing new, with the unique "tin cry" sound given by tin being plastically deformed as a common one. The above examples all share the common trait of the generation of elastic stress waves within a medium due to plastic deformation. The term *acoustic emission* typically refers to elastic wave propagation in the ultrasonic frequency range (~20–2000 kHz). Unlike ultrasonic non-destructive techniques (NDT), which are a means of active scanning (*i. e.* generation, transmission, and collection of signal), acoustic emission is mostly a passive means of scanning, much akin to holding a microphone or other sensor and "listening" for various phenomena.

Since plastic deformation is a strong source of AE, material deformation-based manufacturing processes have the most potential for acoustic emission-based monitoring. Some of the earliest work on examining the source of AE in plastic deformation was conducted by Fisher and Lally [2.10], who found sharp drops in flow stress during uniaxial loading of single crystal metals and attributed these drops to the propagation of localized deformation bands (possibly slip bands) in the crystal. The formation of these bands was believed to be due to the near simultaneous movement of a large number of dislocations, upon which stress relaxation waves were released and subsequently detected as AE.

Since then, a great deal of research has been performed on identifying sources of AE in manufacturing processes, with Moriwaki [2.11] conducting a comprehensive review of AE sources in machining. Fracture was identified as one possible source, as the propagation of microcracks releases elastic energy due to the generation of new surfaces. Friction or rubbing between two surfaces (such as at the tool/chip

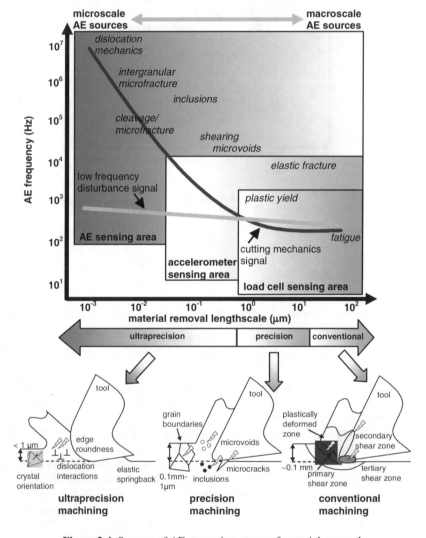

**Figure 2.4.** Sources of AE at varying stages of material removal

interface) is another potential source; surface asperities come into contact and are plastically deformed, and possibly even welded together. As the surfaces slide over one another, these asperities are deformed further and even fractured, hence generating AE.

In precision machining, AE can be attributed to several sources at the tool/chip interface (see Figure 2.4). At the tool/chip interface in conventional machining, AE is typically generated in the primary shear zone (shear region ahead of the tool), secondary shear zone (region of contact between tool and chip), and the tertiary shear zone (contact between the machined surface and flank). While at the conventional machining scale where the DOC is typically on the order of tens of microns and higher, AE is largely due to rubbing and friction at the tool/chip interface, including both the secondary and tertiary shear zones. At the precision scale and below, however, it is believed that the majority of AE signal generation is generated through the interaction of the tool tip with microstructural features of the workpiece, such as voids, inclusions, grain boundaries, and bulk dislocation interactions in the shear zone (see Figure 2.4).

Several studies have focused on the effect of various machining parameters on the resultant AE signal. Liu established a good match between modeled and experimental AE root-mean-square (RMS) signal for diamond turning, and found a strong correlation between chip thickness, cutting speed, and AE RMS [2.12]. Likewise, further work by Chen established an exponential decrease in specific AE RMS (analogous to specific cutting energy, energy density, or energy/chip volume) vs. chip thickness [2.13]. AE sensitivity down to chip thicknesses of 0.01 μm was found, with a sharp difference in specific AE RMS between worn and sharp tools. The increase in specific cutting energy with decreasing chip thickness and tool wear was believed to be due to the ratio of tool edge radius to uncut chip thickness; as the uncut chip thickness decreases below the edge radius of the tool, a transition in the cutting mechanism takes place where an effective negative rake angle is established, which shifts the cutting mode to a more energy-consuming one dominated by rubbing and burnishing. This effective negative rake angle also helps explain the increase in specific cutting energy for worn tools, as worn tools have a larger edge radius and require more energy than sharp tools to initiate chip formation.

The following sections will now cover several case studies of the use of acoustic emission in a series of manufacturing operations covering the range of the Taniguchi curve.

## 2.4 AE-based Monitoring of Grinding Wheel Dressing

Currently, a conventional grinding process requires a complex sequence of tasks in order to start production on a workpiece, including equipment setup, machining of initial test samples, calibration of the tool, and frequent control and correction of the process parameters. AE can serve as a means for the complete automation of all these steps [2.14]. However, in grinding, the two main limitations of AE-based monitoring solutions are the oscillation of the RMS level and signal saturation. Despite these limitations, AE can be very effective for contact detection between moving surfaces. The following section will focus on a Fast AE RMS analysis and mapping technique for wheel condition monitoring in grinding.

### 2.4.1 Fast AE RMS Analysis for Wheel Condition Monitoring

A new method for process monitoring based on short-term analysis of AE RMS patterns is proposed. Since AE RMS changes occur after a considerable period of time, short time evaluation could be a solution for a reliable AE application. Figure 2.5 shows a proposed system where the acoustic emission obtained from the contact between diamond tool and grinding wheel (or grinding wheel and workpiece in grinding monitoring) is converted to an RMS level and acquired by the computer by using an A/D conversion board [2.14]. The sampling rate range varies from 60–500 kHz, depending on the chosen resolution in the circumferential direction. A sampling rate of 2 kHz was used in this particular case to map each rotation of the grinding wheel, which corresponds to a resolution of about 0.5 mm/sample. To be able to measure the contact of the diamond tool with each abrasive grain, the RMS calculation must be performed using a very fast time constant. The time constant was calculated as the average time spent for two consecutive hits between abrasive grains and the diamond tool. This calculation was done for a 60 mesh, L structure aluminum oxide wheel. The average distance between grains measured was about 0.38 mm with a grinding speed of 45 m/s, so a time constant of 10 µs was found.

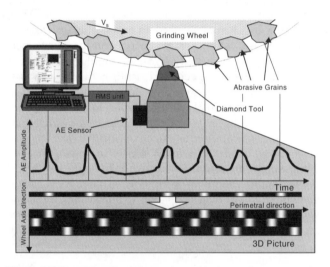

**Figure 2.5.** Procedure for AE signal graphical mapping construction

In AE signal processing, the RMS calculation is normally done after having the raw signal filtered. Due to the very small time constant used in the RMS calculation, it was established that a high pass filter with a cut-off frequency of 100 kHz be used. A specific AE signal processing unit has been developed for this system. The data acquisition is made in data groups where each one-dimensional array of data corresponds to a full rotation of the grinding wheel. The AE signal acquisition starts in each rotation triggered by a magnetic sensor positioned in the wheel hub where a reference pin is installed. A graphical mapping of the AE signal with a color intensity scale is then created. During the wheel dressing operation, the mapping is constructed in real time by adding columns in the array as the dresser travels along the wheel surface.

The system can be used for different evaluation procedures, including

*Dressing evaluation*: During the dressing operation the interaction between dresser and grinding wheel can be acoustically mapped. Lack of contact between dressing tool and grinding wheel will appear as dark areas in the map.

*Topographic mapping*: In this case the map is similar to that obtained for the dressing operation but using a dressing depth of cut close to zero or with a value close to the undeformed chip thickness for the operation. In this case the map shows the active surface of the grinding wheel, which means the surface that will actually be in contact with the workpiece during grinding.

*Grinding evaluation*: During a plunge grinding operation the interaction between the grinding wheel and the workpiece can be evaluated. In this case a different map is obtained where one axis is the grinding time and the other shows the average acoustic energy in the whole wheel length along its perimeter.

### 2.4.2 Grinding Wheel Topographical Mapping

Figure 2.6 shows an output from the acoustic mapping system when used during a topographic mapping procedure. The vertical and horizontal directions are the wheel perimeter and width respectively, with a spatial resolution of 2 samples/mm. The depth of interaction between diamond tool and grinding wheel used was 1 micron (in the range of elastic contact). The color intensity shows the acoustic emission RMS value measured from the interaction between dressing tool and the abrasive grains. Darker areas correspond to lower acoustic energy detected by the sensor. The L shaped mark was created in the wheel surface in order to evaluate the system performance. Dark areas show worn regions of the wheel where the dressing tool had lower interaction with the abrasive grains. The main vertical worn band on the left side was the result of a grinding operation made with the workpiece shown in the photo of Figure 2.6a. In this figure, the "L" mark produced on the wheel surface and a magnified view of its representation in the AE graphical mapping is shown.

It was observed that the AE mapping system could generate an image similar to the surface topography present on the grinding wheel surface, even using a very small depth of interaction with the diamond tool. A spark-out dressing was also tested (depth of interaction = 0) and the result was nearly the same. The use of equal depths of interaction may lead to reading error in the case of any thermal deformation in the machine structure.

The same grinding wheel was dressed in order to remove the damage produced by the grinding operation. The dressing depth of cut was 2 microns per pass. This small value was chosen in order to evaluate the depth of the damage produced by the grinding operation. The maps of three consecutive dressing strokes made in an unbalanced grinding wheel are shown in Figure 2.6b. The wheel was dressed and then unbalanced using an automatic balancing system in manual mode. Each dressing operation was performed with a dressing in-feed of 1 micron. After six dressing strokes, the grinding wheel eccentricity disappeared, and the real unbalancing displacement level was calculated to be about 6 microns. The

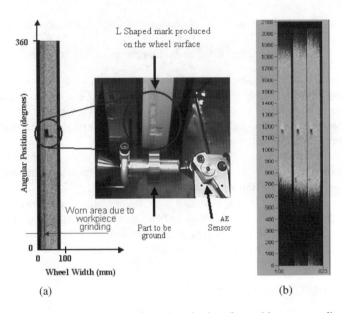

**Figure 2.6.** (a) L-shaped mark produced on the wheel surface with corresponding AE map during dressing, (b) dressing of unbalanced grinding wheel (first three strokes, 2 microns per pass)

unbalancing device installed in the machine demonstrated a vibrational amplitude of only 1 micron (corresponding to the green zone for grinding). This strong difference is caused by the fact that the vibration sensor measures the vibration amplitude in the machine bearings while the acoustic mapping system measures the actual wheel eccentricity caused by the lack of balancing. The sequence shows exactly the position of the heavier side of the grinding wheel and the reduction in the eccentricity after dressing. This is another important piece of information provided by the mapping system, since most balancing devices for grinding machine use vibration sensors as feedback. It was observed that the minimum vibration level does not lead to the best concentricity of the grinding wheel. Therefore, the information provided by the mapping system could be used as a feedback signal when balancing CBN wheels during touch dressing operations where this problem is more critical [2.14].

### 2.4.3 Wheel Wear Mechanism

Figure 2.7 presents two experiments to demonstrate the influence of the wheel wear behavior in the AE map. Each stripe in the graph represents a single grinding cycle or a single workpiece in a production line of automotive components. The workpiece material was Inconel and the grinding wheel a very hard grade, low friability aluminum oxide specification (DA 80 R V). The second experiment was a plunge grinding operation of hardened AISI 4340 steel using a soft white aluminum oxide grinding wheel (AA 60 G V).

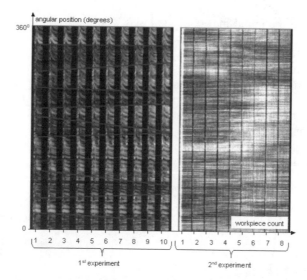

**Figure 2.7.** Mapping of AE from a grinding process showing two types of wear mechanisms

The image composition of the several mappings in Figure 2.7 shows two distinct types of wheel wear behavior. In the first experiment the consistent pattern around the wheel shows that it is not losing grains. In the second experiment the transformation indicates that the wheel is indeed losing grains. These results were confirmed by checking the workpiece size plot (stable for the first case and an increasing tendency in the second case), workpiece temperature (grows for the first experiment and constant for the second) and power plots (power grows in the first experiment and remains constant for the second one) [2.14].

## 2.5 AE-based Monitoring of Face Milling

The fast AE RMS graphic mapping system used for the AE monitoring work for grinding was also used to monitor a face milling process. Previous work by Diei and Dornfeld established a high degree of sensitivity of the AE signal to the individual chip formation mechanism in face milling, with a significant variation observed in the AE signal (processed through time-difference signal processing techniques) with respect to different stages of the chip formation process, with significant increases in AE signal observed upon the initiation of chip formation (due to initial impact of the insert with the workpiece) and exit of the insert from the workpiece [2.15][2.16] (see Figure 2.8a).

In the face milling monitoring work, a ROMI Discovery 3-axis CNC milling center was used with an AE sensor attached directly to the workpiece [2.17]. The AE RMS signal was collected during a typical machining operation (6 insert cutter, 200 RPM, 0.1 mm/insert feed, 0.2 micron axial DOC) in a continuous fashion similar to the techniques and DAQ system parameters used in the grinding monitoring technique. Each vertical data trace in the AE RMS intensity mapping in Figure 2.8b corresponds to one spindle rotation, with a clear spike in AE signal occurring due to the initial impact of the insert with the workpiece (similar to that seen in Figure 2.8a). Successive spindle rotations are shown along the horizontal axis, for a total of approximately 200 revolutions in the intensity mapping. The AE RMS signal for each cutting insert can be clearly seen (numbered 1–6) with a separation of 60 degrees between inserts. After several tens of spindle revolutions, AE signal due to the rubbing of inserts on the trailing edge of the cutter can be observed, although the AE signal due to rubbing of inserts 2 and 6 can barely be seen (see Figure 2.8b). This AE mapping technique serves as a means for tool condition monitoring in face milling, particularly for tool contact, breakage, and insert positional precision within the cutter [2.17].

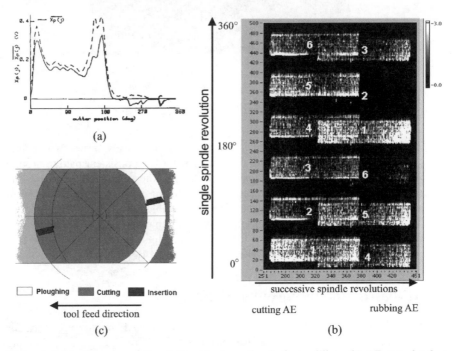

**Figure 2.8.** (a) AE response for single chip formation in face milling, (b) AE map for face milling operation, (c) tool motion relative to workpiece

## 2.6 AE-based Monitoring of Chemical Mechanical Planarization

Chemical mechanical planarization (CMP) is one of the key enabling technologies in the semiconductor manufacturing industry today for the fabrication of extremely smooth and flat surfaces on a variety of semiconductor substrate materials. In order to meet the requirements of current lithography tools which require extremely stringent tolerances for flatness and planarity, CMP is capable of planarizing a 300 mm (current industry standard) diameter wafer achieving surface roughness on the order of 1–2 nm $R_a$ and global planarity well below 0.5 μm. However, CMP has also

become one of the key bottleneck or roadblock issues in semiconductor manufacturing today [2.18]. The decreasing line widths of semiconductor devices require
new materials, such as copper and the so-called low-$k$ dielectrics, which further
challenge the process. Preferential polishing rates of adjacent materials, or surface
features resulting from previous manufacturing steps, often lead to defects such as
dishing, which frustrate efforts to obtain planarity. The abrasive slurry can also
induce defects such as surface contamination, scratches, slurry residue, *etc.*, hence
predicating the need for a reliable means of monitoring the CMP process.

### 2.6.1 Precision Scribing of CMP-treated Wafers

As a means of investigating the process physics of CMP, and to further evaluate the
sensitivity of AE to the material removal physics taking place, a series of scribing
tests on a CMP-processed oxide wafer were conducted [2.19]. In the CMP process
for oxide planarization, the bulk of material removal takes place on a "chemically
weakened" layer consisting of a highly hydrated and loosely bound network of silica
on the order of a few nanometers in thickness (see Figure 2.9). The second layer is
a "plastically compressed layer" around 20 nm deep from the chemically weakened
layer, depending on process conditions. Unlike the chemically weakened layer, this
layer is represented by a plastically compressed network of silica that has higher
density. A bulk layer also exists below the plastically compressed layer that is not
affected by the CMP process. Because of the variation in the material properties of
each layer, it was initially postulated that the AE RMS signal obtained during
material removal of these distinct layers would differ.

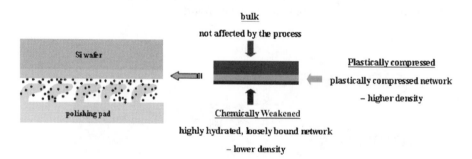

**Figure 2.9.** Three distinct layers exhibited in oxide CMP

In order to assess the mechanical properties of the oxide layer, a scribing
operation using a single point diamond turning machine was used (see Figure 2.10).
The setup involved mounting a polished oxide wafer at a slight angle of ~2° and
scribing across surface, with the tool engaging with an increasing depth of cut
(DOC) as the tool traversed over the surface. An AE sensor was attached to the front
side of the wafer and used to monitor these scratch tests.

During the scribing operation, two transitions (one from the chemically
weakened layer to the plastically compressed layer, and the other from the
plastically compressed layer to the bulk) appear as distinct features in the AE RMS

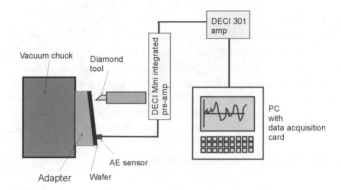

**Figure 2.10.** Experimental setup for CMP oxide wafer scribing operation

signal. Figure 2.11 shows the AE signals from the scratch test of post-CMP wafers. As postulated, the variation in the AE signal reflects the three distinct layers and the transitions from one layer to another. The first part in Figure 2.11 shows an "air-cut" region where the tip is disengaged from the chemically weakened layer, and the very onset of contact between the tool and wafer. Since the chemically weakened layer consists of a loosely-bound network of silica and is only a few nanometers deep, the AE RMS signal from this layer differs only slightly from that of the air-cut region. As the tip starts to touch the wafer, within 10 ms, the AE signals burst and increase over time for about 70 ms, which corresponds to a DOC of 30 nm, and confirms that the plastically compressed layer was engaged at this time. Beyond this part, the AE signals monotonically increase, without significant deviation or burst signal, meaning that the tool is cutting in the bulk layer of the oxide.

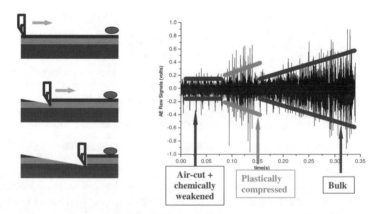

**Figure 2.11.** Variation in AE RMS signal during scribing of CMP-treated oxide wafer

### 2.6.2 AE-based Endpoint Detection for CMP

In semiconductor manufacturing, the use of CMP in thin-film polishing for a fixed time is commonly used. However, due to fluctuations in the process, such as material-removal-rate (MRR) inconsistency, pad degradation, and non-uniformity issues, over- or under-polishing can occur. An endpoint detection system is required to insure that only the desired thickness of material is polished during the CMP process, thus offering many manufacturing advantages such as improved process yields, closer conformance to target requirements, and higher throughput.

In order to evaluate the feasibility of AE as an in-situ endpoint detection technique, two different sets of wafers were polished with a desktop CMP machine. The first set of wafers consisted of stacked films of oxide (5,000 Å) at the bottom, tantalum (5,000 Å) in the middle, and copper (1,500 Å) at the top as an example of the copper damascene process. The wafer was polished with a conventional IC 1000 polyurethane pad and alumina-based slurry with 2.5% $H_2O_2$. The second set of wafers consisted of film stacks ranging from oxide (2,000 Å) at the bottom to nitride (1,000 Å) at the top, representing a shallow trench isolation process, and was polished with fixed abrasives and ph-adjusted deionized water (pH = 11.5). During the CMP process, both AE signals and frictional force data were collected with a DAQ system, and the experimental data for the copper damascene test wafers are plotted against time in Figure 2.12a.

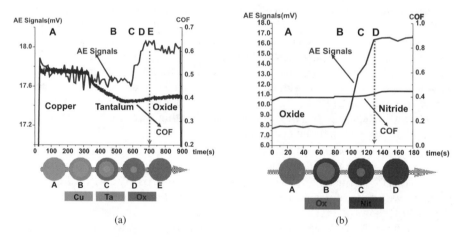

**Figure 2.12.** AE RMS endpoint detection for (a) Cu damascene CMP process, (b) STI CMP process

The endpoint was triggered at the edge of the wafer where copper was first cleared because of higher material removal at the edge. The phenomenon of the edge of the wafer clearing first also occurred in the second polishing step that removes the remaining copper and barrier layer. As shown in Figure 2.12a, AE RMS signals clearly show the transition from tantalum to oxide, indicating the ideal endpoint whereas frictional force signals constantly increase as polishing time continues, making it difficult to detect the desired endpoint with frictional force alone. A similar trend was also observed for the STI test wafers during polishing, with a very sharp transition observed in the AE RMS signal when the oxide completely cleared (see Figure 2.12b). These transitions are believed to be directly related to the variation in material properties of each of the films, with harder materials (such as nitride) demonstrating an increase in AE RMS signal during polishing due to the increase in energy required for initial material removal, whereas less-hard materials such as copper exhibit lower AE RMS signal during polishing [2.19]. However, the frictional force does not exhibit the same sharp transition as the AE RMS signal, demonstrating the superior S/N ratio of AE for the material removal process in CMP when compared to frictional force.

## 2.7 AE-based Monitoring of Ultraprecision Machining

### 2.7.1 Monitoring of Precision Scribing

To examine the effect of crystallographic orientation during ultraprecision machining, a series of nearly-parallel slow speed scribes on the surface of an oxygen-free high-conductivity (OFHC) polycrystalline copper workpiece (250 micron average grain size) was taken. A 0.274 mm nose radius single crystal diamond tool was chosen as the tool of choice since the tool tip was considerably smaller than the average grain size of the workpiece. The workpieces were clamped onto the spindle of the lathe, and the spindle was alternately rotated and locked to allow for slow-speed scratches. A scratch speed of 0.7 mm/s was used, and the infeed (depth of cut) setting was set at a constant value of 10 microns throughout the experiment. After each scribe, the spindle was unlocked and rotated slightly (~1 degree) to produce a radial pattern of scribes (although the scribe pattern can be approximated as a raster scanning pattern over this small angle of rotation). A similar DAQ system to that shown in Figure 2.10 was used, with both cutting force and AE RMS signal collected during each scribing operation.

After collecting data from a series of scribes (typically on the order of 15 quasi-raster pattern scribes), the data was reduced, and a color intensity mapping function in MATLAB® was used to plot the cutting force and AE signal as a function of position, with the color map representing the respective magnitude of the signal. Figure 2.13 shows a graphical representation of the cutting force and AE RMS signal for a series of 15 scratches, along with a micrograph of the workpiece surface before scratching. Figure 2.14 shows the cutting force and AE RMS for the fifth scribe.

Both the cutting force and AE RMS signal reproduce a crude representation of the grain structure of the material. The variation in force and AE RMS signal is largely due to the fact that each grain has a particular crystallographic orientation, so as the tool passes from one grain to another, a new slip system in the grain is activated, which changes the amount of applied stress (and cutting force) required to initiate deformation. If the cutting speed and tool cross-section are constant, then the AE RMS is simply proportional to the energy (and cutting force) required to initiate deformation. The activation of different slip systems as a function of grain orientation causes the energy of the resulting AE RMS signal to fluctuate accordingly. Of particular note is the good match found between the mappings for the force and AE RMS signal, indicating that advances in load cell technology can allow for improved sensitivity to process mechanisms at the precision scale.

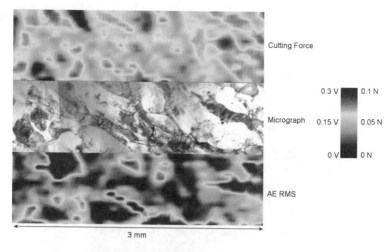

**Figure 2.13.** Force/AE response for 'quasi-raster' scratch pattern

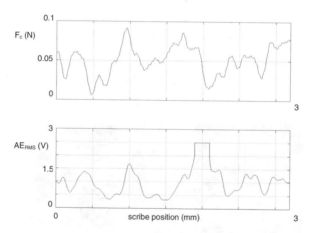

**Figure 2.14.** Individual force/AE response for single scribe (fifth scribe)

## 2.7.2 Monitoring of Ultraprecision Turning of Single Crystal Copper

The sensitivity of AE to crystallographic orientation was also tested for single crystal materials. Figure 2.15a shows a sample trace of AE RMS data for a single revolution of cutting on a <100> workpiece. An approximate sinusoidal variation in AE RMS can be seen in this data set. Each trace of AE RMS data is then collected in a data array in LABVIEW, and is represented graphically as an intensity plot in Figure 2.15b. This intensity plot shows AE RMS signal as a function of spindle revolution, with subsequent traces of data for each spindle revolution progressing from left to right for a total of about 80 revolutions for a single face turning pass of the single crystal workpieces. Signal intensity or voltage levels are plotted according to a color intensity map, with a black color corresponding to zero signal, signal saturation at 3.5 V corresponding to a white color, and intermediate signal values appearing as shades of grey.

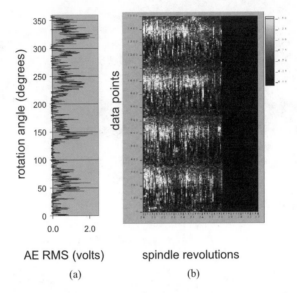

**Figure 2.15.** (a) AE RMS data trace for single revolution of face turning of <100> workpiece, (b) Cartesian intensity plot of AE RMS data for face turning

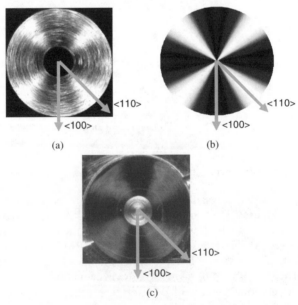

**Figure 2.16.** (a) Experimental AE RMS polar map for <100> workpiece, (b) Taylor factor-based theoretical AE RMS polar map for <100> workpiece, (c) chemically etched surface of <100> workpiece after machining

A Cartesian–polar transform allows for the data in Figure 2.15 to be replotted on a polar intensity map as shown in Figure 2.16a. In the polar intensity map, the AE RMS voltage is plotted against physical position on the workpiece, giving an indication of the variation of cutting energy as a function of crystallographic orientation. Figures 2.16b and 2.16c show the theoretical variation in Taylor factor for a <100> crystal as a function of orientation, and an image of the chemically etched surface of the <100> workpiece after machining, respectively. A good correlation can be seen between the theoretical and experimental polar mappings, and variation in the surface finish of the chemically etched workpiece.

Figure 2.17 shows SEM images of the machined surface before chemical etching. Figure 2.17a shows the surface corresponding to areas that demonstrated low AE RMS signal levels of approximately 0.5V, and Figure 2.17b shows the surface of areas with AE RMS values of approximately 2V. Dornfeld *et al.* postulated that this variation in AE signal corresponded to the variation in the orientation-dependent Taylor factor (representative of the yield stress, and consequently, specific energy required to initiate plastic deformation in a single crystal material) of the workpiece [2.20]. These regions of high and low AE RMS signal correspond to regions that demonstrate high and low Taylor factors (ranging from 2.4 to 3.7), respectively. Crystallographic orientations that have a relatively

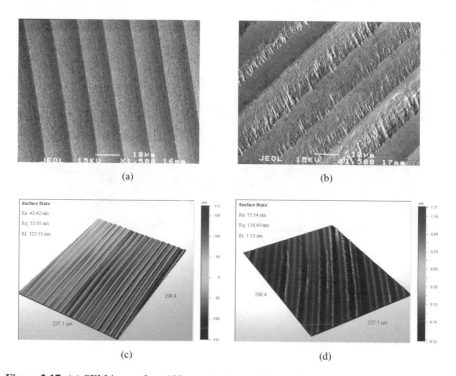

**Figure 2.17.** (a) SEM image for <100> workpiece, <100> cutting direction, "smooth" region, (b) image for <100> workpiece, <110> cutting direction, "rough" region, (c) Wyko image for <100> workpiece, <100> cutting direction, "smooth" region, (d) Wyko image for <100> workpiece, <110> cutting direction, "rough" region

high Taylor factor yield machined surfaces that are significantly rougher than areas with a relatively low Taylor factor. Figure 2.17c shows a relatively smooth surface finish of 42.4 nm $R_a$ (measured with a Wyko surface interferometer) while Figure 2.16d shows the rougher surface with a surface finish of 75.6 nm $R_a$.

### 2.7.3 Monitoring of Ultraprecision Turning of Polycrystalline Copper

Figure 2.18a shows an AE polar map for a polycrystalline OFHC copper workpiece. Although this piece was supposed to have an average grain diameter of 250 µm, several large grains were still observed on the workpiece. The AE polar map in Figure 2.18a shows good correlation with the chemically-etched workpiece surface, shown in Figure 2.18b. While small grains below 100 µm cannot be resolved in the AE polar map due to DAQ limitations, the larger grains appear very clearly in both the chemically-etched surface and AE polar map. Because these large grains can serve as defects that affect the homogeneity of the surface finish, the AE polar mapping technique, in addition to serving as a tool contact sensor, provides a very convenient means of detecting potential trouble spots or defective areas on the workpiece [2.20]. A perfectly homogenous and isotropic workpiece would most likely result in little variation in the AE signal and polar map, so variations in the AE signal can serve as useful feedback in a fully-automated manufacturing environment.

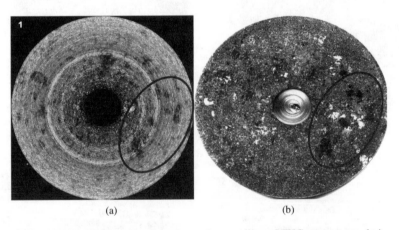

(a)                                          (b)

**Figure 2.18.** (a) AE RMS polar map for polycrystalline OFHC copper workpiece, (b) micrograph of chemically-etched workpiece

## 2.8 Conclusions

As current manufacturing trends aim for smaller and finer form and features, a reliable means of process monitoring with the appropriate sensor technology is necessary to properly characterize and monitor the manufacturing operation. As demonstrated by endpoint detection in CMP and crystallographic orientation detection in ultraprecision machining, AE is capable of meeting these requirements, especially for material removal process parameters such as material removal rates and very small uncut chip thicknesses. The use of AE as an in-situ process monitoring and characterization tool can serve as a means of closely linking the manufacturing and quality control stages together, and with further development, in a fully automated manufacturing environment, the quality control stage can be entirely eliminated with optimal in-situ monitoring and process control.

# References

[2.1]    Taniguchi, N., "Current Status In, And Future Trends Of, Ultraprecision Machining and Ultrafine Materials Processing," Annals of the CIRP, **32**(2), 1983.

[2.2]    Dornfeld, D. A., Lee, Y., Chang, A., "Monitoring of Ultraprecision Machining Processes," International Journal of Advanced Manufacturing Technology, **21**, pp. 571–578, 2003.

[2.3]    Moriwaki, T., "Experimental Analysis of Ultra-Precision Machining," International Japan Society of Precision Engineering, **29**(4), pp. 287–290, 1995.

[2.4]    Lee, Y., "Monitoring and Planning for Open Architecture Manufacturing of Precision Machining Using Acoustic Emission," Ph.D. Dissertation, Mech. Eng. Department, University of California at Berkeley, 1991.

[2.5]    D. A. Dornfeld W. Konig and G. Kettler, "Present State of Tool and Process Monitoring in Cutting", Proceedings of New Developments in Cutting, VDI Berichte NR 988, Dusseldorf, pp. 363–376, 1993.

[2.6]    Tlusty, J. andrews, G. C., "A Critical Review of Sensors for Unmanned Machining," CIRP Annals, **32**(2), 1983.

[2.7]    Tonshoff, H. K., Wulfberg, J. P., Kals, H. J. J., Konig, W., "Developments and Trends in Monitoring and Control of Machining Processes," CIRP Annals, **37**(2), 1988.

[2.8]    G. Byrne, D. A. Dornfeld, I. Inasaki, W. Konig and R. Teti, "Tool Condition Monitoring (TCM) – The Status of Research and Industrial Application", CIRP Annals, **44**(2), 1995.

[2.9]    K. Iwata, "Sensing Technologies for Improving the Machine Tool Function", Proceedings of the 3rd International Machine Tool Engineer's Conference, JMTBA, Tokyo, pp. 87–109, 1988.

[2.10]   Fisher, R. M., Lally, J. S., "Microplasticity Detected by an Acoustic Technique," Canadian Journal of Physics, **45**, pp. 1147, 1967.

[2.11]   Moriwaki, T., "Application of Acoustic Emission Measurement to Sensing of Wear and Breakage of Cutting Tool", Bulletin Japan Society of Precision Engineering, **17**, pp. 154–160, 1983.

[2.12]   Liu, J. J., "Monitoring the Precision Machining Process: Sensors, Signal Processing, and Information Analysis," Ph.D. Dissertation, Mechanical Engineering Department, University of California at Berkeley, 1991.

[2.13]   Chen, X., Tang, J., Dornfeld, D. A., "Monitoring and Analysis of Ultraprecision Metal Cutting with Acoustic Emission," Mechanical Engineering Congress and Exposition, ASME, Atlanta, GA, 1996, pp. 387.

[2.14]   Oliveira, J. F. G., Dornfeld, D. A., 2001, Application of AE Contact Sensing in Reliable Grinding Monitoring, CIRP Annals, **50**(1), 2001.

[2.15]   Diei, E. N., Dornfeld, D. A., "A Model of Tool Fracture Generated Acoustic Emission During Machining," ASME Trans., J. Eng. Ind., **109**(3), 1987, pp. 227–234; also appears in Sensors and Controls for Manufacturing, Kannatey-Asibu, E. and Ulsoy, A. G., eds., ASME, New York, 1985.

[2.16]   Diei, E. N., Dornfeld, D. A., "Acoustic Emission from the Face Milling Process - the Effects of Process Variables," ASME Trans., J. Eng. Ind., **109**(2), 1987, pp. 92–99.

[2.17]   Ansenk, J., Broens, N., "Investigation on Acoustic Emission Mapping Applications in Milling," OPF Report, University of Sao Paulo, Sao Carlos, Brazil, 2004.

[2.18]   R. DeJule, "CMP challenges below a quarter micron", Semiconductor International, pp. 54–60, 1997.

[2.19]   Hwang, I., "In-Situ Process Monitoring and Orientation Effects-Based Pattern Design for Chemical Mechanical Planarization", Ph.D. Dissertation, Mech. Eng. Department, University of California at Berkeley, 2004.

[2.20]   Dornfeld, D. A., Oliveira, J. F. G., Lee, D., Valente, C. M. O., "Analysis of Tool and Workpiece Interaction in Diamond Turning Using Graphical Analysis of Acoustic Emission," CIRP Annals, **52**(1), 2003.

# 3

# Tool Condition Monitoring in Machining

Mo A. Elbestawi, Mihaela Dumitrescu and Eu-Gene Ng

McMaster University
Hamilton, ON L8S 4L7, Canada
Email: elbestaw@mcmaster.ca

## Abstract

Condition monitoring and diagnosis systems which are capable of identifying machining system defects and their location are essential for unmanned machining. Unattended (or minimally manned) machining would result in increased capital equipment utilization, thus substantially reducing manufacturing costs. A review of tool monitoring systems and techniques and their components is presented. The proposed algorithm for MPC fuzzy neural networks is a fast, effective, and simple method for dealing with multi-sensor, multi-class, overlapped classification problems. Two case studies are presented on Multiple Principle Component fuzzy neural networks for tool condition monitoring in turning and drilling. Experimental application of this method yielded a success rate up to 97%. A case study for online detection of drill chipping is also included. The online detection methodology employs vibration signals to detect tool chipping, based on the particle frequency being excited. The Continuous Wavelet Transformation method has proven effective, however, it is not able to generate an online monitoring classification script capable of analyzing its output map.

## 3.1 Introduction

The need for continuous improvements in product quality, reliability, and manufacturing efficiency has imposed strict demands on automated product measurement and evaluation. Modern day manufactured products demand ever-higher precision and accuracy. Automated process monitoring is thus crucial in successfully maintaining high quality production at low cost.

Automated tool condition monitoring implies identifying the tool conditions without interrupting the manufacturing process operation, under minimum human supervision. Unattended or minimally manned machining would result in increased capital equipment utilization, thus substantially reducing the manufacturing costs.

An "Intelligent Sensor System" was defined by [3.1] as an integrated system consisting of sensing elements, signal conditioning devices, signal processing algorithms, and signal interpretation and decision-making procedures. This system acts not only as a signal-collecting device, but also as a sorting and analyzing machine. In the absence of a human operator, the system should sense signals indicating the process status and its changes, interpret incoming sensed information, and decide on the appropriate control action. We define a system as *Automated/ Intelligent Monitoring System* if the system possesses abilities of sensing, analyzing, and knowledge learning, and error correction, which are essential to machining tool condition monitoring. An automated/intelligent machining process and tool condition monitoring system should be able to emulate as closely as possible the sensing, recognizing, responding, and learning abilities of human operators. To emulate the human monitoring action, an automated tool condition monitoring system has four components:

*Sensing Technique*: Typically, indirect sensing techniques such as cutting forces, vibrations, and acoustic emission are used. Different types of sensors and sensory data from different locations are combined to yield maximum useful information.

*Feature Extraction*: Ideally, sensory signals contain the necessary information required to discriminate between different process and tool conditions. However, the sensed signals are usually noisy and have to be further processed to yield useful features that are highly sensitive to the tool conditions but insensitive to noises.

*Decision Making*: Decision-making strategies process the incoming signal features and perform a pattern association task, mapping the signal feature to a proper class (tool condition). This processing task can be done sequentially or in parallel depending on the monitoring system architecture.

*Knowledge Learning*: In order to make a correct decision, learning algorithms have to be provided. Such algorithms tune system parameters by observing the sample features corresponding to different tool conditions. Like human operators, automated monitoring systems should have the ability to learn from their experiences (past work) as well as the new information generated from the machining process.

Key concerns with both signal processing and decision-making algorithms, jointly known as monitoring methods, include reliable and fast identification or response to an abnormal event occurring under normal process conditions [3.2].

This chapter, building on work done by [3.3] and [3.4], presents two novel approaches to the development of an automated tool condition monitoring system – the Multiple Principle Component (MPC) Fuzzy Neural Networks, and an online monitoring technique for the detection of drill chipping. Both subjects are exemplified with case studies.

## 3.2 Research Issues

The major research goals in tool condition monitoring are to develop self-adjusting and integrated systems capable of monitoring under various working conditions with minimum operator supervision.

The purpose of automated tool condition monitoring in machining is to relate the process signals to the tool conditions, and detect or predict the tool failure. Automated tool condition monitoring involves the act of identifying the characteristic changes of the machining process based on the evaluation of process signatures without interrupting normal operations. Basically, a monitoring process has two parts:

- *Sensing* – the process of obtaining cutting process signals from sensors. Appropriate signals used for tool condition monitoring are force, torque, vibration, temperature, acoustic emission, electric current, *etc.*
- *Monitoring* – composed of *signal processing* and *decision making,* can be divided into model-based and feature-based methods. Both methods use sensor signals from the cutting process for the system input.

Any automated machining process and tool condition monitoring system should involve the following:

1. *Multi-sensor System* – uses more than one sensor for monitoring machining processes and tool conditions, yielding an extended survey of sensitive features, as most process variables influence one another.
2. *Automated Feature Extracting System* – automatically generates monitoring features through learning. The signals sensed from multiple sensors are analyzed, compacted, and selected by the system to generate the most sensitive features to the monitoring subjects. The extracted features are also further refined or reselected by the monitoring system.
3. *Learning and Decision Making System* – builds up flexible and comprehensive monitoring strategies and automatically generates control parameters. The concentrated information from the learning procedure is stored in the system for classification purposes and can be modified by knowledge updating procedures. With increasing experience, the system will become more and more reliable and promote the monitoring/control functions. These strategies should be robust and valid for a reasonable range of cutting conditions.

Significant research work performed in this research field focused on analytical forecast, dynamic structure identification, monitoring techniques, and adaptive control approaches; [3.5]–[3.8] published research papers on the development of modern monitoring techniques for machining. Critical reviews on sensors for machining monitoring were published in [3.6] and [3.9]. Applications include geometric corrections, machine diagnosis, surface finish controls, tool condition monitoring, and machining process monitoring.

Researcher [3.5] describes the monitored conditions in machining processes, as listed in Table 3.1, and identifies five monitoring tasks: machine, tool, process, tool condition, and workpiece. The monitored functions were also classified into two groups: time critical and non-time critical. The former requires a system response within a range of milliseconds while the latter may take seconds or even minutes.

### 3.2.1 Sensing Techniques

Metal cutting is a dynamic process. The sensor signals can be considered as the output of the dynamic system in the form of time series. Consequently, process and tool condition monitoring can be conducted based on system modeling and model evaluation. One of the most widely used models is the linear time-invariant system, such as state space models, input–output transfer function models, Auto-regressive (AR) models, Auto-regressive and Moving Average (ARMA) models, and the

Dynamic Data Systems (DDS) methodology. When a suitable model is identified, monitoring can be performed by detecting the changes of the model parameters and/or the changes of expected system responses.

Current sensing techniques can be divided into two basic types:

- *Direct Technique* – the most accurate measure for determining drill failure; however, the trade-off is production stoppage. In this method, a direct analysis of the drill tip or workpiece surface is performed at the end of the drill cycle. Basic analysis methods include optical measurements, surface finish using contact probes (profilometers), chip size measurements, *etc.* The main disadvantage of this method is that any significant deterioration occurring in between measurements goes unnoticed, allowing for potential damage to the "machine tool–tooling–workpiece" (MTW) system.
- *Indirect Technique* – utilizes correlated variables with process signals to monitor for a specific drill failure signature. This technique can be applied continuously while drilling, and therefore can be used in an online monitoring algorithm. As online monitoring for chipping is the focus of this case study, the most common methods used to correlate a variable with drill chipping are presented in Table 3.1.

**Table 3.1.** Monitored conditions in machining processes [3.5]

|  | **Time Critical** | **Non-time Critical** |
|---|---|---|
| Machine | * CNC-control<br>* collision | * accuracy<br>* thermal deformation |
| Tool | * tool fracture<br>* tool approach | * tool wear<br>* tool presence |
| Process | * chatter<br>* force/torque/power<br>* chip forming | * coolant |
| Tool condition | * dressing | * tool compensation |
| Workpiece | * dimension in process<br>* shape in process<br>* roughness in process | * raw stock dimension<br>* workpiece material<br>* surface integrity |

### *Spindle Current Monitoring*

Researchers [3.10]–[3.12] suggest that monitoring both spindle and feed motor current for fluctuations caused by variations in cutting thrust and torque can successfully detect drill failure. Various signal processing and decision-making algorithms have been experimented with to determine whether current is the best monitoring index for the detection of drill chipping.

The most basic method is simple threshold monitoring of the stator current, as applied in [3.11]. The current signal is acquired from the motor stator and verified against a specified threshold. Signals whereby the motor stator current exceeds the threshold due to increasing torque are categorized as belonging to a catastrophically fractured drill. The actual stator current is not overly sensitive to motor torque variations due to its limited bandwidth, however, the response time for severe tool failure detection is acceptable. Insensitivity to minor chipping of the drill bit due to this limited bandwidth makes simple threshold monitoring unreliable for the purpose of this study. A more complex algorithm with a reliable detection technique is thus required. [3.10] and [3.12] propose using wavelet transforms to analyze the spindle and feed motor currents. The AC servo motor spindle current is analyzed using the Continuous Wavelet Transforms (CWT) to detect failure of the drill bit, and the AC servo feed motor current is decomposed using a Discrete Wavelet Transform (DWT) algorithm to additionally verify drill failure. As instantaneous spikes in both the feed and spindle motor current signals occur during chipping, detection by an algorithm that senses changes in frequencies over short time periods is required. Since the power of wavelet transforms lies in the ability to detect changes in frequencies, and locate them in the time domain, they are ideal in this situation.

### Acoustic Emissions

Monitoring the Acoustic Emission (AE) signal during the drilling process is another way of detecting drill state. The advantage of using AE is that the signal propagates through the entire workpiece, and therefore placing a sensor next to the drilling location is not crucial, allowing for ease of implementation in any drilling process. Additionally, the frequency range (9.7–10.5 kHz) of the AE signal is usually much higher than the natural frequencies excited by the drilling process and therefore, cannot be misinterpreted for vibration generated by normal drilling.

Parameters such as energy, RMS of the signal, and a measure of counts and hits were selected as monitoring indices for this technique. It was concluded that energy was the only reliable index that may accurately be correlated with drill lip height variations, making this a usable detection strategy for drill lip wear.

Deterrents to using AE include the high costs associated with implementing AE sensors and the further testing required for reliable monitoring of chipping [3.10]–[3.13].

### Vibration Signature Analysis

Detailed analysis of the vibration signal generated by drilling provides a great deal of information about the state of the drill. Vibration signals contain information on the natural frequencies of the workpiece, fixture, machine, and spindle, and are transformable into various domains to determine signatures of uneven drill wear or possible chipping.

The most basic vibration analysis technique transforms the time domain signal to the frequency domain using the Fast Fourier Transform (FFT) in order to detect unusual changes in vibration frequencies possibly due to wear or chipping. However, effective application of this to online monitoring is difficult and therefore alternative analysis techniques have been researched.

In [3.14], both frequency and time monitoring indices were applied in the form of the Cepstrum and Kurtosis values to detect drill wear and chipping. Cepstrum analysis has been widely accepted as a powerful technique for monitoring wandering during drilling. This concept was tested in [3.14] to assess its ability to detect drill failure, and was successful for small diameter drills, but has not been researched for large diameter drills (>6 mm).

As spikes are also generated by inconsistencies in the material, a method of eliminating false alarms is required. [3.14] recommends calculating a triggering value, the Ratio of Absolute Mean Value (RAMV), to initialize the Kurtosis value. The amplitude of spikes generated by chipping are much greater than those generated by material inconsistencies, therefore application of a triggering value can be used to differentiate and prevent false alarms. This method is not dependent on cutting conditions, but has only been tested on small diameter drills (<6 mm).

### Sensor Fusion

In most cases, signals from only one sensor are typically insufficient to give enough information for machining and tool condition monitoring.

Using several sensors at different locations simultaneously has been proposed for data acquisition [3.8][3.15]–[3.19]. Signals from different sources are integrated to provide the maximum information needed for monitoring and control tasks. A schematic diagram showing the use of multiple sensors in monitoring systems is shown in Figure 3.1.

*Sensor Fusion* generally covers all the issues of linking sensors of different types together in one underlying system architecture [3.16]. The strategy of integrating the information from a variety of sensors will increase the accuracy and resolve ambiguities in the knowledge about the environment. The most significant advantageous aspects of sensor fusion are its enriched information for feature extraction and decision-making strategy, and its ability to accommodate changes in the operating characteristics of the individual sensors due to calibration, drift, failure, *etc*. The type and number of sensors used for tool conditioning monitoring are chosen according to the type of monitoring tasks. Table 3.2 illustrates types of sensors commonly used in monitoring systems.

The use of multiple sensors for machining process and tool condition monitoring gives extended information about the process. As most process variables influence one another, more than one model is typically needed to analyze the sensor signals. Some models considered are: dynamic structure models for cutting force, such as an empirical cutting force model [3.20]; dynamic models for tool wear, such as diffusion wear models and adhesive wear models [3.21]; empirical models [3.22]; linear steady-state models for tool wear and cutting forces [3.23]–[3.25]; and parametric models including AR (auto-regressive) for chatter [3.26][3.27], AR for tool wear [3.28], and AR for tool breakage [3.29].

Emphasis is placed on high reliability and fast response in tool condition monitoring systems to ensure the manufacture of high quality parts in an efficient manner. Early detection of tool deterioration improves process productivity and reliability. Therefore, monitoring systems must be developed with the above criteria in mind. Computing time and adaptive learning are also both important factors to consider in developing monitoring systems [3.30].

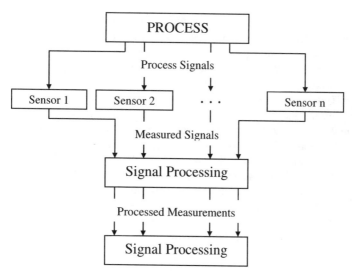

**Figure 3.1.** Multiple sensors in monitoring

**Table 3.2.** Sensors used for tool condition monitoring [3.6]

| Sensor | Dimensional | | Cutting Force | | Feed Force | Spindle Motor | Acoustic Emission |
|---|---|---|---|---|---|---|---|
| Task | Touch Trigger | Non Contact | Dynamo meters | Process | | Torque Power | |
| Dimensional check of the blank, machined surface, or thermal drifts | *** | | | | | * | |
| Tool Wear | * | | *** | *** | | *** | * |
| Tool Breakage: | | | | | | | |
| in drilling | * | ** | | | ** | | |
| in turning | * | | *** | * | | | *** |
| in milling | * | | ** | * | | ** | ** |
| Preload in rolling bearings | | | | *** | | | |
| Friction in guideways resistance to feed | | | | | | *** | |
| Adaptive Control | | | *** | ** | | ** | |

## 3.2.2 Feature Extraction Methods

Monitoring methods can be sub-divided into two basic types: model-based and feature-based methods [3.2].

*Model-based methods* involve finding a model that fits the process and monitoring specific parameters in that model to detect changes. Changes in the expected system response can be interpreted as changes in the process, signifying an abnormal cycle. Model-based monitoring methods are also referred to as failure detection methods [3.31], and [3.32].

*Feature-based monitoring methods* use mapping processes to identify the process and tool conditions and to relate the tool conditions to the sensor features. Such techniques include pattern recognitions, expert systems, neural networks, and fuzzy classifications. The feature-based methods consist of two phases: *learning* and *classification*. Learning, also called training, is the procedure for establishing the system structure and classification rules. The knowledge for decision making is obtained from the learning samples as well as from instructions. Knowledge *updating* or *continuous learning* implies that the system is retrained with new available information. Monitoring tasks are done in classification phase. The structure and knowledge base built in the learning phase are used for decision making during monitoring.

### 3.2.3 Decision-making Methods

Decision-making strategies are one of the major issues in the development of automated machining process and tool condition monitoring. They range from simple threshold limit values being exceeded and triggering abnormal conditions, to systems requiring substantial training and learning to determine the process characteristics of each cutting condition and using these to detect out-of-specification conditions. A decision-making process in monitoring is based on the relationship between the process/tool conditions and the feature-bearing signals (monitoring indices).

Currently used feature-based monitoring methods include pattern recognition, fuzzy systems, decision trees, expert systems and neural networks.

Among the large number of decision-making methods that have been developed, statistical pattern recognition, neural networks and fuzzy classification are very interesting aspects in the development of automated/intelligent tool condition monitoring in machining. They have been applied successfully to many monitoring tasks in turning, milling, drilling, and other metal cutting processes.

The pattern recognition technique has been applied to recognize cutting states and to monitor tool conditions in machining for decades. The simplest and most popular algorithm is the linear classifier.

A linear model was used in applications of linear classifiers in [3.33], [3.34] and [3.35]. Features for classifying cutting states included cutting speed, feed and the power spectrum in different frequency bands. Features used for tool condition monitoring are usually feed rate, depth of cut, cutting force, cutting torque, sums of the magnitudes of spectral components at certain frequencies, and other signal features. Experiments have shown that the number of features and different combinations of features have a great effect on correct classification rates. An error-correction procedure was used to obtain weight vector. Arbitrary initial weight vectors were selected and adjustments were made whenever the classification scheme responded incorrectly to any pattern. The success rates of classification with these schemes were 77% and greater.

Other pattern recognition algorithms for tool condition monitoring in machining include the class-mean scatter criterion, the class variance criterion, and Fisher's weighted criterion [3.36][3.37]. The class-mean scatter criterion maximizes class separation and minimizes within-class variance. The class variance criterion maximizes the difference between the within-class variance of each class. Fisher's weight criterion maximizes class separation and minimizes the within-class variance between each pair of classes. This methodology was applied in order to detect tool wear and breakage in turning operations using acoustic emission spectral information under fixed cutting conditions. The tool wear sensing results had performances ranging from 84 to 94%. Research [3.38][3.39] divided pattern recognition systems into statistical and distribution-free methods.

*Statistical methods* are useful if the probability distribution of samples is fairly Gaussian in nature. As the name implies, they are based on a Bayes estimation of the probability that sample $x$ belongs to the process condition $h_i$. The initial $h_i$ value for the process condition is set during the learning phase by finding the minimum of Equation (3.1) for the process based on a number of learnt samples. Each $q_j$ is then calculated using Equation (3.2) during the classification phase for every sample $x$ and compared to find to which process condition $h_j$ the current sample belongs. The use of a cost, $C_{\alpha j}$, and probability density function $p_j$ helps to determine by how much the sample belongs to a particular condition, $h_j$.

$$h_j^* = \arg\min\left\{ q_j(x) \right\}$$

(3.1)

$$q_j(x) = \sum_{\alpha=1}^{n} p_j C_{\alpha j} f_j\left( {x}/{\Omega_j} \right)$$

(3.2)

*Distribution-free methods* are used if the samples are not Gaussian but instead can effectively be compared to a set of pattern vectors $p_j = [p_{j1}, p_{j2}, ..., p_{jm}]$ to determine which vector the sample vector $x$ best resembles. The pattern vectors are seen as the process conditions and are determined during the learning phase based on experimentation with preliminary data.

Monitoring problems using pattern recognition occur when there is a large grey area between process conditions or patterns, such as in drill wear, leading to development of fuzzy systems.

Bayes' rule is essential for classifications in statistical decision making. The function used to relate the monitoring indices and the tool conditions can be linear, polynomial, or in other forms.

## 3.3 Neural Networks for Tool Condition Monitoring

The tasks of an automated tool condition monitoring system involve the ability to recognize the tool condition by analyzing measured cutting process parameters such as forces and vibrations. This ability is based on the accumulation of useful information from related laws of physics and operators' experiences. In building automated/intelligent tool condition monitoring systems, some basic functions have to be considered:

1. fusion of multiple sensors;
2. learning or training strategies for the monitoring system;
3. knowledge updating techniques; and
4. description of the imprecision in tool conditions for various cutting conditions.

With the increasing need for effective and robust automated machining process and tool condition monitoring, a significant amount of research work has been performed to find decision-making strategies. The principal constituents of soft computation include fuzzy logic for imprecision in the acquired data, neural networks for learning, and probability reasoning for uncertainty. These three components are usually overlapped. "Soft computation" is easily implemented by fuzzy neural networks.

### 3.3.1 Structure of MPC Fuzzy Neural Networks

*Learning* refers to the processes which build the monitoring system in a given structure with information from the learning data. In addition, some logic rules are also created, which determine the data processing and govern the relationship between the processing elements. During the learning phase, a limited amount of data is used to adjust the parameters of the monitoring system. The trained monitoring system uses the stored knowledge to classify the data terms of tool conditions.

If the sampled data for training the system are labeled with the class to which the sample belongs, the decision making is performed with *a priori* knowledge. This is called pattern classification, or supervised classification, common in automated tool condition monitoring in machining. When the training samples are collected, the tool conditions related to each training sample are provided to give the necessary information.

Knowledge updating, or self-learning, essential for an automated tool condition monitoring system, refers to processes in which the structure and parameters of a monitoring system are modified according to the new information about the classification. Classification results should be checked online to ensure the system gives correct results. If the results are not correct, the system should be retrained or modified.

Neural networks are computing systems made up of a number of simple, highly interconnected processing elements that provide the system with the capability of self-learning. Using neural networks, simple classification algorithms can be used and the system parameters are easily modified. One major characteristic of building neural networks is the training time. Training times are typically longer when complex decision regions are required and when networks have more hidden layers. As with other classifiers, the training time is reduced and the performance improved if the size of a network is optimally tailored.

### 3.3.2 Construction of MPC Fuzzy Neural Networks

Multiple Principal Component (MPC) fuzzy neural networks are constructed based on the idea of "soft computation". Neural networks, fuzzy logic and statistical reasoning are employed. Simple classification procedures can be implemented at individual processing elements (neurons). The interconnections between neurons in the network communicate the information and make it possible to solve complex classification problems. Statistical reasoning is used in the learning procedure for the feature extraction and partition strategies.

For conventional neural networks, each of the processing elements (for input, output, and hidden layers) is always connected to every single processing element in the neighboring layers. Decision tree classifiers are hyperplane classifiers that can be regarded as a type of partially connected neural network since each node in the tree is connected to only its "father" and "sons", requiring comparatively less classification computations and can be implemented using parallelism from the decision region by performing simple, easily understood operations on the neural network. In more sophisticated implementations, multi-layered neural networks, consisting of nonlinear connections between inputs and outputs, are employed.

As an alternative to conventional neural networks, a partially connected, fuzzy neural network approach can be used for automated tool condition monitoring in machining. Different from matrix-type decision making, a tree structure is used for reducing unnecessary connections between elements in the input and the output layer. The fuzzy classifications are used in the neural networks to provide a comprehensive solution for certain complex problems. The neural network that utilizes fuzzy classification is shown in Figure 3.2.

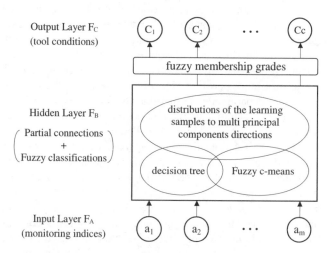

**Figure 3.2.** The Multiple Principal Component (MPC) fuzzy neural network

The input layer, $F_A = (a_1, a_2, ..., a_m)$, has m processing elements, one for each of the m dimensions of the input pattern $x_m$. The hidden layer of the network, $F_B$, consists of the neurons that use the fuzzy classification to separately address subsets of the original data set while invoking necessary information from other neurons. The probability distribution and the membership function are used for interconnections within the hidden layer and the connections to the output layer. The neurons of the output layer, $F_C$, represent the degrees to which the input pattern xk fits within the each class. There are two possible ways that the outputs of $F_C$ class nodes can be utilized. For a soft decision, outputs are defined with fuzzy grades. For a hard decision, $F_C$ nodes with the highest membership degree are located (a "winner-take-all" response). Connections within the hidden layer are not from one element to every one in the neighboring layers. The structure depends on the training data and is created through the training process. These partial connections result in simpler and faster training and classification.

### 3.3.3 Evaluation of MPC Fuzzy Neural Networks

A pattern classifier should possess several properties: online adaptation, nonlinear separability, dealing with overlapping classes, short training time, soft and hard decisions, verification and validation, tuning parameters, and nonparametric classification [3.40]. MPC fuzzy neural networks for automated tool condition monitoring have been developed by considering these requirements. The training and classification algorithms are based on the theories of neural networks, fuzzy logic, and probability reasoning.

For fuzzy classification, neural networks are effective in dealing with nonlinear separable and/or class-overlapping classification problems, which are common in tool condition monitoring in machining, especially for monitoring variations in cutting conditions. The partial interconnections within fuzzy neural networks make the training time very short compared to that of fully connected networks, such as back-propagation neural networks. The calculations necessary for classification are also significantly reduced since not all neurons in the hidden layer are used while a sample is being processed. Soft and hard decisions are optional for different applications. The maximum partition algorithm is based on the distributions of the learning samples and no parameter estimations are needed.

This method functions similarly, in the partition of training samples, to the linear fuzzy equation algorithm proposed by [3.41]. The linear fuzzy equation method uses a matrix to describe the relationship between the monitoring indices and the tool conditions. The proposed MPC fuzzy neural networks use a tree structure similar to that in the fuzzy decision tree described in the work of [3.42]. Because the decision tree is more flexible than a matrix approach, it has better performance in tool condition monitoring in machining. In constructing the fuzzy decision tree, the maximum partition generates nodes holding the samples from only one tool condition. The other samples are put into other nodes. This means each partition leads to a final decision at a leaf node of the tree. The maximum partition in the MPC fuzzy neural networks chooses a better partition so that a new-born neuron can hold samples from different tool conditions. A neuron can lead to other neurons in the hidden layer as well as neurons in the output layer. The consequence of this structure is simplicity in the interconnection and the short routines in the classification. Experimental tests using the same set of data showed that the MPC fuzzy neural networks gave a better success rate than the fuzzy decision tree algorithm [3.43].

For online adaptation (online learning) and tuning parameters, a classifier should have as few parameters to tune as possible. Both back-propagation neural networks and the proposed MPC fuzzy neural networks have very few tuning parameters. The structure of the MPC fuzzy neural network is, however, easily modified with new learning samples. Unlike back-propagation neural networks, which require complete retraining of the system with both old and new learning data, MPC fuzzy neural networks need only to partially change their neurons and connections when new learning information is added. Both supervised and unsupervised classification algorithms are easily implemented with the available learning samples.

The proposed classification algorithm for MPC fuzzy neural networks is based on the three components of "soft computation". In the learning procedure, partitions of the learning samples are carried out by examining distribution of the learning samples in multiple principal component directions. Like other statistical reasoning methods, "maximum partition" relies on the distribution of the data. If the data are scattered in feature space, a cluster centre is impossible to recognize. So the basis of classification is that the learning patterns are distributed around their centre(s). This is also believed to be true for all kinds of pattern classifiers. The major difference between the proposed algorithm and other statistical reasoning algorithms is that parameter estimation is not required.

To ensure good distribution of the learning data, the training samples have to be representative of the whole span of the feature space. In tool condition monitoring, all applicable tool and cutting conditions have to be considered during the training phase. On the other hand, if a poor distribution is encountered, then a modified feature extraction procedure has to be implemented. Information about new phenomena can be added to the monitoring system by knowledge updating.

### 3.3.4 Fuzzy Classification and Uncertainties in Tool Condition Monitoring

During machining, cutting conditions (*e. g.* cutting speed, feed, depth of cut) and tool conditions (*e. g.* tool wear) significantly affect process parameters such as cutting forces and vibrations, which are usually used as the input signals to a monitoring system. Deterministic models which attempt to describe the relationship between tool conditions and the various measured parameters are typically valid for only a limited range of cutting conditions. Fuzzy classification can be used to describe the uncertainties and overlapping relationships of tool conditions and monitoring indices. Briefly, the fuzzy expression for a tool condition $A$ can be defined by

$$A = \{x \mid \mu_A(x)\} \tag{3.3}$$

where $x$ is the value of $A$, and $\mu_A(x)$ is a fuzzy measure, also known as the membership function. $\mu_A(x)$ is a monotonous function, and $0 \leq \mu_A(x) \leq 1$. The function increases in accordance with the decrease in the uncertainty of $A$. If $B$ is also a fuzzy set and is more uncertain than $A$, then:

$$\mu_A(x) > \mu_B(x) \tag{3.4}$$

This might be interpreted as "the membership grade of small tool wear is greater than that of large tool wear". The fuzzy representation of tool conditions in machining has advantages. The concept of fuzzy decision making in machining tool condition monitoring is illustrated in Figure 3.3, where, $A_H$ and $B_H$ are categories classified by the hard decision, while $A_F$ and $B_F$ are classified by the fuzzy decision.

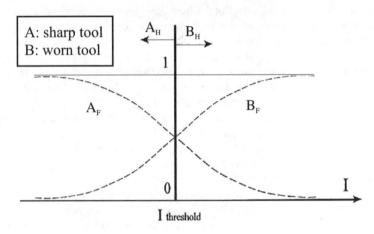

**Figure 3.3.** Soft boundaries in fuzzy classification

## 3.4 Case Studies

### 3.4.1 Experimental Tests on MPC Fuzzy Neural Networks for Tool Condition Monitoring

The proposed MPC fuzzy neural network (FNN) for automated tool condition monitoring in machining was tested for turning and drilling.

Several sensors were used to measure the process signals. Force, torque, vibration, and spindle motor power signals were fused using principal component analysis to produce highly sensitive features. These features were taken as the inputs to fuzzy neural networks for automated tool condition monitoring. The tool conditions considered in the monitoring tests included sharp tool, tool breakage, chipping and different states of tool wear. The classified tool conditions were obviously the outputs of the fuzzy neural networks.

Experiments were conducted to test the performance of MPC fuzzy neural networks in supervised classification, unsupervised classification, and knowledge updating. The learning and testing samples were obtained from cutting tests conducted over a reasonable range of cutting conditions.

## *Turning Experiments*

The five different tool conditions for the experiments are summarized in Table 3.3.

**Table 3.3.** Definition of the tool conditions in turning tests

| Tool Condition | Sharp Tool (SHP) | Breakage (BRK) | Slight Wear (SLW) | Medium Wear (MDW) | Severe Wear (SVW) |
|---|---|---|---|---|---|
| Tool Features | wear < 0.1 mm; no chipping | chipping > 0.04 mm$^2$ | 0.1 mm < wear < 0.16 mm | 0.16 mm < wear < 0.3 mm | 0.3 mm < wear < 0.6 mm |

## *Signal Conditioning and Feature Selection*

A schematic representation of the signal processing system is shown in Figure 3.4.

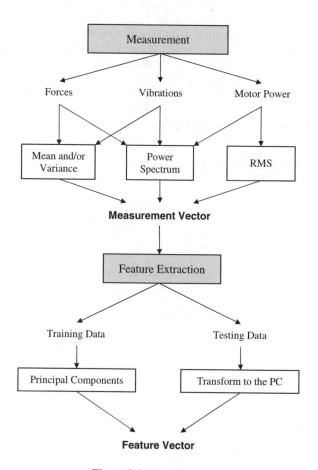

**Figure 3.4.** Signal processing

A total of 97 features were used to generate a measurement vector **x**. Of these features, nine were obtained from time domain records (length 2048) of the force, vibration, and power signals. From each time domain record of the forces, both the mean and the variance were calculated. The variances of the acceleration signals were also calculated. From the power records, the root-mean-square values were determined. In addition, 88 features were obtained from the power spectra of the force, vibration, and power signals.

### Experimental Results

#### Tests for Supervised Classification

Four sets of experimental data (namely A, B, C, and D) were generated from the cutting tests, under conditions presented in Table 3.4. For comparison, the results obtained using well-known feed-forward neural networks trained using back-propagation (BPNN) with the same sets of data are also given.

As shown in this table, any decision-making method performs better when using the same samples for both learning and classification. In general, the results show that using the proposed MPC fuzzy neural network for tool condition monitoring in machining, along with the integration of multi-sensor information, resulted in higher success rates than those obtained using BPNN. An important advantage of the proposed method is its good classification performance over a reasonable range of cutting conditions, even though the classification samples used were obtained under different cutting conditions from those of the training samples.

**Table 3.4.** Results of supervised classification with different tests

| Test | Training Data | Classification Data | Classification | Success Rate of FNN | Success Rate of BPNN |
|------|--------------|--------------------|----------------|--------------------|--------------------|
| #1 | Group A | Group A | Self-classification | 94.7 % | 89.3 % |
| #2 | | Group B | Different Record Same Cutting Condition | 89.3 % | 80.0 % |
| #3 | Group B | Group B | Self-classification | 94.7 % | 84.0 % |
| #4 | | Group A | Different Record Same Cutting Condition | 84.0 % | 70.7 % |
| #5 | Group C | Group C | Self-classification | 96.0 % | 86.7 % |
| #6 | | Group D | Different Cutting Condition | 80.0% | 69.3 % |

Figure 3.5 shows the detailed results of classifying the three stages of tool wear by two different neural networks. In detecting tool wear, the fuzzy neural networks give better separation between the three wear stages than the back-propagation trained neural networks. Another advantage of the proposed MPC fuzzy neural

networks over the back-propagation neural networks is the short training time. This is a critical issue when the system is used for online monitoring where self-improvement and self-adjustment are needed.

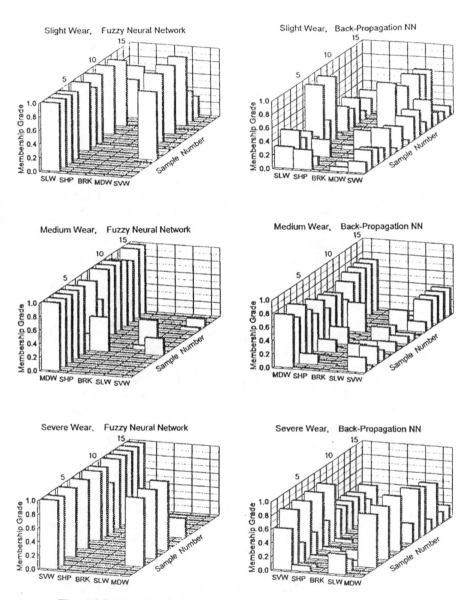

**Figure 3.5.** Classification results of tool conditions by two neural networks

For classifying tool wear, the MPC fuzzy neural network gives significantly better estimations. The results are detailed in Table 3.5.

**Table 3.5.** Detailed results of classification using two neural networks

| Test | Network | Sharp | | Breakage | | Slight W | | Medium W | | Severe W | | Total | |
|------|---------|----|------|----|------|----|-------|----|-------|----|-------|----|-------|
| #2 | FNN | 15 | 100% | 15 | 100% | 11 | 73.3% | 14 | 93.3% | 12 | 80.0% | 67 | 89.3% |
| | BPNN | 14 | 93.3% | 15 | 100% | 7 | 46.7% | 15 | 100% | 9 | 60.0% | 60 | 80.0% |
| #4 | FNN | 13 | 86.7% | 15 | 100% | 13 | 86.7% | 9 | 60.0% | 13 | 86.7% | 63 | 84.0% |
| | BPNN | 13 | 86.7% | 15 | 100% | 7 | 46.7% | 7 | 46.7% | 11 | 73.3% | 53 | 70.7% |
| #6 | FNN | 15 | 100% | 13 | 86.7% | 9 | 60.0% | 12 | 80.0% | 11 | 73.3% | 60 | 80.0% |
| | BPNN | 14 | 93.3% | 13 | 86.7% | 5 | 66.7% | 10 | 66.7% | 10 | 66.7% | 52 | 69.3% |

### Tests for Unsupervised Classification

The proposed MPC fuzzy clustering neural networks for tool condition monitoring were tested with experimental data obtained over a range of cutting conditions. Two sets of experimental data, named A and B, were randomly selected. Each set of data contained 75 samples obtained at the five tool conditions defined in this study (*i. e.* 15 samples for each condition).

To reduce the dimension of the original data vectors, samples were transformed into their principal component directions before being used as inputs to the networks. To investigate the effect of cluster numbers on the clustering results, three- and five-class tests were designed. The five-class tests included all the tool conditions mentioned earlier. In the three-class tests, the three stages of tool wear were combined into a single class, so the tool conditions were simply sharp tool, tool breakage, and tool wear. The result showed that the larger the number of clusters, the more levels are required in the hidden layer, and the more complicated the corresponding network is.

Figure 3.6 shows the fuzzy clustering neural networks built during the clustering of the data in set A. Each neuron in the networks receives information from the input layer as well as from other neurons at the preceding level in the hidden layer. The outputs of a neuron are sent to either the output layer and/or other related neurons at the next level. These connections are jointed with the fuzzy membership grades.

### Tests for Knowledge Updating

Two sets of experimental data, A and B, were generated from the cutting tests. Each set of data contained 75 labeled samples belonging to the five tool conditions and transformed into the six-dimensional monitoring feature space before being used for learning and testing.

Three experiments were designed to provide a comparison of the classification results using the monitoring systems trained by the different learning procedures. The results of these experiments are given in Table 3.6 and the comparison is illustrated in Figure 3.7.

The experimental results indicate that the proposed MPC fuzzy neural network for automated tool condition monitoring has the abilities of self-learning and knowledge updating. New knowledge in the retraining data is easily added into the system by a simple retraining procedure, and the classification results with the retrained system are improved.

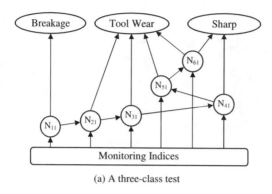

(a) A three-class test

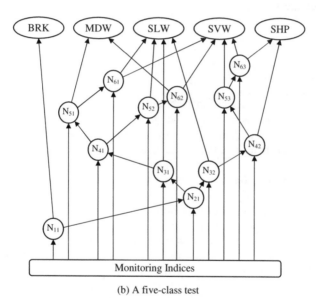

(b) A five-class test

**Figure 3.6.** Classification results of tool conditions by two neural networks

**Table 3.6.** Classification results for knowledge updating

| Test | Training Set | Retraining Set | Classification Results (with 30 testing samples in each tool condition) | | | | | | |
|------|--------------|----------------|------|------|------|------|------|------|------|
| | | | SHP | BRK | SLW | MOW | SVW | Total | |
| 1 | A | none | 27 | 30 | 26 | 25 | 24 | 132 | 88.0 % |
| 2 | A&B | none | 29 | 28 | 27 | 30 | 26 | 140 | 93.3 % |
| 3 | A | B | 27 | 30 | 30 | 27 | 21 | 135 | 90.0 % |

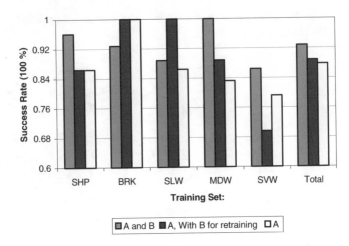

**Figure 3.7.** Comparison of the classification results with retraining

### Drilling Experiments

#### Definition of Tool Conditions

In drilling, tool wear changes along the cutting edge from the margin to the chisel edge due to the complex geometry of the drill bit and the cutting process. At the drill point, wear occurs at the flute (crater wear), the clearance face (flank wear), the chisel edge, and the margin. Flank wear was the main consideration in this study. The tool conditions in the drilling experiments comprised four categories: *Sharp tool* (SHP), *Chipping* (CHP), *Small wear* (SMW) and comprised *Large wear* (LGW).

#### Signal Conditioning and Feature Selection

The same signal processing scheme was used for the drilling tests as for the turning tests. The measurements were taken randomly during steady cutting processes for one second. The measurement vectors were further treated by principal component analyses to generate feature vectors for learning and classifications.

#### Experimental Results

Two sets of experimental data were obtained from the drilling tests. Both A and B were randomly generated within the experimental data obtained within the considered range of cutting conditions. After the principal component analysis was completed, three features were extracted from the experimental data to form the feature vectors. All the samples for learning and classification were transferred into these directions. Three tests, designed to evaluate the performance of the MPC fuzzy neural networks for tool condition monitoring during the drilling process, were carried out.

The performance with different learning data and system retraining was also tested and the comparison is illustrated in Figure 3.8.

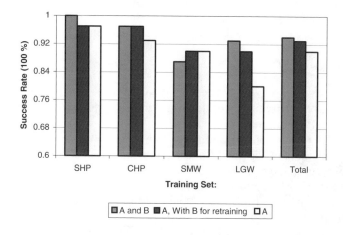

**Figure 3.8.** Comparisons of classification results for drilling tests

Table 3.7 gives details of these experimental results. The results show the good performance of the proposed MPC fuzzy neural networks for automated tool condition monitoring in drilling tool condition monitoring.

**Table 3.7.** Experimental results in drilling tests

| Test | Training Set | Retraining Set | Classification Results (with 30 testing samples in each tool condition) | | | | | |
|------|--------------|----------------|------|------|------|------|------|------|
| | | | SHP | CHP | SMW | LGW | Total | |
| 1 | A&B | none | 30 | 29 | 26 | 28 | 113 | 94.2 % |
| 2 | A | none | 30 | 26 | 24 | 25 | 105 | 87.5 % |
| | B | none | 29 | 28 | 27 | 24 | 108 | 90.0 % |
| 3 | A | B | 30 | 27 | 25 | 27 | 109 | 90.8 % |
| | B | A | 29 | 29 | 27 | 27 | 112 | 93.3 % |

## 3.4.2 Online Monitoring Technique for the Detection of Drill Chipping

The objective of this case study is to identify a detection strategy based on correlation of the vibration signal monitored online and the drill status, and to determine if vibration signals generated during uninterrupted drilling contain a signature correlated to chipping of the cutting edges. Two analysis techniques were used in this research; experimental testing was conducted to see which technique is better suited for reliable and computationally effective detection.

### Experimental Procedure

Chipping of the drill results in sharp transients occurring in the vibration signal. These transients resemble those of a resonating system responding to an impulsive excitation at frequencies independent of cutting conditions. In order to successfully correlate this event with chipping, and to determine if this is the signature of drill chipping, careful experimentation duplicating this phenomenon is necessary.

*Time Domain Analysis*

Preliminary analysis involved detecting repeated spikes in the unprocessed raw vibration data of the chipped drill, using kurtosis value calculation successfully verified by [3.14] with small diameter drills.

Amplitude time graphs obtained by plotting the vibration signal followed by calculation of the kurtosis values did not provide concrete evidence sufficient to detect chipping of large diameter drills. Figure 3.9 shows small spikes appearing randomly that may be correlated to material inconsistencies or chip accumulation and not drill chipping. To provide a comparison, the graph of the vibration signal using a sharp drill is illustrated in Figure 3.9(a) and Figure 3.10(a), and the graph of a drill cycle where chipping occurred is illustrated in Figure 3.9(b) and Figure 3.10(b). Figure 3.9(b) for the chipped drill shows minor spikes, but nothing distinct enough to use in a detection strategy. Figure 3.10(b) has large spikes occurring in the signal, whereby an amplitude threshold may be applied to classify drill condition. However, these spikes are primarily due to the belated total destruction of the drill's cutting edge and extreme wandering cases. As the goal of this case study is to detect minor chipping that may also lead to potential wandering and an out-of-specification hole condition without false alarms, it is essential to develop other analysis techniques providing unambiguous and reliable results.

*Frequency Domain Analysis*

Transformation of the data to the frequency domain provides information not readily noticeable in the time domain. It provides a visualization of the frequencies excited by the drilling process itself. The use of the Fast Fourier Transform (FFT) function in MATLAB® was used to plot the graphs in Figure 3.11 and Figure 3.12.

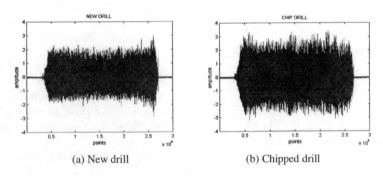

(a) New drill                              (b) Chipped drill

CUTTING PARAMETERS: N= 970 rpm, f=9 in/min, $f_s$=24,242 Hz
DRILL: 2 flute Chamdrill
WORKPIECE: H13 tool steel

**Figure 3.9.** Time domain signal (Test #1)

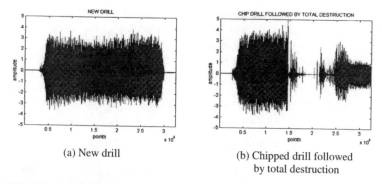

(a) New drill

(b) Chipped drill followed
by total destruction

CUTTING PARAMETERS: N= 970 rpm, f=8 in/min, $f_s$=24,242 Hz
DRILL: 2 flute Chamdrill
WORKPIECE: H13 tool steel

**Figure 3.10.** Time domain signal (Test #2)

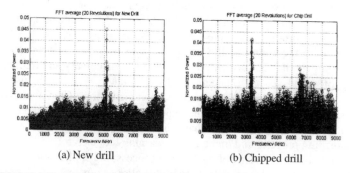

(a) New drill

(b) Chipped drill

CUTTING PARAMETERS: N= 970 rpm, f=9 in/min, $f_s$=24,242 Hz
DRILL: 2 flute Chamdrill
WORKPIECE: H13 tool steel

**Figure 3.11.** Frequency spectra (Test #1)

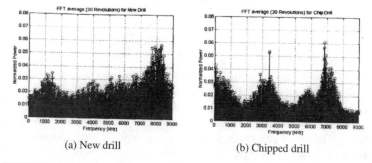

(a) New drill

(b) Chipped drill

CUTTING PARAMETERS: N= 970 rpm, f=8 in/min, $f_s$=24,242 Hz
DRILL: 2 flute Chamdrill
WORKPIECE: H13 tool steel

**Figure 3.12.** Frequency spectra (Test #2)

The FFT has been calculated based on an average of 20 drill revolutions, in order to damp out minor spikes attributed to inconsistencies in the workpiece and chip accumulation, arriving at better averages of the dominant frequencies over a fixed number of revolutions. Segments of 20 revolutions were sampled over the entire drill cycle. For each test case, the frequency spectra for a sharp drill is illustrated in Figure 3.11(a) and Figure 3.12(a) and that of a chipped drill is illustrated in Figure 3.11(b) and Figure 3.12(b) for comparison.

*Online Monitoring System Design Constraints*

The ability to provide a fast response within a few drill revolutions of the chipping event was a crucial requirement, as further chipping of the drill may lead to damage of the spindle motor, bearings, machine structure, or workpiece.

Since the system is designed to be used in a manufacturing environment where production stoppages are seen as economic losses, false alarms must be minimized and therefore reliability and consistency in detection are extremely important. The system hardware must be robust to handle the 'wear and tear' placed on it by harsh large-scale manufacturing environments. Flexibility and generalization are essential as not all drilling processes are the same. Therefore a learning cycle to train the system to adapt to the unique drilling characteristics of each process can contribute to providing this flexibility. Ease of implementation is necessary and minimizing the complexity of initial set-up is beneficial. Development of a simple user-friendly interface to guide the operator permits easy implementation of the application, and encourages more manufacturing companies to evaluate and implement this application. Since the system is online, chip detection must occur in real time, halting the drilling process, and immediately warning the operator.

To enable online monitoring of the drilling process, the application uses a special handling technique to buffer data. This technique allows to be held in a pre-specified amounts of data buffer, saved to smaller arrays, and analyzed while continuing to acquire further data – this feature allows the application to analyze data segments and determine if the drill is chipped without waiting until the end of the drill cycle.

## 3.5 Conclusions

### A. System Architecture

Three major components of "soft computation" are involved in the construction of the proposed fuzzy neural networks. The combination of fuzzy logic with neural networks has a sound technical basis because these two techniques approach the design of intelligent machines from different angles. Fuzzy neural networks employ the advantages of both neural networks and fuzzy logic. Neural networks offer good performance dealing with sensor information in parallel at a low computational level. The large number of interconnections within the networks gives the capabilities of exchanging information efficiently and managing nonlinearity. Fuzzy logic gives a means for representing, manipulating, and utilizing data and information that possesses non-statistical uncertainties.

## B. Supervised Classification

A supervised classification algorithm has been developed for the MPC fuzzy neural networks. "Maximum partition" is proposed to give better division of the involved classes. All the monitoring indices are examined against all the tool conditions to select the "pivot indices" for the maximum partition. Statistical methods are used to locate the class centers and the fuzzy membership is used for generating "soft" boundaries between the classes. Neurons in the network are generated with pivot indices, class centers, fuzzy membership grades and connections to other related neurons.

In classification, the fuzzy membership grades of the input pattern are calculated at the neurons. The membership grades define the path directions at each neuron for classification. The final results are given by fuzzy membership grades which measure the uncertainty in the classification.

## C. Unsupervised Classification

Unsupervised classification is based on principal component analysis and fuzzy membership of the potential clusters. Two major procedures for clustering at a neuron in the network are principal component analysis in multiple directions and cluster combinations. A fuzzy membership function is defined for the optimum combinations. The fuzzy neural network is built and these procedures are repeated in the learning procedure.

## D. Knowledge Updating

Knowledge updating is easily implemented due to structure of MPC fuzzy neural networks. The proposed knowledge updating algorithm deals with two issues: tuning old neurons and adding new neurons. The updating algorithm is developed with the assumption that the system has been trained and it keeps only the necessary information for classification. New information in the new training data is combined with the stored information learned previously. The old training data are not required. The main structure of the system is not destroyed and only some parameters and minor connections are modified.

## E. Sensor Fusion and Feature Extraction

Several sensors are used for selection of the monitoring features. The principal cutting mechanics are studied to choose the measurement of process signals which are the most sensitive to changes in machining tool conditions. Signals in both the time domain and frequency domain are used. The signals from multi-sensors of different types at different locations are fused by principal component analysis to produce highly information-bearing features (monitoring indices).

The monitoring indices are further analyzed in the learning procedure of the fuzzy neural networks. They are used selectively at each individual neuron in the form of the pivot index. The monitoring features are extracted in both the signal processing and learning procedures.

## F. Experimental Tests in Turning and Drilling

Experiments for testing the proposed MPC fuzzy neural networks for automated tool condition monitoring in machining were performed for turning and drilling.

The experimental results showed the good performance of the system with these tests. The success rates of self-learning were 94–96%. Those of learning with different time records under the same cutting conditions were 84–89%. Learning with samples under different cutting conditions gave 80% success rate. The proposed MPC fuzzy neural networks performed better than the back-propagation neural networks in theses tests. In unsupervised classification, the system gave success rates of 80–97%, depending on the number of classes (tool conditions) pre-defined. Knowledge updating improved the performance of the system by 50% when comparing the success rates of self-learning (all the samples used for learning) and half-learning (half the samples used for learning).

The proposed algorithm for the MPC fuzzy neural network is a fast, effective, and simple method for dealing with multi-sensor, multi-class, overlapped classification problems.

### G. Online Monitoring Techniques for Drill Chipping

An online monitoring technique capable of detecting drill chipping, based upon specified drill length and certain other process parameters was presented. Initial development work determined and verified the chipping signature using a signature based on a particular frequency being excited corresponding to the natural frequency in the torsional and longitudinal directions. This impulsive excitation signature of sharp spikes was used to determine techniques capable of reliably identifying the spikes among vibration data, leading to use of the technique in an online monitoring system.

Application of the DWT technique failed to reliably capture this signature due to poor frequency resolution at high frequency, and worse time resolution due to the removal of data points at lower frequencies. The CWT has been proven effective by some researchers, however, limitations arose in its ability to create an online monitoring classification script capable of analyzing its output map. Filtering and RAMV analysis methods were successfully applied to the acquired experimental data, used as the core of the monitoring system developed.

## References

[3.1]   Dornfeld, D. A., 1986, "Acoustic Emission Monitoring for Untended Manufacturing," *Proceedings of Japan/USA Symposium on Flexible Automation*, JAACE.

[3.2]   Du, R., Elbestawi, M. A. and Wu, S. M., 1995, "Automated Monitoring of Manu-facturing Processes, Part 1: Monitoring Methods," *ASME Journal of Engineering for Industry*, **117**, pp. 121–132.

[3.3]   Li, S., 1995, "Automated Tool Condition Monitoring in Machining Using Neural Networks," *PhD Thesis*, McMaster University, Hamilton, ON, Canada.

[3.4]   Ruparelia, S., 2002, "Online Monitoring Technique for detection of Drill Chipping", *M.Eng. Thesis*, McMaster University, Hamilton, ON, Canada.

[3.5]   Tonshoff, H. K. and Wulfsberg, J. P., 1988, "Developments and Trends in Monitoring and Control of Machining Processes," *Annals of the CIRP*, **37**(2), pp. 611–622.

[3.6]   Tlusty, J. and Andrews, G. C., 1983, "A Critical Review of Sensors for Unmanned Machining," *Annals of the CIRP*, **32**(2), pp. 563–572.

[3.7]  Iserrmann, R., 1984, "Process Fault Detection Based on Modeling and Estimation Methods – A Survey," *Automatica*, **20**(4), pp. 387–404.

[3.8]  Dornfeld, D. A., 1990, "Neural Network Sensor Fusion for Tool Condition Monitoring," *Annals of the CIRP*, **39**(1), pp. 101–105.

[3.9]  Dornfeld, D. A., 1992, "Monitoring of Machining Process – Literature Review," Presented at CIRP STC"C" Meeting, Paris, France.

[3.10]  Du, X., Dong, S. and Yuan, Z., 1999, "Discrete Wavelet Transform for Tool Breakage Monitoring," *International Journal of Machine Tools and Manufacture*, **39**, pp. 1935–1944.

[3.11]  Liu, H. S., Lee, B. Y. and Tang, Y. S., 1996, "Monitoring of Drill Fracture from the Current Measurement of a Three-Phase Induction Motor," *International Journal of Machine Tools and Manufacture*, **36**(6), pp. 729–738.

[3.12]  Xiaoli, L., 1999, "On-Line Detection of the breakage of Small Diameter Drills using Current Signature Wavelet Transform," *International Journal of Machine Tools and Manufacture*, **39**, pp. 157–164.

[3.13]  Everson, C. and Cheraghi, S., 1999, "The Application of Acoustic Emission for Precision Drilling Process Monitoring," *International Journal of Machine Tools and Manufacture*, **39**, pp. 371–387.

[3.14]  El-Wardany, T. I., Gao, D. and Elbestawi, M. A., 1996, "Tool Condition Monitoring in Drilling Using Vibration Signature Analysis," *International Journal of Machine Tools and Manufacture*, **36**(6), pp. 687–711.

[3.15]  Ruokangas, C. C., Blank, M. S., Matin, J. F. and Schoenwald, J. S., 1986, "Integration of Multiple Sensors to Provide Flexible Control Strategies," *Proceedings of the 1986 IEEE International Conference on Robotics and Automation*, pp. 947–953.

[3.16]  McClelland, S., 1988, "Tackling the Problem of Sensor Fusion," *Sensor Review*, **8**(2), pp. 89–91.

[3.17]  Chryssolouris, G. and Domroese, M., 1988, "Sensor Integration for Tool Wear Estimation in Machining," *Sensors and Control for Manufacturing*, ASME PED- **33**, pp. 115–123.

[3.18]  Chryssolouris, G. and Dornroese, M., 1989, "An Experimental Study of Strategies for Integrating Sensor Information in Machining," *Annals of the CIRP*, **38**(1), pp. 425–428.

[3.19]  Chryssolouris, G., Domroese, M. and Beaulieu, P., 1992, "Sensor Synthesis for Control of Manufacturing Process," *ASME J. of Engineering for Industry*, **114**, pp. 158–174.

[3.20]  Endres, W. J., Sutherland, J. W., DeVor, R. E. and Kapoor, S. G., 1990, "A Dynamic Model of the Cutting Force System in the Turning Process," *Monitoring and Control for Manufacturing Process*, ASME PED. **44**, pp. 193–212.

[3.21]  Kannatey-Asibu, E. Jr., 1985, "A Transport-Diffusion Equation in Metal Cutting and its Application to Analysis of the Rate of Flank Wear," *ASME Journal of Engineering for Industry*, **107**, pp. 81–89.

[3.22]  Koren, Y., Ko, T. R., Ulsoy, A. G. and Danai, K., 1991, "Flank Wear Estimation Under Varying Cutting Conditions," *ASME Journal of Engineering for Industry*, **113**, pp. 300–307.

[3.23]  Koren, Y., 1978, "Flank Wear Model of Cutting Tools Using Control Theory," *ASME Journal of Engineering for Industry*, paper No. 77-WA/PROD-20.

[3.24]  Koren, Y., Danai, K., Ulsoy, A. G. and Ko, T. R., 1987, "Monitoring Tool Wear Through Force Measurement," *Proceedings of the 15th NAMRI Conference*, pp. 463–468.

[3.25]  Matsumoto, Y., Tjinag, N. and Foote, B., 1988, "Tool Wear Monitoring Using a Linear Extrapolation," *Sensors and Control for Manufacturing*, ASME-PED **33**, pp. 83–88.

[3.26]  Yang, W.Q., Hsieh, S.H. and Wu, S.M., 1982, "Adaptive Modelling and Characterization for Machining Chatter," *Measurement and Control for Batch Manufacturing*, ASME, pp. 135–144.

[3.27]  Tsai, S.Y., Eman, K.F. and Wu, S.M., 1983, "Chatter Suppression in Turning," *Proceedings of the 11th NAMRC*, pp. 399–402.

[3.28]  Liang, S.Y. and Dornfeld, D., 1989, "Tool Wear Detection Using Time Series Analysis of Acoustic Emission," *ASME Journal of Engineering for Industry*, **111**, pp. 199–205.

[3.29]  Takata S. and Sata, T., 1986, "Model Referenced Monitoring and Diagnosis Application to the Manufacturing System," *Computers in Industry*, **7**, pp. 31–43.

[3.30]  Du, R., Elbestawi, M.A. and Wu, S., 1995, "Automated Monitoring of Manufacturing Processes, Part 2: Applications," *Journal of Engineering for Industry*, **117**, pp. 133–141.

[3.31]  Isermann, R., Ayoubi, M., Konrad, H. and Reib, T., 1993, "Model Based Detection of Tool Wear and Breakage for Machine Tools," *Proceedings of the IEEE International Conference System Manufacturing Cybernetics*, **3**, pp. 72–77.

[3.32]  Lee, B.Y., Liu, H.S. and Tarng, Y.S., 1998, "Modeling and Optimization of Drilling Process," *Journal of Material Processing Technology*, **74**, pp. 149–157.

[3.33]  Zhang, Y.Z., Liu, Z.F, Pan, L.X., Liu, Y.J. and Yang, W.B., 1982, "Recognition of the Cutting States for the Difficult-to-Cut Materials Application of the Pattern Recognition Technique," *Annals of the CIRP*, **31**(1), pp. 97–101.

[3.34]  Marks, J. and Elbestawi, M.A., 1988, "Development of Machine Tool Condition Monitoring System Using Pattern Recognition," *Sensors and Controls for Manufacturing*, ASME-PED-**33**, pp. 89–98.

[3.35]  Liu, T.I. and Wu, S.M., 1988, "On-Line Drill Wear Monitoring," *Sensors and Control for Manufacturing*, ASME-PED-**33**, pp. 99–104.

[3.36]  Emmell, E. and Kannatey-Asibu, E. Jr., 1987, "Tool Failure Monitoring in Turning by Pattern Recognition Analysis of AE Signals," *Sensors for Manufacturing*, ASME-PED. **30**, pp. 39–57.

[3.37]  Emmell, E. and Kannatey-Asibu, E. Jr., 1988, "Tool Failure Monitoring in Turning by Pattern Recognition Analysis of AE Signals," *ASME Journal of Engineering for Industry*, **112**, pp. 137–145.

[3.38]  Everitt, B., 1980, *Cluster Analysis*, Richard Clay Ltd.

[3.39]  Kandel, A., 1982, *Fuzzy Techniques in Pattern Recognition*, John Wiley & Sons.

[3.40]  Simpson, P.K., 1992, "Fuzzy Min-Max Neural Networks – Part 1: Classification," *IEEE Transactions on Neural Networks*, **3**(5), pp. 776–786.

[3.41]  Du, R.X., Elbestawi, M.A. and Li, S., 1992, "Tool Condition Monitoring in Turning Using Fuzzy Set Theory," *International Journal of Machine Tool and Manufacturing*, *32*(6), pp. 781–796.

[3.42]  Li, S., Elbestawi. M.A. and Du, R.X., 1992, "A Fuzzy Logic Approach for Multi-Sensor Process Monitoring in Machining," *Sensors and Signal Processing for Manufacturing*, ASME PED-**55**, pp. 1–16.

[3.43]  Li, S. and Elbestawi, M.A., 1994, "Tool Condition Monitoring in Machining by Fuzzy Neural Networks," *Dynamic Systems and Control*, ASME-DSC. **55**(2), pp. 1019–1034.

# 4

# Monitoring Systems for Grinding Processes

Bernhard Karpuschewski[1] and Ichiro Inasaki[2]

[1] Otto-von-Guericke-University Magdeburg
Post box 4120, D-39016 Magdeburg, Germany
Email: karpu@mb.uni-magdeburg.de

[2] Keio University
3-14-1 Hiyoshi, Kouhoku-ku, Yokohama-Shi, Japan
Email: inasaki@sd.keio.ac.jp

**Abstract**
This chapter is dedicated to the description of monitoring systems for grinding processes. Grinding is by far the most important abrasive process with geometrically non-defined cutting edges and plays a prominent role to generate the final surface quality of machined parts. The monitoring systems will be discussed in terms of their ability to measure process quantities during manufacturing, or on the grinding wheel or the workpiece. Monitoring of peripheral units like dressing systems will also be discussed. After the description of different technical solutions an outline of adaptive control and intelligent grinding systems is provided.

## 4.1 Introduction to Grinding Processes

Grinding, being the most prominent of all abrasive processes, is widely applied to achieve high accuracy and high quality mechanical, electrical and optical parts. This is due to the fact that modern grinding technology can meet the demands of not only high precision machining but also high material removal rate. For some difficult-to-machine materials, such as engineering ceramics, grinding is often the only economic manufacturing solution.

## 4.2 Need for Monitoring During Grinding

The behavior of any grinding process is very much dependent on the tool performance. The grinding wheel should be properly selected and conditioned to meet the requirements of the parts to be machined. In addition, its performance may change significantly during the grinding process, which makes it difficult to predict the process behavior in advance (Figure 4.1). Conditioning of the grinding wheel is necessary before the grinding process is started. It is necessary after the wheel has reached its useful life, to restore the wheel configuration and the surface topography to the initial state. This peripheral process needs sufficient sensor systems to minimize the auxiliary machining time, to assure the desired topography and to keep the amount of wasted abrasive material during conditioning to a minimum.

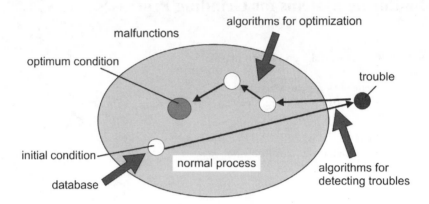

**Figure 4.1.** Roles of a monitoring system

Sensor systems for a grinding process should also be capable of detecting any unexpected malfunctions in the process with high reliability so that the production of substandard parts can be minimized. Some major troubles in the grinding process are chatter vibration, grinding burning and surface roughness deterioration. These troubles have to be identified in order to maintain the desired workpiece quality.

In addition to problem detection, another important task of the monitoring system is to provide useful information for optimizing the grinding process in terms of total grinding time or total grinding cost. Optimization of the process will be possible if degradation of the process behavior can be followed with the monitoring system. The information obtained from any sensor system during the grinding process can also be used to establish a database as part of an intelligent system [4.1].

## 4.3 Monitoring of Process Quantities

As with all manufacturing processes, it is most desirable to measure quantities of interest as directly and close to their origin as possible. Every grinding process is determined by a large number of input quantities, which may all have an influence on process quantities. Due to the interaction of tool and workpiece, material removal is initiated and the zone of contact is generated. Only during this interaction can process quantities be detected. The measurement of these quantities is of major interest to characterize the process. The most common sensors used in both industrial and research environments are force, power and acoustic emission sensors [4.2][4.3]. Figure 4.2 shows the set-up for the most popular integration of sensor systems in surface and outer diameter grinding.

## Force

The first attempts to measure grinding forces go back to the early 1950s and were based on strain gauges. Although the system performed well capturing a substantial data on grinding, the most important disadvantage of this approach was the quantity of significant reduction in total stiffness during grinding. Thus research was carried out to develop alternative systems. With the introduction of piezo-electric quartz force transducers, a satisfactory solution was found. Figure 4.2 shows different locations for these sensors during grinding. In surface grinding most often the sensor platform is mounted on the machine table carrying the workpiece. In inner (ID) or outer (OD) diameter grinding, this solution is not available due to the rotation of the workpiece. In this case, either the whole grinding spindle head is mounted on a platform or the workpiece spindle head and sometimes also the tailstock are put on a platform. For OD grinding, it is also possible to use ring type piezo-electric dynamometers. With each ring all three perpendicular force components can be measured; they are mounted under preload behind the nonrotating center points.

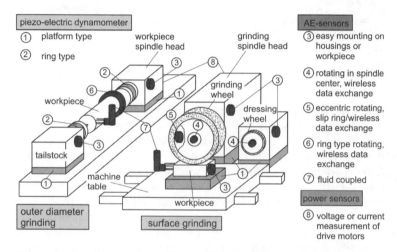

**Figure 4.2.** Possible positions of force, AE and power sensors during grinding

To complete possible mounting positions of dynamometers in grinding machines, dressing forces can also be monitored by the use of piezo-electric dynamometers, *e. g.* the spindle head of rotating dressers can be mounted on a platform. Besides these general solutions, a lot of special set-ups have been used for non-conventional grinding processes like inner diameter (ID) cut-off grinding of silicon wafers or ID grinding of long small bores using rod-shaped tools.

Generally speaking, the application of dynamometers can be regarded as state of the art. The only problems are high investment and missing overload protection.

## Power

The measurement of power consumption of a spindle drive can be regarded as technically simple. During grinding, the amount of power used for the material removal process is always only a fraction of the total power consumption. Nevertheless, power monitoring of the main spindle is widely used in industry to define specific thresholds to avoid overload of the whole machine tool due to bearing wear or any errors from operators or automatic handling systems. There are also attempts to use the power signal of the main spindle in combination with the power consumption of the workpiece spindle, to avoid grinding burn.

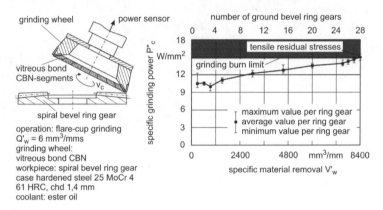

**Figure 4.3.** Power monitoring in spiral bevel gear grinding to avoid grinding burn

A typical result is shown in Figure 4.3 for a grinding process on spiral bevel ring gears, introducing a vitreous bond CBN grinding wheel [4.4]. Monitoring of grinding power revealed a constant moderate increase in material removal $V'_w$. At a specific material removal rate of 8100 mm$^3$/mm, grinding burn was detected for the first time, using nital etching. The macro- and micro-geometry of the 28th workpiece was still within tolerance, so the tool life criterion was taken as the surface integrity state. For this type of medium or large-scale production in the automotive industry, using grinding wheels with lifetime power monitoring is a very effective way to avoid thermal damage of the workpiece and also to get rid of the environmentally harmful etching process. These results reveal that power monitoring can be a suitable technique to avoid surface integrity changes during grinding. The most promising application is seen for superabrasives, because the slow wear increase of the grinding wheel can be clearly determined with this dynamical limited method.

*Acceleration*

In abrasive processes, the major application for acceleration sensors is related to balancing systems for grinding wheels, and detection of chatter. Large grinding wheels without a metal core may have significant inbalance at the circumference. With the aid of acceleration sensors the vibrations generated by this inbalance are monitored during rotation of the grinding wheel at cutting speed. Different systems are in use to compensate this inbalance, *e. g.* hydro compensators using coolant to fill different chambers in the flange, or mechanical balancing heads that move small weights to specific positions. Although these systems are generally activated at the beginning of a shift, they are able to monitor the change of the balance state during grinding and can continuously compensate the inbalance. In Figure 4.4 shows an example of acceleration measurement at this point [4.5]. The purpose of the research was to improve the poor dynamic stability of an inner diameter grinding process by investigating different measures. In the case shown, a carbon fiber reinforced plastic (CFRP) mandrel was used instead of steel material to improve the damping behavior. Although there were still significant amplitudes detected at high depths of cut, the behavior of this new mandrel was superior to that of the common steel material, at least with small material removal rates.

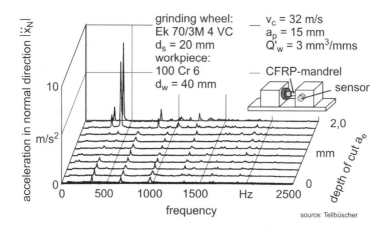

**Figure 4.4.** Acceleration in the normal direction at the spindle housing during ID grinding

### *Acoustic Emission*

Systems based on acoustic emission must be regarded as very attractive for grinding processes. Figure 4.2 points out the possible mounting positions for AE sensors on different components of a grinding machine. The spindle drive units, the tool, the grinding wheel or the workpiece can be equipped with a sensor. In addition, fluid coupled sensors are in use without any direct mechanical contact with the mentioned components. The time domain course of the root-mean-square value $U_{AE,RMS}$ is one of the most important quantities used to characterize the process state.

Besides time domain analyses the AE signal can be investigated in the frequency domain. Different effects, like wear or chatter vibrations, have different influences on the frequency spectrum, thus it should be possible to use frequency analysis to separate these effects. Figure 4.5 shows the result of a frequency analysis of the AE signal when outer diameter plunge grinding with a vitreous bond CBN grinding wheel [4.6]. As a very special feature, the AE sensor is mounted on the grinding wheel core and the signals transferred via a slip ring to the evaluation computer; thus grinding as well as dressing operations can be monitored. The results reveal that no significant peak can be seen after dressing and first grinding tests. After a long grinding time specific frequency components emerge in the spectrum which continue to increase in power throughout the test. The detected frequency is identical with the chatter frequency, which can be confirmed by additional measurements. AE signals were used as input data for a neural network to automatically identify the occurrence of chatter vibrations in grinding [4.6]. From the very first AE applications in grinding, researchers have tried to correlate the signal to the occurrence of grinding burn. The work of Klumpen and Saxler is directly related to the possibility of grinding burn detection with AE sensors [4.7][4.8]. One fundamental result is that all process variations, which finally generate grinding burn, like increasing material removal or infeed or reduced coolant supply, lead to an increase of the acoustic emission activity.

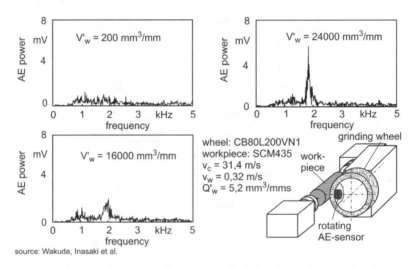

**Figure 4.5.** Acoustic emission frequency analysis for chatter detection in grinding

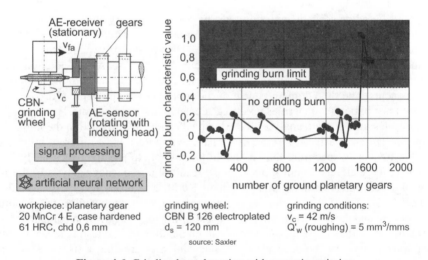

**Figure 4.6.** Grinding burn detection with acoustic emission

An analysis of the AE signal in the time domain is shown in Figure 4.6. Based on investigations and theoretical considerations, it was concluded that the AE sensor must be mounted to the workpiece to be most sensitive to grinding burn detection. This is of course a major drawback for practical applications. An industrial test was conducted during gear grinding of planetary gears with an electroplated CBN grinding wheel. Using artificial neural networks, a dimensionless grinding burn characteristic value was deduced from the AE values of different frequency ranges in the time domain. The large effort required for training the artificial neural network and the problems related to sensor mounting at the workpiece must be seen as limiting factors for wide industrial application. However, the results have clearly shown that AE systems can be regarded as suitable process quantity sensors in grinding to monitor surface integrity changes.

Due to the general advantages of AE sensors and their variety, almost any process with bonded abrasives can be investigated with the use of acoustic emission. Surface grinding as well as ID and OD grinding, centerless grinding, flexible disc grinding, gear profile grinding, ID cut-off grinding of silicon wafers, honing and grinding with bond abrasives on tape or film type substrates have all been the subject of AE research [4.3].

*Temperature*

In any abrasive process, mechanical, thermal and even chemical effects are usually superimposed in the zone of contact. Grinding of any kind generates a significant amount of heat, which may cause deterioration of the dimensional accuracy of the workpiece, undesirable changes of the surface integrity state or lead to increased wear of the tool. Figure 4.7 shows the most popular temperature measurement devices [4.3]. The preferred method for temperature measurement in grinding is the use of thermocouples. The second metal in a thermocouple can be the workpiece material itself; this set-up is called the single wire method.

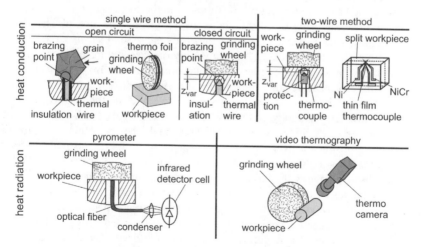

**Figure 4.7.** Temperature measurement systems used in grinding

A further distinction is made according to the type of insulation. Permanent insulation of the thin wire or foil against the workpiece by use of sheet mica is known as open circuit. The insulation is interrupted by the individual abrasive grains, thus, measurements can be repeated or process conditions varied until the wire is worn or damaged. Also the grinding wheel can be equipped with the thin wire or a thermo foil, if the insulation properties of abrasive and bond material are sufficient. Figure 4.8 shows the realization of a so-called "intelligent" grinding wheel with four integrated thermocouple sensors using a telemetry system for wireless data transmission [4.9].

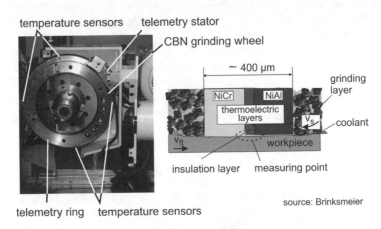

**Figure 4.8.** CBN grinding wheel with integrated temperature sensors

In the closed circuit type permanent contact of the thermal wire and the workpiece by welding or brazing is achieved. The most important advantage of this method is the possibility to measure temperatures at different distances from the zone of contact until the thermocouple is finally exposed to the surface. For the single wire method, it is necessary to calibrate the thermocouple for every different workpiece material. This disadvantage is overcome by the use of standardized thermocouples, where the two different materials are assembled in a ready-for-use system with sufficient protection. A large variety of sizes and material combinations are available for a wide range of technical purposes. With the two-wire method, it is again possible to measure the temperature at different distances from the zone of contact. A special variation of the two-wire method is the use of thin film thermocouples. The advantage of this method is an extremely small contact point to resolve temperatures in a very small area and the possibility to measure a temperature profile for every single test depending on the number of evaporated thermocouples in simultaneous use. In all cases, either single or two-wire methods, the major disadvantage is the large effort required to carry out these measurements. Because of the necessity to install the thermocouple as close to the zone of contact as possible, it is always a technique where either grinding wheel or workpiece have to be specially prepared. Thus all these methods are used only in fundamental research; industrial use for monitoring is not possible due to the partial destruction of major components.

Besides these heat conduction based methods, the second group of usable techniques is related to heat radiation. Infrared radiation techniques have been used to investigate the temperature of grinding wheel and chips. By the use of a special infrared radiation pyrometer, with the radiation transmitted through an optical fiber, it is even possible to measure the temperature of working grains of the grinding wheel just after cutting [4.10]. Also the use of coolant is possible and its effects can be evaluated. In all cases, these radiation-based systems need careful calibration, taking into account the properties of the material to be investigated, the optical fiber characteristics and the sensitivity of the detector cell. But again for most investigations, preparation of the workpiece is necessary, as shown in Figure 4.7 bottom left. The second heat radiation based method is using thermography. For this type of measurement, the use of coolants is always a severe problem, because the initial radiation generated in the zone of contact is significantly reduced in the mist or direct flow of the coolant until it is detected in the camera. Thus major applications of this technique are limited to dry machining. Brunner was able to use a high speed video thermography system for OD grinding of steel to investigate the potential of dry or minimum quantity lubrication (MQL) grinding [4.11].

## 4.4 Sensors for the Grinding Wheel

The grinding wheel state is of great importance for good results. The tool condition can be described by the characteristics of the grains. Wear can lead to flattening, breakage and even pullout of whole grains. Moreover, the number of cutting edges and the ratio of active/passive grains are of importance. Also the bond of the grinding wheel is subject to wear. The hardness and composition influence significantly the described variations in the grains. In all cases, wheel loading generates negative effects due to insufficient chip removal and coolant supply. All these effects can be summarized as grinding wheel topography, which changes during the tool life between two dressing cycles. As a result of the grinding process, the diameter of the grinding wheel reduces thoughout its life. In most cases, dressing cycles have to be carried out without any information about the actual wheel wear. Commonly, grinding wheels are dressed without reaching their end of tool life in order to prevent workpiece damage, e. g. workpiece burn. Many attempts have been made to describe the surface topography of a grinding wheel and to correlate surface quantities to workpiece results. Here, only those methods will be discussed which are capable at being used in the grinding machine during rotation of the tool. If only the number of active cutting edges is of interest, some techniques previously introduced can be used. Either piezo-electric dynamometers or thermocouple methods have been used to determine the number of active cutting edges.

In Figure 4.9, other methods suitable for dynamic measurement of the grinding wheel are shown. Most of the systems are not able to detect all micro- and macro-geometrical quantities, but can only be used for special purposes.

### Sensors for Macro-geometrical Quantities

The majority of sensors are capable of measuring macro-geometrical features. Any kind of mechanical contact of a sensor with the rotating tool causes serious

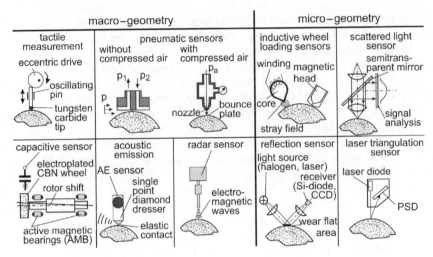

**Figure 4.9.** Sensors for grinding wheel topography measurement

problems, because the abrasives always tend to grind the material of the touching element. Only by realizing short touching pulses with small touching forces and by using a very hard tip material like tungsten carbide, is it possible to achieve satisfactory results. For instance, such a system with an eccentric drive to provide oscillation of the pin to measure the radial wheel wear at cutting speeds of up to 35 m/s has been used. Another group of sensors for the measurement of grinding wheels is based on pneumatic systems. Although this method is, in principle, also not able to detect micro-geometrical features of a grinding wheel due to the nozzle diameter of 1 mm or more, they are important for determining macro-geometry.

Systems with compressed air supply and those without have to be distinguished. These systems are capable of measuring distance changes, namely, radial wear, with a resolution of 0.2 μm. This feature and the comparatively easy set-up and moderate costs are the main reasons that pneumatic sensors have found acceptance in industrial applications.

One possibility to register the macro-geometry of an electroplated CBN grinding wheel at very high circumferential speed is to apply a capacitive sensor. Using the rotor shift option of active magnetic bearings (AMB), it is possible to compensate any inbalance of these non-dressable tools [4.12].

The AE signal can be used to determine the macro-geometry of the grinding wheel. In [4.13], a system was proposed consisting of a single point diamond dresser equipped with an AE sensor to detect exactly the position of the grinding wheel surface. This concept was further improved for rotating dressing tools [4.14].

Another principle used to determine radial wheel wear is based on a miniature radar sensor [4.15]. Usually, radar techniques in speed and traffic control have a maximum accuracy in the centimeter range. Sensors used for grinding work on an interferometric principle. With an emitting frequency of 94 GHz and a wavelength $\lambda = 3.18$ mm, this sensor has a measuring range of 1 mm and a resolution of 1 μm. Its main advantages are robustness against any dust, mist or coolant particles and the possibility to measure on any solid surface.

## Sensors for Micro-geometrical Quantities

Besides the systems for macro-geometrical features, other sensor are able to give information about the micro-geometry. The loading of a grinding wheel with conductive metallic particles enables micro-geometrical wear to be detected by using sensors based on inductive phenomena. Also a conventional magnetic tape recorder head may be used to detect the presence and relative size of ferrous particles in the surface layer of a grinding wheel.

The limitations of the techniques described thus for suggest the use of optical methods. They appear promising because of their frequency range and lack of dependence on the surface material. A scattered light sensor has been used to determine the reflected light from a grinding wheel surface using CCD arrays. Tests have shown that the number of pulses change during grinding, thus a possible scheme for monitoring the wear state was proposed [4.3]. A highly technical optical method has been proposed based on laser triangulation. Figure 4.10 shows some results of macro-geometrical measurements.

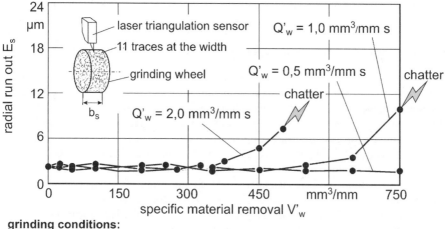

**grinding conditions:**
grinding wheel: EKW 80K5 V62
$q = 60$, $v_c = 30$ m/s $= v_{mea}$
spark out time: 10 s

multi-point diamond dresser
workpiece: 100 Cr 6 (ball bearing steel)

**Figure 4.10.** Optical macro-geometrical grinding wheel topography measurement

The change in radial run out as function of material removal rate at 30 m/s has been documented at three different material removal rates. For the smallest material removal rate, no change is detectable from the initial value after dressing. For increasing material removal rate, the radial run out increases beyond a specific threshold material removal rate. In the latter cases, the increasing radial run out leads to chatter vibrations, with visible marks on the workpiece surface. Obviously, the system is capable of detecting significant macro-geometric changes due to wear of the grinding wheel.

The examples shown for grinding wheel sensors reveal that the majority of systems are related to macro-geometrical features. However, many attempts have been made to develop optical systems for the measurement of micro-geometrical quantities. The overall limitation of these techniques will always be the dirty conditions in the working space of a grinding machine with coolant and process residues in direct contact with the object to be measured. In many cases, it is thus preferable to directly measure the manufactured workpiece itself.

## 4.5 Workpiece Sensors

Two essential quality aspects determine the results of a grinding process on a workpiece. On the one hand, geometrical quality demands have to be fulfilled: these are dimension, shape and waviness as essential macro-geometrical quantities. Roughness is the main micro-geometrical quantity. On the other hand, increasing attention is also paid to the surface integrity state of a ground workpiece due to its significant influence on the functional behavior. The physical properties are characterized by changes in hardness and residual stresses on the surface and in subsurface layers, and by changes in the structure and the likely occurrence of cracks.

### *Contact-based Workpiece Sensors for Macro-geometry*

The determination of macro-geometrical properties of workpieces during manufacturing is the most common application of sensors in abrasive processes, especially grinding. For decades, contacting sensors have been used to determine dimensional changes in workpieces during manufacturing. A large number of in-process gauges for any kind of operation are available. In ID- or OD-grinding, the measuring systems can be either comparator or absolute measuring heads, with the capability of automatic adjustment to different part diameters.

The contacting tips are usually made of tungsten carbide, combining the advantages of wear resistance, moderate costs and sufficient frictional behavior. The repeatability is in the order of 0.1 μm [4.16]. Internal diameters can be gauged starting from 3 mm size. If constant access to the dimension of interest during grinding is possible, these gauges are often used as signal sources for adaptive control (AC) systems. The conventional technique for measuring round parts rotating around their rotational axis can be regarded as state of the art. The majority of automatically operating grinding machines are equipped with these systems. Even geometry measurement during eccentrical movement of a crankshaft pin is now possible [4.17]. This is true for grinding (Figure 4.11 top left) as well as for a finishing process (Figure 4.11 bottom left).

The detection of waviness on the circumference of rotation symmetrical parts during grinding is more complex due to the demand for a significantly higher scanning frequency. A system was developed with three contacting pins at nonconstant distances to detect the development of waviness on workpieces during grinding as a result of regenerative chatter [4.16] (Figure 4.11 top right). Only by using this set-up was it possible to identify the real workpiece shape, taking into account the vibration of the workpiece center during rotation.

The last example of contact-based macro-geometry measurement in a machine tool is related to gear grinding (Figure 4.11 bottom right). For manufacturing small batch sizes or single components of high value, it is essential to fulfil the "first part good part" philosophy. For these reasons, several gear grinding machine tool builders have decided to integrate an intelligent measuring head in their machines to measure the characteristic quantities of a gear, such as flank modification, pitch or root fillet. Of course the measurement can only be done if the manufacturing process is interrupted. But still the main advantage is a significant time saving. Any removal of the part from the grinding machine tool for checking on an additional gear measuring machine will take a longer time. Also the problem of a loss of precision due to rechucking is overcome, because the workpiece is rough machined, measured and finished in the same set-up. These arguments are generally true for any kind of high value part with small batch size and complex grinding operations. Thus it is not surprising that also in the field of aircraft engine manufacturing new radial grinding machines are equipped with the same kind of touch probe system in the working space. Geometrical quality data are acquired on the machine tool before the next grinding operation in the same chuck position is started [4.18].

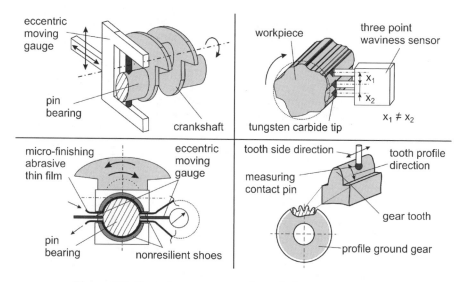

**Figure 4.11.** Contact sensor systems for workpiece macro-geometry

### Contact-based Workpiece Sensors for Micro-geometry

The determination of micro-geometrical quantities on a moving workpiece using contacting sensor systems is a challenging task. Permanent contact of any stylus with the surface is not possible, because the dynamic demands are much too high. Only intermittent contact can be used to generate a signal, which should be proportional to the roughness. Different sensor systems have been proposed, such as oscillating diamond tips on a parallel spring set-up, or rotating wheels with an inbuilt touching pin, but they did not lead to industrial success [4.3].

## Noncontact-based Workpiece Sensors

All the above-mentioned restrictions of contacting sensor systems on a workpiece surface gave a significant push to the development of noncontact sensors. As for grinding wheels, again optical systems seem to have great potential. In Figure 4.12, different optical systems as well as two other noncontacting sensor principles are introduced. As a very fast optical system to measure macro-geometrical quantities, a laser scanner is shown. The diameter is a function of the speed of the polygonal mirror and the time the laser beam does not reach the covered photo diode. Conicity can be evaluated by axial shifting of the workpiece [4.19]. For the determination of macro- and micro-geometrical quantities, a different optical system has to be applied. The basis of a scattered light sensor for the measurement of both roughness and waviness is the angular deflection of nearly normal incident rays. The set-up of a scattered light sensor is shown in Figure 4.12, bottom left. A beam-splitting mirror guides the reflected light to an array of diodes. A commercially available system was introduced in the 1980s [4.20] and used in a wide range of tests.

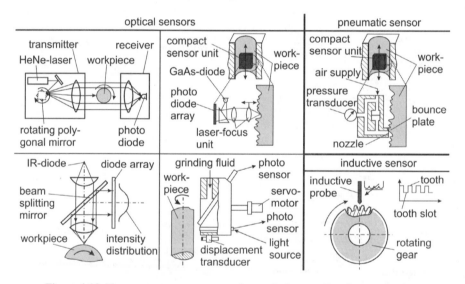

**Figure 4.12.** Non-contact sensor systems for workpiece quality characterization

The optical roughness measurement quantity of this system is called the scattering value $S_N$ and is deduced from the intensity distribution. It was possible to establish a correlation between optical and stylus roughness measurements [4.3].

A different optical sensor is based on a laser diode (Figure 4.12, top middle). The sensor is equipped with a gallium-arsenide diode, which is commonly used in a CD player. The beam is focussed on the surface with a lens system and the reflected light is registered on an array of four photodiodes. This system can be used as an autofocus system; the focus-lens is moved until the best position for minimum diameter of the signal from the four photodiodes is reached. Another optical approach for in-process roughness measurement is based on the use of optical fiber sensors [4.21]. The workpiece surface is illuminated through fiber optics and the intensity of the reflected light is detected and evaluated (Figure 4.12 bottom middle).

Besides these systems, other optical techniques for online measurement of surface topography have been proposed, including speckle patterns. Although the measurement speed suggests that these systems may be installed in production line conditions, practical use as sensors in the machine tool working space is not realistic.

In summary, due to the problems related to coolant supply, it must be stated that these conditions do not allow the use of optical systems during grinding as reliable and robust industrial sensors. Only optical sensor applications measuring during interruptions of the coolant supply either in the working space of the machine tool or its direct surroundings have gained importance in industrial production.

In addition to optical sensors, two other principles are also used for noncontact workpiece characterization. A pneumatic sensor as shown in Figure 4.12 top right was designed and used for the measurement of honed cylinders [4.22]. The measurement is based on the nozzle-bounce plate principle. Correlation with stylus measurements is possible. The main advantages of this system are the small size, the robustness against impurities and coolant, and the fact that an area but a trace is evaluated. So, in principle, movement of the sensor during measurement is not necessary.

The final noncontact workpiece sensor described is based on an inductive sensor. The sensor is used in gear grinding machines to identify the exact position of tooth and tooth slot at the circumference of the pre-machined and usually heat-treated gear (Figure 4.12, bottom right). The gear is rotating at high speed and the signal obtained is evaluated in the control unit of the grinding machine. This signal is used to index the gear in relation to the grinding wheel to define the precise position to start grinding and to avoid any damage of a single tooth. It is also possible to detect errors in tooth spacing. Gears with intolerable distortions after heat treatment can be identified and rejected to avoid overload of the grinding wheel, especially when using CBN as abrasive.

### Contact-based Workpiece Sensors for Surface Integrity

The best way to investigate the influence of any grinding process on the physical properties of the machined workpiece would be to directly measure the generated surface. But until now only very few sensors have been available to meet this demand. In the following, two techniques will be explained, which have the greatest potential for this purpose, an eddy-current and a micro-magnetic method (Figure 4.13).

The principle of eddy-current measurement for crack detection is based on the fact that cracks at the workpiece surface disturb the eddy current lines, which are in the measuring area of a coil with alternating-current excitation. An eddy-current sensor was used to monitor cracks generated during profile surface grinding of turbine blade roots. Cracks in the workspace of the machine tool were detected by moving a bridge with the sensor over the ground surface [4.23]. Crack detection is possible, but the measurement speed is lower than the grinding table speed, thus measurements have to be performed after grinding. Furthermore only cracks in the surface of the ground workpiece can be identified.

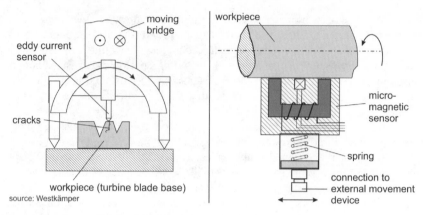

**Figure 4.13.** Workpiece surface integrity sensors

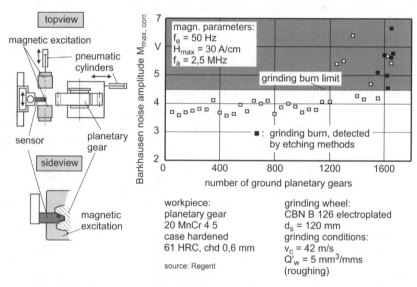

**Figure 4.14.** Micro-magnetic surface integrity characterization of ground planetary gears

The measuring principle of micro-magnetic techniques is based on the fact that residual stresses, hardness values and the structure in subsurface layers influence the magnetic domains of ferromagnetic materials. The existence of compressive stress in ferromagnetic materials reduces the intensity of the so-called Barkhausen noise, whereas tensile stresses increase the signal. A magnetization device and a signal receiver have to be adapted to the shape of the workpiece. In [4.24], a detailed investigation of the potential of the two-parameter micro-magnetic approach to characterize surface integrity states of workpieces with different heat treatments was described. Figure 4.14 shows the results of a large industrial test on planetary gears ground with electroplated CBN-grinding wheels [4.3][4.25]. The geometrical fitting of the micro-magnetic sensor and excitation is shown left in top and side view. It can be seen that all gears with identified grinding burn by nital etching are also recognized with the micro-magnetic set-up. But in addition, gears with high $M_{max, corr.}$ values are identified, which do not show any damage under nital etching. A possible explanation for this is the different penetration depth of the methods. Nital etching gives information about only the very top layer of the workpiece. Micro-magnetic measurements can reveal deeper damage, depending on the frequency used. It has to be further mentioned that the measuring time required to scan all flanks of one gear is significantly higher than the grinding time. With intelligent strategies or an increased number of sensors in parallel, this time can be shortened for suitable random testing. Furthermore, initial tests on in-process measurements of surface integrity changes based on this micro-magnetic sensing have been conducted in the laboratory for outer diameter and surface grinding [4.25].

## 4.6 Sensors for Peripheral Systems

Primary motion between tool and workpiece characterizes the grinding process, but also supporting processes and systems are of major importance. In this chapter, monitoring of the conditioning process and the coolant supply will be discussed.

### Sensors for Monitoring the Conditioning Process

The condition of the grinding wheel is very important for achieving the desired results from the grinding process. Thus the grinding wheel has to be prepared for the desired purpose using a suitable conditioning technology. The major problem in any conditioning operation is the possible difference between nominal and real conditioning infeed. There are four main reasons for these deviations. The unknown radial grinding wheel wear after removal of a specific workpiece material volume must be regarded as a significant factor. Also, the changing relative position of grinding wheel and conditioning tool due to thermal expansion of machine components is relevant. As a third reason, infeed errors related to friction of the guideways or control accuracy have to be considered, although their influence is declining in modern grinding machines. The last reason to mention is wear of the conditioning tool, which of course depends on the individual type. For rotating dressers, only after regular use for several weeks can first wear effects be registered.

Due to the immense importance of grinding wheel topography, monitoring of the conditioning operation has been a subject for research for many decades. Starting with the monitoring of a static single point diamond dresser on conventional abrasives, complex contact detection has been extended to rotating dressers for superabrasives like CBN [4.3][4.21][4.26][4.27]. The great hardness and wear resistance of these grinding wheels require a different conditioning strategy and monitoring accuracy compared to conventional abrasives. The conditioning intervals due to the superior wear resistance can amount to several hours. The dressing infeed should be limited to the range between 0.5 and 5 μm instead of 20 to 100 μm for conventional wheels in order to save wheel costs. For vitreous bonded CBN grinding wheels, it was proposed to use very small dressing infeeds more frequently in order to avoid additional sharpening. This strategy, known as "touch dressing", required the establishment of a reliable contact detection and monitoring system for dressing superabrasives. In most cases, rotating dressing tools are used. The set-up of a conditioning system with a rotary cup wheel, which is often used on internal grinding machines, is shown in Figure 4.15.

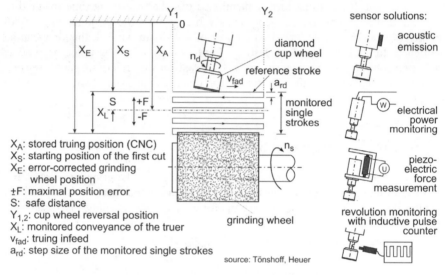

**Figure 4.15.** Dressing monitoring with rotating diamond tools

The conditioning cycle consists of four stages: fast approach, contact detection, defined infeed and new initiation. Setting parameters such as number of strokes, dressing infeed and dressing feed rate are stored in the control unit of the grinding machine. Besides AE techniques, other methods can also be used. Heuer investigated the possibility of using either the required power of the dressing tool spindle or a piezo-electric force measurement for monitoring [4.27]. The latter technique was available, because a piezo-electric actuator was installed as a high precision positioning system for the infeed of the dressing tool. With this additional equipment, infeed amounts of 0.25 μm were realized. Thus this system may also upgrade older machine tools with less accuracy. In a modern machine tool, the built-in x-axis provides the infeed. A further technique is measurement of the rotational speed change of the high frequency dressing spindle, which gives a maximum number of revolutions of 60,000 min$^{-1}$. After contact detection using any of the mentioned systems, the conditioning program is continued until the desired number of strokes and infeed is reached. Depending on the type of system, it is possible to monitor the course of the signal over the whole width of the grinding wheel. Thus uneven macro-geometrical wear of the wheel can be registered, if the measured signal does not exceed the defined threshold reference level over the whole width. This strategy also assures perfect macro-geometrical shape after conditioning. After finishing the conditioning process, the final position of the machine x-axis is stored to initialize the next operation.

The use of AE sensors for contact detection during conditioning and dressing operations can be regarded as state of the art. Many different systems are available. New grinding machine tools with self-rotating conditioning tools are usually equipped with an AE system. Systems with dressing spindle rotational speed monitoring have found acceptance in industry, because this system is regarded as very robust and is not influenced by coolant supply or bearing noise, which is still regarded as the major limitation of AE systems.

### *Sensors for Coolant Supply Monitoring*

Grinding operations are characterized by relatively large contact areas. The large number of cutting edges generates a considerable amount of heat in the contact zone. Thus, reduction of friction and cooling of the interacting parts are often necessary to avoid thermal damage. Therefore in almost all cases, coolants are used to reduce heat and to provide lubrication. These are the main functions of any coolant supply. The removal of chips and process residues from the workspace of the machine tool, the protection of surfaces and human compatibility should also be provided. Modern coolant compositions also try to fulfil the contradictory demands of long-term stability and biological recyclibility. With the wider use of superabrasives like CBN, the possibility of high speed grinding and highly efficient deep grinding, a closer view of the coolant supply began. Brinksmeier and Heinzel made a very systematic approach to investigate coolant-related influences and to optimize the relevant parameters and designs. For development of a suitable shoe nozzle design, a special flow visualization technique was used, as shown in Figure 4.16 [4.28][4.29].

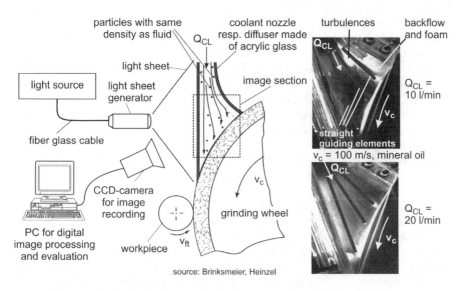

**Figure 4.16.** Flow behavior monitoring by means of particle image velocimetry

Tracer particles with almost the same density are added to the transparent fluid. All important parts of the nozzle are made from acrylic glass and a CCD-camera is used to record the flow images perpendicular to the light sheet plane. Although only a qualitative result is available, this technique offers the possibility to systematically study and improve the whole design of coolant nozzles. As an example on the right side of Figure 4.16, the flow behavior of a nozzle with straight guiding elements is shown at two different flow rates. The coolant is mineral oil and the grinding wheel rotates at a speed of 100 m/s. For the smaller flow rate of 10 l/min inhomogeneous flow behavior can be observed. Turbulences, backflow and foam between top and center guiding elements and at the entry side of the grinding wheel are visible. A doubling of the flow rate leads to steady flow behavior.

Besides this use of an optical monitoring method to optimize the design of coolant nozzles, special sensors for pressure investigations in combination with force measurement were introduced [4.29]. This method is suitable to investigate the influence of different coolant compositions. The efficiency of additives can be evaluated, if the coolant pressure force component is known and can be subtracted from the total normal force to emphasize the effect on cutting, friction and deformation component.

The use of special sensor systems for coolant supply investigations is a relative new field of activities. Initial results have shown that these sensors can contribute to a better understanding of the complex thermo-mechanical interaction in the zone of contact. Also, direct industrial improvements such as coolant nozzle optimization or additive efficiency evaluations for grinding can be performed.

## 4.7 Adaptive Control Systems

In this section, the use of the above-described sensor systems for automated feedback of their signals to the control unit of the grinding machine tool will be discussed. The main problem of any grinding operation is the unstable process behavior. For example, in inner diameter (ID) grinding, the limited stiffness of the small grinding spindle leads to significant deflections, accompanied by fast changes of grinding wheel topography, and thus cutting performance. These changing process conditions require fast measurement of relevant quantities for online control of the operation through feedback of necessary information. To explain this idea the situation in inner and outer diameter grinding is described [4.30]. In a so-called conventional grinding process, the infeed motion of the grinding wheel is a setting parameter in the control unit of the machine tool. No measurement of effective process quantities and no feedback of signals is used. This is a pure path-related speed control (Figure 4.17, left). The initial cut changes dependent on the oversize variations. Due to the elasticity of the system grinding wheel, workpiece and machine tool, the normal force in the contact zone increases very slowly. The force level at the end of the roughing process is different, leading to different levels of deflection and resulting in different size deviations even after finishing and spark-out. In comparison to that, it is the purpose of an adaptive control system to keep one or more process quantities at a defined level throughout the entire grinding process. In Figure 4.17 right, the example of a force-controlled adaptive grinding process is shown. The air grinding time is reduced to a minimum, because the grinding wheel approaches the workpiece surface with higher infeed speed. After contact is detected, the infeed speed is switched to reach the nominal value of force in a very short time. The force is then kept constant until the end of the roughing position of the grinding wheel is reached. Regardless of the different grinding times, which depend on the oversize, a constant defined process condition is assured. Different variations of force profile during finishing are possible to the end of the cycle, e. g. a linear decrease to a defined force level without spark-out [4.30].

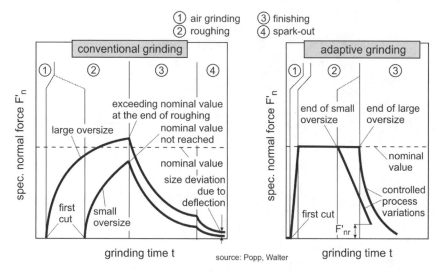

**Figure 4.17.** Specific grinding normal force for conventional and adaptive grinding

In general, AC systems relate to either a geometrical or a technological desired condition, but mixed conditions are possible. Geometrical systems are designed to achieve a desired workpiece geometry by measuring the actual diameter of the part and by controlling the infeed. Sensor systems, as explained earlier, belong to the group of tactile macro-geometry sensors. Technological-oriented systems usually use the feed speed of the grinding wheel as the regulated quantity and grinding forces or spindle power as the regulating quantity. With this concept, the quality of the workpiece as well as economic needs can be met.

## 4.8 Intelligent Systems for Abrasive Processes

Any kind of sensor technology for monitoring purposes is applied to ensure the availability of manufacturing systems and to guarantee a desired output from the process. Examples of sensor applications in the preceding sections have revealed a large variation in these applications. The emphasis was put on different technical solutions to obtain the desired process, tool or workpiece information. All the sensors described can be categorized into two different kinds of control loops

(Figure 4.18 [4.3]). Sensors for measuring process quantities as well as sensors for any kind of in-process measurement are related to the machine internal control loop. Sensors which are not installed in the working space of the machine tool or which only measure outside the process time belong to the post-process control loop. In the superior control loop, direct use of sensors is not scheduled. At this highest level, any kind of intelligent information system is used to either download control tasks to the sensor systems together with threshold values or to collect measuring data for further processing. [4.31] surveys different approaches to the use of these artificial intelligence methods in grinding.

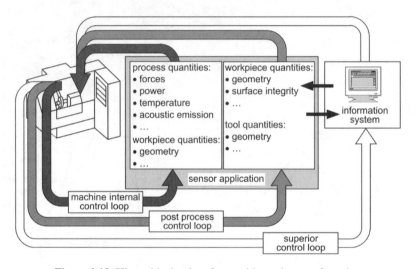

**Figure 4.18.** Hierarchical order of control loops in manufacturing

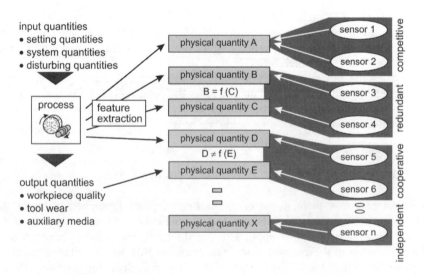

**Figure 4.19.** Multi-sensor approach to grinding monitoring

Although sensor systems appear at a lower level in the hierarchy of control loops, they must be regarded as an essential part of an overall intelligent system. The major task is to get as much information as possible from the current process. With single phenomenon monitoring, this aim can often not be met, so the application of multiple sensors in one process is one of the many activities necessary to achieve an intelligent system [4.32] (see Figure 4.19).

In general, the sensor system itself is the same as for a single-purpose monitoring task. A major difference can be seen in sensor choice. Some systems use several sensors to detect the same type of phenomenon, *e. g.* power and force monitoring to avoid overload of the main spindle. Although the sensors are not necessarily similar in operation, their field of application is enabling some redundancy. This strategy is chosen to increase the reliability of sensor systems, which is one of the features most desired by industrial users. The other approach is to combine the data from different types of sensors to increase the flexibility of the monitoring system. In this case, the term sensor fusion is also used, but usually it is not a fusion of sensors but a fusion of data from different sensor sources.

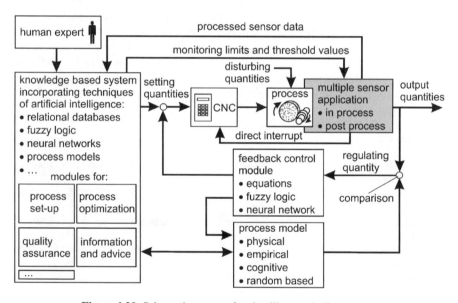

**Figure 4.20.** Schematic set-up of an intelligent grinding system

In Figure 4.20, an attempt is made to show a range of sensor applications in an intelligent grinding system. The term "intelligent" is not clearly defined. It is most often related to the application of artificial intelligence (AI) techniques such as knowledge-based systems, neural networks or fuzzy logic. Processing of sensor data is usually still done in a conventional way, using the latest available equipment for A/D conversion, filtering, sampling and further calculations. One of the major tasks is to provide parallel processing of different sensor signals as fast as possible to enable extremely short response times in the event of troubles. Different hardware solutions are available but will not be discussed in detail. In principle, there is a tendency to change from stand-alone computers to PC-based solutions, because this is also the kernel of modern CNC machine tool control units. In addition, a lot of work is directed towards a modular set-up of monitoring systems, which should help to implement these systems in a machine tool environment. The sensor system is connected to other modules of the set-up. First, monitoring limits and threshold values have to be fixed depending on the experience of any expert. Databases as part of a knowledge-based system can support this essential step. The measured and processed sensor data are transmitted to succeeding modules such as grinding databases and control modules. Depending on the type of application, direct feedback to the process according to adaptive control strategies is possible. There might be a direct connection to the CNC machine tool control unit to initiate an interrupt in the case of exceeding a specific limit, such as spindle maximum power. Also, integrating the sensor data in databases can support tasks in the superior control loop, such as planning and quality control. Special attention has to be paid to the use of process models. These models are used either to gain knowledge about the process, or to simulate or optimize the process input parameters. All sensor data from in-process or post-process application can be compared to the output of the different models mentioned to optimize and train them. These models directly support the process set-up and optimization modules of the knowledge-based system. Also the output of the feedback control unit may be used to improve the accuracy of the process model. Above all these parts of a schematic intelligent grinding system, the human expert is responsible for the configuration of the system including all hard- and software development.

# References

[4.1]    Tönshoff, H. K., Friemuth, T., Becker, J. C., 2002, "Process monitoring in grinding", *Annals of CIRP*, **51**(2), pp. 551–571.
[4.2]    Byrne, G., Dornfeld, D. A. *et al.*, 1995, "Tool condition monitoring (TCM): The status of research and industrial application", *Annals of the CIRP*, **44**(2), pp. 541–567.
[4.3]    Karpuschewski, B., 2001, "Sensoren zur Prozeßüberwachung beim Spanen", *Habilitation Thesis*, University of Hannover.
[4.4]    Haupt, F., Seidel, T., Karpuschewski, B., 1997, "Zahnflankenschleifen bogen-verzahnter Kegelradsätze mit CBN-Schleifscheiben", VDI-Z **139**(9), pp. 62–65.
[4.5]    Inasaki, I., Karpuschewski, B., Lee, H.-S., 2001, "Grinding chatter – origin and suppression", *Annals of the CIRP*, **50**(2), pp. 515–534.
[4.6]    Wakuda, M., Inasaki, I. *et al.*, 1993, "Monitoring of the grinding process with an AE sensor integrated CBN wheel", *Journal of Advanced Automation Technology*, **5**(4), pp. 179–184.
[4.7]    Klumpen, T., 1994, "Acoustic Emission (AE) beim Schleifen, Grundlagen und Möglichkeiten der Schleifbranddetektion", *Dissertation*, RWTH Aachen.
[4.8]    Saxler, W., 1997, "Erkennung von Schleifbrand durch Schallemissionsanalyse", *Dissertation*, RWTH Aachen.
[4.9]    Brinksmeier, E., Heinzel, C., Meyer, L., 2005, "Development and application of a wheel based process monitoring system in grinding", *Annals of the CIRP*, **54**(1), pp. 301–304.
[4.10]   Ueda, T., Hosokawa, A., Yamamoto, A., 1985, "Studies on temperature of abrasive grains in grinding – Application of infrared radiation pyrometer", *Journal of Engineering for Industry, Trans. of the ASME*, **107**, pp. 127–133.

[4.11]   Brunner, G., 1998, "Schleifen mit mikrokristallinem Aluminiumoxid", *Dissertation*, University of Hannover.

[4.12]   Westkämper, E., Klyk, M., 1993, "High-speed I. D. grinding with CBN wheels", *Production Engineering*, I (1), pp. 31–36.

[4.13]   Gomes de Oliveira, J., Dornfeld, D. A., Winter, B., 1994, "Dimensional characterization of grinding wheel surface through acoustic emission", *Annals of the CIRP*, **43**(1), pp. 291–294.

[4.14]   Karpuschewski, B., Wehmeier, M., Inasaki, I., 2000, "Grinding monitoring based on power and acoustic emission sensors", *Annals of CIRP*, **49**(1), pp. 235–240.

[4.15]   Westkämper, E., Hoffmeister, H.-W., 1997, "Prozeßintegrierte Qualitätsprüfung beim Profilschleifen hochbeanspruchter Triebwerksbauteile", in *Arbeits- und Ergebnisbericht 1995–97 des Sonderforschungsbereiches 326*, University of Hannover and Technical University of Braunschweig, pp. 345–401.

[4.16]   Tönshoff, H. K., Foth, M., 1988, "In-process detection and reduction of workpiece waviness caused by vibration in external plunge grinding", *3rd International Grinding Conference*, Oct. 4–6, Fontana/Wisconsin, SME Technical Paper MR88-620.

[4.17]   Tönshoff, H. K., Karpuschewski, B. *et al.*, 1998, "Grinding process achievements and their consequences on machine tools – challenges and opportunities", *Annals of CIRP*, **47**(2), pp. 651–668.

[4.18]   Rio, R., 1998, "Jet-engine CNC data collection takes off", *American Machinist*, July, pp. 66–70.

[4.19]   Tönshoff, H. K., Brinksmeier, E., Karpuschewski, B., 1990, "Information system for quality control in grinding", *4th International Grinding Conference*, October 9.-11., Dearborn, Michigan, Paper MR90-503.

[4.20]   Brodtmann, R., Gast, T., Thurn, G., 1984, "An optical instrument for measuring the surface roughness in production control", *Annals of the CIRP*, **33**(1), pp. 403–406.

[4.21]   Inasaki, I., 1985, "Monitoring of dressing and grinding processes with acoustic emission signals", *Annals of the CIRP*, **34**(1), pp. 277–280.

[4.22]   Maskus, P., 1992, "Prozeßintegrierte Qualitätsprüfung und Prozeßregelung beim Honen", *Dissertation*, Technical University of Braunschweig.

[4.23]   Lange, D., 1996, "Sensoren zur Prozeßüberwachung und Qualitätsprüfung", 8. *Internationales Braunschweiger Feinbearbeitungskolloquium*, 24.-26.4., pp. 16/1–16/19.

[4.24]   Karpuschewski, B., 1995, "Mikromagnetische Randzonenanalyse geschliffener einsatzgehärteter Bauteile", *Dissertation*, University of Hannover.

[4.25]   Regent, C., 1999, "Prozeßsicherheit beim Schleifen", *Dissertation*, University of Hannover.

[4.26]   Meyen, H. P., 1991, "Acoustic Emission (AE) –Mikroseismik im Schleifprozeß", *Dissertation*, RWTH Aachen.

[4.27]   Heuer, W., 1992, "Außenrundschleifen mit kleinen keramisch gebundenen CBN-Schleifscheiben", *Dissertation*, University of Hannover.

[4.28]   Brinksmeier, E., Heinzel, C., Wittmann, M., 1999, "Friction, cooling and lubrication in grinding", *Annals of the CIRP*, **48**(2), pp. 581–598.

[4.29]   Heinzel, C., 1999, "Methoden zur Untersuchung und Optimierung der Kühlschmierung beim Schleifen", *Dissertation*, University of Bremen.

[4.30]   Walter, A., 1995, "Prozeßinterne Optimierregelung für das Innenrundschleifen mit unscharfer Logik", *Dissertation*, University of Hannover.

[4.31]   Rowe, W. B., Yan, L. *et al.*, 1994, "Applications of artificial intelligence in grinding", *Annals of the CIRP*, **43**(2), pp. 1–11.

[4.32]   Inasaki, I., 1995, "Monitoring technologies for an intelligent grinding system", VDI-Berichte Nr. 1179, VDI-Verlag Düsseldorf, pp. 31–45.

# 5

# Condition Monitoring of Rotary Machines

N. Tandon and A. Parey

Industrial Tribology, Machine Dynamics & Maintenance Engineering Center
Indian Institute of Technology
Hauz Khas, New Delhi 110016, India
Emails: ntandon@itmmec.iitd.ernet.in, anandparey@hotmail.com

**Abstract**
Condition monitoring of machines provides knowledge about the condition of machines. Any deterioration in machine condition can be detected and preventive measures taken at an appropriate time to avoid catastrophic failures This is achieved by monitoring such parameters as vibration, wear debris in oil, acoustic emission etc. The changes in these parameters help in the detection of the development of faults, diagnosis of causes of problem and anticipation of failure. Maintenance/corrective actions can be planned accordingly. The application of condition monitoring in plants results in savings in maintenance costs, and improved availability and safety. The techniques covered in this chapter are performance, vibration, motor stator current, shock pulse, acoustic emission, thermography and wear debris monitoring. The instrumentation required, method of analysis and applications with some examples are explained. Signal processing techniques to gain more benefits of vibration monitoring are covered. Wear debris monitoring methods include magnetic plugs, ferrography, particle counter and spectrographic oil analysis.

## 5.1 Introduction

Condition-based maintenance strategy is used to carry out machinery maintenance dependent upon the actual condition of the machinery. Maintenance may be performed at irregular intervals rather than at fixed intervals as in the case of preventive periodic maintenance. The main function of condition monitoring is to provide a knowledge of the condition of the machine and its rate of change, so that preventive measures can be taken at an appropriate time. Knowledge about the condition of machinery can be obtained by selecting a parameter that indicates machine condition deterioration. The value of this parameter can be measured periodically or continuously. In some machines, the deterioration in condition develops so fast that there may be only a few seconds available between fault detection and total failure. In such cases continuous or online monitoring with automatic shutdown is recommended. In the case of periodic monitoring, called offline condition monitoring, the trend in the value of the selected parameter can be plotted as shown in Figure 5.1. Most machines usually give enough "lead time" before total failure occurs and provide sufficient warning of incipient machine failure. Where considerable experience of a machine and its failures is available, a safe period of further running can be estimated when the level of the parameter starts increasing. The maintenance work can then be planned accordingly. The machine should be stopped when the alarm level is reached. The alarm level should not be too late and at the same time should be set sufficiently high to avoid false alarms. The parameter measured can also be analyzed to indicate the most likely cause of the problem. In general, a condition monitoring system is used to detect faults in machines, anticipate breakdown and diagnose the problem.

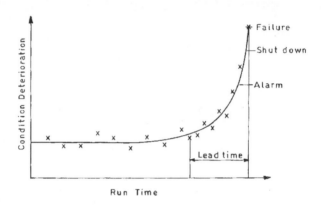

**Figure 5.1.** Trend monitoring showing lead time before failure

Critical machines or components to be monitored are selected on the basis of loss of productive earning capacity (*i. e.* whose breakdown gives rise to high costs), failure damage, and safety. The plant that is an obvious candidate for condition monitoring is where at some part of the manufacturing process, the stoppage of one machine affects the entire output. This is usually true for capital intensive plants having few employees, and in which the final output mainly depends upon the troublefree operation of the machinery. A measurement technique/parameter which is sensitive to change in the machine condition and is least influenced by operating condition changes and interference from other machines is selected for condition monitoring. The method/parameter should be able to give sufficiently early warning. The parameter may be pressure, temperature, flow rate, friction, corrosion, oil contaminants, exhaust emission, vibration, noise *etc.* Often more than one parameter is monitored so that the machinery problem is effectively detected as well as diagnosed. Parameters which are needed to estimate the performance of a machine such as pressure, flow, speed *etc.* are called "primary" parameters or signals. The changes of primary signals is sometimes slow. On the other hand, variation of secondary signals, which appear as disturbing effects (such as vibrations, noise, oil contaminants *etc.*), is usually fast.

Condition monitoring has evolved over the last three decades to become one of the important maintenance strategies in an increasingly diverse range of industries. The advantages of condition monitoring are that its use avoids unexpected catastrophic machinery failures, which may be very expensive and unsafe. The number of overhauls is reduced, resulting in savings in maintenance costs. Spare parts can be planned and procured in time. Plant operation is more efficient with greater plant availability; and consistent product quality can be obtained. Properly applied condition monitoring systems are highly cost effective. The modern trend in condition monitoring systems is towards the use of more and more automation – the use of transducers and instruments which can perform self-tests and automatic calibration; automatic decision-making tools such as expert systems, pattern recognition and neural network techniques for machine diagnostic purpose. There are a large number of techniques and instruments available for condition monitoring of machines. The techniques covered in this chapter are performance, vibration, shock pulse, acoustic emission, thermography and wear debris monitoring.

## 5.2 Performance Monitoring

Performance monitoring determines if the equipment is operating as expected. This is achieved by monitoring performance-indicating parameters. An adavantage of performance monitoring is that it can provide an indication of the efficiency of the system during its operation. The inter-relationship between pressure, temperature, flow, speed *etc.* for all parts of the system should be understood in order to diagnose faults using performance monitoring. For example, pressure drop and through flow in a heat exchanger can be related. Any build up of deposits in the flow lines will be revealed as an increase in the pressure drop needed to maintain a certain flow rate. In a chemical plant, deviation of the required temperature or pressure may affect the reaction of chemicals and result in poor quality product. So monitoring the temperature and pressure and the performance of systems which maintain these is important.

Online performance monitoring has special importance in power plants and in equipment such as pumps and compressors [5.1][5.2]. In gas turbines, combustor flame image is observed and fluctuations in its intensity are monitored on a screen. Passage of wear particles generated by seal rubbing in exhaust ducts can be monitored using exhaust gas debris detectors. Another parameter that can be monitored is the specific steam consumption rate in steam plants.

Efficiency and power of equipment such as pumps, compressors and internal combustion (IC) engines can also be monitored to indicate their performance. In IC engines, abnormal lubricating oil consumption indicates that this oil is being burnt and is passing through the piston rings due to inadequate sealing being provided by failing piston rings. Some other examples of performance monitoring are: temperature monitoring of furnaces; exhaust gas analysis of IC engines; temperature, pressure and steam flow rate in steam turbines; and shape of IC engine indicator diagram.

## 5.3 Vibration Monitoring

Vibration measurements are widely used in industry for monitoring the condition of a variety of machines and their components [5.3]–[5.8]. It is a very powerful technique for detecting and diagnosing faults in machinery. Both online and offline measurements are used. In the case of online measurements, vibration transducers are permanently mounted at the selected location which, in rotating machines, is usually bearings. Normally, root mean square (RMS) velocity or peak to peak displacement of vibration is measured. Sometimes, vibration acceleration is also measured, especially where high frequencies dominate the vibration signal.

Basic vibration measuring systems consist of a transducer (ususally an accelerometer or a proximity probe), signal conditioning element (such as a charge amplifier) and a display or recording unit. Accelerometers, the most commonly used transducers, are seismic type devices that measure absolute vibration of the object to which they are attached. Inside the accelerometer is a mass mounted on a spring damper system. A piezoelectric crystal is either compressed or sheared by this mass producing an electric signal proportional to the acceleration of the vibration being measured. The maximum frequency up to which the accelerometers can be used to measure vibrations is approximately one-third their natural frequency. Conventional accelerometers have high output impedance and so cannot be directly connected to measuring or analyzing instruments. To eliminate this problem, the accelerometer output is fed through an impedance conversion preamplifier called a "charge ampli-fier". This device has high input impedance and low output impedance, suitable for connection to the low input impedance of measuring instrumentation. Currently, vol-tage output ICP® accelerometers are also used widely. These accelerometers convert the low level high impedance charge output of a piezoelectric crystal into low impedance voltage output using an internal amplifier circuit. They require a constant current power supply and two-conductor cable – one lead each for power and signal return. Another type of accelerometers called TEDS (transducer electronic data sheet) are smart sensors which use mixed mode analogue and digital operation to communicate with condition monitoring instruments. These sensors send digital information stored in their electronic data sheet once a triggering protocol has been received. After this, the accelerometer starts giving an analogue signal output proportional to vibration acceleration. TEDS accelerometers have inbuilt identify-cation number, self-test and auto calibration facilities. Micro electromechanical system (MEMS) accelerometers, which are low cost and extremely small in size (can be less than 1 gm and 5×2 mm), are also being used now. These are made by forming a small mechanical element, such as a beam mass, on the surface of a semi-conducting integrated circuit. The mechanical element responds to vibration and the response is converted to an electrical signal. Silicon micromachined capacitive, resistive, as well as piezoelectric accelerometers are made [5.9]. Some of them incorporate a moving inertial mass suspended by springs from the surrounding frame structure. Both the sensor and the signal conditioning circuit are provided on chip. Analogue, as well as digital output MEMS accelerometers are available.

Vibration can also be monitored from a distance using laser Doppler vibrometers (LDV), which measures the vibration velocity of objects using Doppler effect. In the LDV configuration, a light wave is emitted from a stationary source and received by a moving (vibrating) target object, which then retransmits to a stationary observer. The scattering Doppler frequency shift $\Delta f$ is given by:

$$\Delta f = \frac{2u}{\lambda}\cos\theta \tag{5.1}$$

where, $u$ is the velocity of the object, $\lambda$ is the wavelength and $\theta$ is a small angle between source and observer. The wave source is a laser and the observer is a photodetector, which measures the target surface velocity related to the Doppler shift. The LDV system basically consists of two main components – the optical sensor/controller processor and the scan control/data acquisition system. The optical sensor contains the laser, optical elements and other components needed to spatially position the laser beam. The controller processor consists of electronic components needed to process the velocity information provided by the optical sensor. It constitutes the interface between the interferometer and the data acquisition computer.

Eddy current based non-contact "proximity probes" are used to measure vibration displacement. These probes are commonly used to monitor shaft vibration in journal bearings. Inside these probes is a coil on a small rod. The coil is excited by high frequency alternating current, inducing an oscillating magnetic field around the coil. This causes generation of eddy currents in the metallic surface whose vibration is to be measured. The eddy currents generated have their own magnetic field which induces a current in the coil in the direction opposite to the supply alternating current and reduces the amplitude of the supply current in proportion to the gap distance between the probe and the surface. Hence, an electric signal proportional to the gap between the probe tip and the surface, or in other words the vibration displacement of the surface is obtained as output, after suitable conditioning of the signal. Two such probes can be used to obtain "orbit plot" or "Lissajous figures" of shaft motion in journal bearings. A circle is obtained by feeding horizontal and vertical vibration signals to horizontal and vertical amplifiers of an oscilloscope. The orbit plot in effect shows the motion of the centre of the shaft. The shape of the orbit provides information about faults such as misalignment, rubbing *etc.* (Figure 5.2).

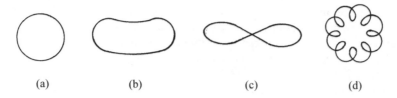

(a)                (b)                (c)                (d)

**Figure 5.2.** Typical orbit plots from two proximity probes for (a) normal condition, (b) misalignment, (c) severe misalignment, (d) rubbing

A circle is obtained when there is no fault. In the case of angular misalignment the orbit shape changes from a circular shape to a banana shape for significant misalignment and figure of eight shape for severe misalignment. Rubbing is a very common phenomenon in most rotating machinery. It generally starts from a light rub and gradually increases to medium, heavy and full rub. For a rotor system without rub a circular shape orbit is obtained, rub creates additional loops in the orbit shape. The number of additional loops increases with advancement of rub. For full rub the orbit rotates in the opposite direction to the rotation of the shaft without any additional loop.

Orbit analysis is basically suited to simple harmonic motion. For the analysis of multifrequency components, it becomes difficult to analyze the orbit diagram. A purified orbit diagram consists of only the specific frequency and other frequencies do not affect it. Another technique to minimize disturbances due to other frequencies is filtered orbit analysis. The signals of horizontal and perpendicular probes are passed through non-phase shifting bandpass filters and orbits are plotted. This method gives orbits in a particular frequency band, minimizing the effect of other frequencies [5.10].

The vibration signal from the accelerometer or proximity probe can be displayed on a vibration meter or on a computer using analogue to digital converters. Fast Fourier transform (FFT) analyzers are used to obtain the frequency spectrum of the vibration signal. In FFT analyzers, the input signal is considered over a finite time called the "frame" or "time window" and is digitized and stored. Discrete Fourier transformation (DFT) of this signal gives the vibration frequency spectrum. The FFT is a computer algorithm for calculating DFT efficiently. FFT analyzers are becoming smaller and smaller in size and have facilities like "hanning weighting" and averaging of a number of spectra. Hanning weighting avoids spurious frequencies which are caused by abrupt discontinuities at each end of the frame. This is achieved by multiplying the input frame by the bell-shaped Hanning function, which tapers the ends of the frame. Instead of using a standalone vibration meter or FFT analyzer, often so called "virtual instrumentation" is employed. This is nothing but computer cards and software which make it possible to digitize the signal and use various types of filters, weightings and FFT available within the software in a computer. One has to choose an appropriate sampling rate for digitizing the vibration signal to avoid spurious frequency components due to "aliasing". According to the Nyquist theorem, the sampling frequency (inverse of sampling rate) should be at least twice the maximum frequency present in the signal.

Development of a fault or deterioration in the condition of machines is indicated by an increase in overall vibration level. Unacceptable overall levels can be established by past experience on a particular machine. In the absence of this, the measured levels can be compared with vibration severity criteria such as VDI 2056 or ISO 2372 to assess the condition of a machine. VDI 2056 is an old (1964) but still used German standard which classifies machines into four different categories depending upon their power and whether their operating speed is below or above foundation natural frequency. The vibration levels specified are in terms of RMS velocity ranging from 0.18 to 45 mm/s. The velocity range for each category of machine is divided into four zones – good, allowable, just tolerable and not permissible. An increase of 20 dB in the level changes the zone from good to not permissible. Maximum starting level for not permissible zone is 18 mm/s. ISO 2372 (1974) [5.11] is similar to VDI 2056. New international standard ISO 10816 (1995–2001) series "Mechanical vibration – Evaluation of machine vibration by measurements on non-rotating parts" is in 6 parts:

10816-1: 1995    Part 1: General guidelines
10816-2: 2001    Part 2: Land-based steam turbines and generators in excess of 50 MW with normal operating speeds of 1500 r/min, 1800 r/min, 3000 r/min and 3600 r/min

| 10816-3: 1998 | Part 3: Industrial machines with normal power above 15 kW and nominal speeds between 120 r/min and 15000 r/min when measured in situ |
| 10816-4: 1998 | Part 4: Gas turbine sets excluding aircraft derivatives |
| 10816-5: 2000 | Part 5: Machines set in hydraulic power generating and pumping plants |
| 10816-6: 1995 | Part 6: Reciprocating machines with power ratings above 100 kW |

Part 7 of this standard: Rotodynamic pumps for industrial applications, including measurements on rotating shafts, is under development. Part 3 is for all general production machinery such as pumps, fans *etc.* ISO 10816-3 separates the machines into four different groups and takes into account if the machine installation is rigid or flexible in each group. Vibration levels for each group of machines are divided in four zones:

A (Green):    Vibration values from machines just put into operation
B (Yellow):  Machines can run in continuous operation without any restriction
C (Orange):  Machine condition is acceptable only for a limited period of time
D (Red):      Dangerous vibration values – damage could occur to the machine

The maximum starting RMS velocity level for zone D is from 11 mm/s. In general, unacceptable levels have been reduced compared to the "not permissible" levels of earlier VDI 2056/ISO 2372 standards. Frequency range is now broadened, and is not limited to 10 to 1000 Hz as specified in ISO 2372. ISO 7919 series, which is for measurements on rotating shafts of non-reciprocating machines, and is in 5 parts. The parts of this series "Mechanical vibration of non-reciprocating machines – measurement on rotating shafts and evaluation criteria" are as follows:

| 7919-1: 1996 | Part 1: General guidelines |
| 7919-2: 2001 | Part 2: Land-based steam turbines and generators in excess of 50 MW with normal operating speeds of 1500 r/min, 1800 r/min, 3000 r/min and 3600 r/min |
| 7919-3: 1996 | Part 3: Coupled industrial machines |
| 7919-4: 1996 | Part 4: Gas turbine sets |
| 7919-5: 1997 | Part 5: Machines set in hydraulic power generating and pumping plants |

Frequency analysis of the vibration signal is usually necessary to diagnose the fault in machines. Vibration generated by a machine component consists of certain frequencies, which do not change during transmission of vibration although their levels may change during vibration transmission from one location to another. Also, mixing of vibration frequencies from different machine components does not result in any loss of information about the frequencies of individual components. These facts are helpful in relating rise in levels of particular frequencies in vibration spectrum to different components of a machine and hence diagnosing the faults. Some of the common faults in rotating machines and the frequencies at which high levels are expected in vibration spectra are given in Table 5.1.

**Table 5.1.** Vibration frequencies associated with common faults in rotating machines

| Fault | Frequencies at Which High Vibration Levels are Expected ($f_r$: Rotational frequency = RPM/60) |
|---|---|
| Unbalance | $f_r$ <br> (Mainly in radial direction) |
| Misalignment | $f_r$, $2f_r$ <br> (Often $2f_r$ is higher than $f_r$ and axial vibration may be higher than radial vibration) |
| Mechanical looseness and rubbing | $0.5f_r$, $f_r$ and a number of their higher harmonics |
| Oil whirl in journal bearings | 40-50% of $f_r$ |
| Damaged rolling bearings | At rolling element pass frequency given by: <br><br> Outer race defect = $\dfrac{N}{2} f_r \left(1 - \dfrac{RD}{PD} \cos \alpha\right)$ <br><br> Inner race defect = $\dfrac{N}{2} f_r \left(1 + \dfrac{RD}{PD} \cos \alpha\right)$ <br><br> Rolling element defect = $\dfrac{PD}{RD} f_r \left[1 - \left(\dfrac{RD}{PD} \cos \alpha\right)^2\right]$ <br><br> Cage defect frequency = $\dfrac{f_r}{2} \left(1 \pm \dfrac{RD}{PD} \cos \alpha\right)$ <br> (+ sign if outer race is rotating, - sign if inner race is rotating) <br> N: Number of rolling elements (balls/rollers) <br> $F_r$: shaft rotational speed, Hz <br> RD: Rolling element diameter <br> PD: Pitch circle diameter <br> $\alpha$: Contact angle |
| Damaged gears | Tooth meshing frequency = $Nf_r$ <br> And sidebands at $Nf_r \pm kf_r$ <br> N: Number of gear teeth <br> K = 1, 2, 3, ..... |
| Damaged fan blades | Blade pass frequency = $Nf_r$ <br> N: Number of blades |
| Pump problems | $Nf_r$ and its harmonics <br> N: Number of vanes |
| Electrically induced vibrations | $f_r$ or $f_e$, $2f_e$ <br> $f_e$: Electric supply frequency <br> (Disappears as soon as power is switched off) |
| Cavitation in pumps [5.12] | $f_r$, $Nf_r$, Broad band peak at frequencies greater than 2000 Hz <br> N: Number of vanes |
| Broken rotor bar in induction motors [5.13][5.14] | $f_r \pm 2sf_e$, $f_e \left[k \left(\dfrac{1-s}{p}\right) \pm s\right]$ <br><br> s: slip <br> $f_e$: Electric supply frequency <br> k/p: 1, 5, 7, 11, 13, ..... (due to normal winding configuration) |

All rotating machines operate under some residual inbalance that gives the synchronous or rotational frequency component $(f_r)$. The machinery may become unbalanced either due to manufacturing errors, mounting errors or non-uniform operating conditions. Mass inbalance increases the $f_r$ component mainly in the radial direction.

Angular misalignment produces $f_r$ and $2f_r$ components in the radial direction and $f_r$ component in the axial direction. (Often $2f_r$ is higher than $f_r$ and axial vibration may be higher than radial vibration.) Parallel misalignment also excites $f_r$ and $2f_r$ components, but no axial component.

Mechanical looseness may exist in horizontal plane or/and vertical plane. Vertical plane looseness excites $0.5f_r$ and $f_r$ component and higher harmonics. The $0.5f_r$ component has generally half the amplitude of the $f_r$ component amplitude. Horizontal mechanical looseness mostly excites $f_r$ and $2f_r$ component.

By observing phase differences across the machine, the possible cause of the $f_r$ component can be identified.

When rub-impact occurs, multiple harmonic components such as $2f_r$, $3f_r$, etc. as well as $\frac{1}{2}f_r$, $\frac{3}{2}f_r$, etc. can be observed. Under some special cases $\frac{1}{3}f_r$, $\frac{2}{3}f_r$, etc. can be observed as well [5.15].

As the shaft start to rotate in a journal bearing, a circular wedge shape of oil film is formed. This wedge shaped oil film drives the shaft ahead of it in forward circular motion. The oil whirl occurs theoretically at $0.5f_r$, but in actual practice it occurs at $0.4f_r$ to $0.46f_r$.

In the case of roller bearings, the presence of defects such as pits also causes generation of high frequencies in the 5 to 18 kHz region due to impacts of the fault with rolling elements. An example of vibration spectra of a good and defective bearing is shown in Figure 5.3. For outer race waviness, the spectrum has components at the outer race defect frequency and its harmonics. In the case of inner race waviness, waviness orders equal to the number of rolling elements and its multiples give rise to spectral components at the inner race defect frequency and its multiples.

In gear spectra, tooth mesh frequency and their harmonics are present due to profile errors. Sidebands are present due to modulation of tooth meshing frequencies. An increase in the number and/or strength of sidebands indicates damaged gears. Some ghost frequencies may also be present due to errors during gear manufacturing.

Blade frequencies are commonly seen in pumps, compressors, fans and turbines. A blade frequency with equally spaced side bands may be due to loose coupling.

Vapor bubbles are formed in a flowing fluid if the local pressure of the liquid droplets, drops below its vapor pressure. Bubbles implode if the fluid pressure increases above the pressure level that can be sustained by the bubbles, causing small pieces of the metallic surface to tear out. In I.C. engines "knocking" generally appears in the 4 to 8 kHz frequency range [5.16].

Changes in vibration spectrum with other parameters, usually speed of rotation, are often obtained as three-dimensional plots. These are called waterfall or cascade plots. Such plots are useful in fault identification and in distinguishing between order-related frequencies and structural natural frequencies in a coast-down analysis.

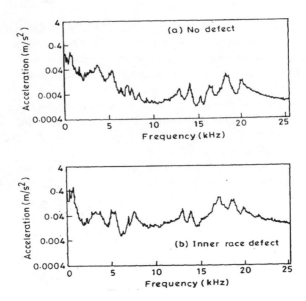

**Figure 5.3.** Vibration spectrum of ball bearing (a) in good condition, (b) with a defective inner race

### 5.3.1 Vibration Signal Processing

Vibration signals acquired from machines for diagnostics purposes may be either deterministic or random. Deterministic signals can be further classified as periodic and non-periodic, whereas random signals can be classified as stationary and non-stationary. Useful information can be extracted from these signals by appropriate signal processing techniques as explained later.

However, these vibration signals often contain a lot of noise. If the noise is too great, useful information will be impossible to extract through signal processing techniques and wrong conclusions may be drawn. In such cases techniques that enhance signal to noise ratio (SNR) are therefore necessary. Adaptive noise cancellation (ANC) is one such technique that enhances SNR [5.7].

### *Statistical Analysis*

The vibration signal acquired from the machine using transducers is basically in the time domain. In order to obtain useful information for diagnostic purposes, various statistical operations can be performed. Crest factor is the ratio of maximum absolute value to the RMS value of the vibration signal, and gives an idea about any impacts present in the signal. An increased value of crest factor over a period of time could be due to the presence of wear or pitting. Kurtosis measures the degree of peakiness of a distribution compared to a normal distribution. It is based on the size of the distribution's tail: the longer the tail, the higher the kurtosis value. In general, even statistical moments give information about spread. Kurtosis is defined as:

$$\text{Kurtosis} = \frac{M_4}{M_2^2} = \frac{\dfrac{1}{N}\sum_{n=1}^{N}(x(n)-\overline{x})^4}{\left[\dfrac{1}{N}\sum_{n=1}^{N}(x(n)-\overline{x})^2\right]^2} \qquad (5.2)$$

where $M_4$ is the fourth-order statistics moment, $M_2$ is the second-order statistics moment, $x(n)$ is the amplitude of the signal at the $n$th sample, $\overline{x}$ is the mean value of the amplitudes and $N$ is number of samples taken in the signal. The kurtosis value of a normal distribution is 3 and for a random signal it is close to $3(\pm 8\%)$. The presence of any impulse increases the kurtosis value from 3, depending upon the severity of the fault. Kurtosis is a better fault indicator than crest factor, because it can detect impulses with minimum repetition period [5.17]. Figure 5.4(a) shows the time domain representation of a gear vibration signal of a gearbox with good gears and Figure 5.4(b), that for gears with spall.

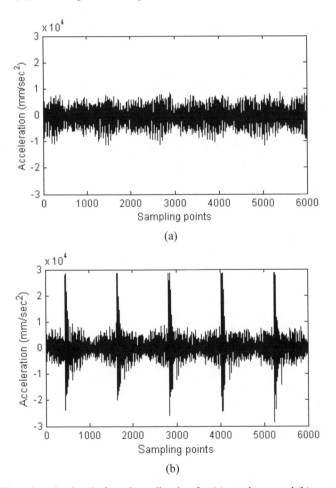

(a)

(b)

**Figure 5.4.** Time domain signal of gearbox vibration for (a) good gear and (b) gear with spall

The gearbox is a single-stage gearbox with spur gears. Pinion and gear have 20 and 21 teeth respectively. The pinion runs at 1000 rpm. The sampling rate is 20000 points per second, *i. e.* 1200 points for one revolution of the pinion. The crest factor and kurtosis value of these signals are given in Table 5.2.

**Table 5.2.** Crest factor and kurtosis values for gear vibration shown in Figure 5.4

| Gear Condition → | Good Gear | Gear with Spall |
|---|---|---|
| Fault Indicator ↓ | | |
| Crest Factor | 2.5371 | 5.9671 |
| Kurtosis | 2.7382 | 12.282 |

One of the limitations of the kurtosis method is that the kurtosis value falls to 3 when the damage is well advanced. This may be due to the higher relaxation time of the impulsive response than the impulsive repetition period. Monitoring overall RMS values can be more useful in such cases. Another limitation is that it can give variable and misleading results if measurements are taken on machines in an unloaded condition. Studying kurtosis values in different frequency bands can be more useful. In the kurtosis meter four frequency bands are normally used: 3 Hz–5 kHz, 5–10 kHz, 10–15 kHz and 15–20 kHz.

### High Frequency Resonance Technique

The high frequency resonance method is an important signal processing technique, which helps in the identification of rolling element bearing defects by extracting characteristic defect frequencies, which may be buried in noise and may not be identifiable in the direct spectrum. Each time a defect in rolling element bearings makes contact with another bearing surface an impulse vibration is generated. The impact excites resonance of the bearing element, housing or bearing structure. These resonances are excited periodically at the defect frequency and are amplitude modulated at defect the frequency. A signal indicative of bearing condition can be recovered by demodulating the resonance. The signal is first bandpass filtered around the resonance frequency. The frequency generated by misalignment, inbalance, gears *etc.* are thus eliminated, leaving a narrow band carrier resonant frequency, amplitude modulated at the resonance frequency. An envelope detector demodulates the signal. The envelope detection process involves passing the bandpass filtered signal through a rectifier and then through a lowpass filter, which removes the resonance frequency and smooths the pulses.

The limitation of this technique is that, with advanced damage, the defect frequency may become submerged in the rising background level of the spectrum. This may happen due to the reduced severity of impacts, which are generated so frequently that the leading edge of one impact is buried in the decay of the previous impact.

### Time Synchronous Averaging

The time domain signal acquired for machine diagnostics is complex in nature. The signal consists of different frequencies, stemming from various rotating components of a mechanical system, *e. g.* bearings and gears of a gear train. To diagnose a particular element, if the frequency of that element can be extracted from the complex signal, any changes in the signal can easily be seen. By time domain averaging, a periodic waveform can be extracted from the noisy signal. This can be achieved by repeating three steps: sampling, storing and ensemble averaging. First, sampling length, which depends on the frequency of the signal to be extracted is found. It requires a trigger pulse at a frequency the same as the frequency of interest, to start sampling. This first sample is stored, the next sample is obtained and is stored after performing ensemble averaging. The next sample is taken and ensemble averaged with the previously averaged signal, as shown in Figure 5.5.

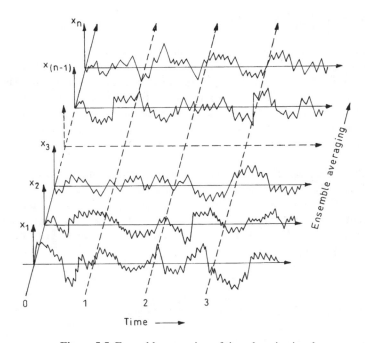

**Figure 5.5.** Ensemble averaging of time domain signal

After repeating this process several times, the frequency of interest can be extracted. This method improves signal to noise ratio in proportion to the square root of the number of averages taken. This technique is suitable for constant rotational speed only. It requires many averages to extract a particular frequency, making it a time consuming process.

### *Cepstrum Analysis*

Cepstrum analysis is a technique used to enhance the understanding of a spectrum. Cepstrum is the power spectrum of the logarithm of the power spectrum, so in a broader sense it is a technique, introducing the spectrum of the log of a spectrum. Mathematically:

$$C(\tau) = \left| \text{FT}\left( \ \log \left| \text{FT}(x(t)) \right|^2 \right) \right|^2 \tag{5.3}$$

where, FT is the forward Fourier transform of a signal $x(t)$.

The power cepstrum is defined as the inverse transform of the logarithm of the power spectrum:

$$C_p(\tau) = \text{FT}^{-1}\left( \ \log \left| \text{FT}(x(t)) \right|^2 \right) \tag{5.4}$$

The complex cepstrum is defined as the inverse Fourier transform of the logrithm of the spectrum:

$$C_c = \text{FT}^{-1}\left[ \log \left\{ \text{FT}\left( x(t) \right) \right\} \right] \tag{5.5}$$

In cepstrum analysis terminology, frequency is written as quefrency, harmonics as rahmonics *etc*. The power cepstrum and complex cepstrum have different applications as they contain different information about the spectrum. The power cepstrum gives magnitude information, whereas the complex cepstrum gives both magnitude and phase information about the spectrum. It is possible to reconstruct a useful component of the spectrum and time domain signal using a complex cepstrum, whereas the power cepstrum can reconstruct only the useful component of the spectrum.

Gearbox vibration spectra normally contain sidebands due to modulation of the tooth meshing frequencies and their harmonics. The power cepstrum can be used for detecting the presence and/or growth of sidebands in gearbox vibration spectra indicating their mean spacing over the entire spectrum, and is thus applicable to both detection and diagnosis of faults. If the complex cepstrum is uesd in sideband analysis, consistency will be lost due to its phase dependence.

Generally, signals are acquired at a certain distance from the source; therefore all signals are a convolution of source and path information. The deconvolution property of the complex cepstrum can separate source and path information. The complex cepstrum of the source and the transmission path exist in separate regions. By using a window, the time domain signal of the source or the impulse response function of the transmission path can be separated.

The important characteristic of the cepstrum is that it clearly identifies periodic peaks in the spectrum, which appears as a distinct peak in the cepstrum. Figure 5.6(a) shows the cepstrum of the signal shown in Figure 5.4(a). Two periodic signals, one for the pinion and another for the gear, are clearly visible in cepstrum (1200 points corresponds to 0.06 seconds, which is equivalent to 16.667 Hz, *i. e.* the rotational frequency of the pinion; and 1260 points corresponds to 0.063 seconds, which is equivalent to 15.873 Hz, *i. e.* the rotational frequency of the gear). Figure 5.6(b) shows the cepstrum of the signal shown in Figure 5.4(b). There is a change in the amplitude of periodic peaks of the gear and pinion due to the defected gear pair. The ratio of the difference between the normalized value of the defective pinion peak and the normalized value of the defective gear peak to the sum of the normalized value of defective pinion and gear peaks gives a defect indicator. The normalized values of defective pinion and gear peaks are obtained by dividing them by their respective peaks for no defect. The defect indicator gives a positive

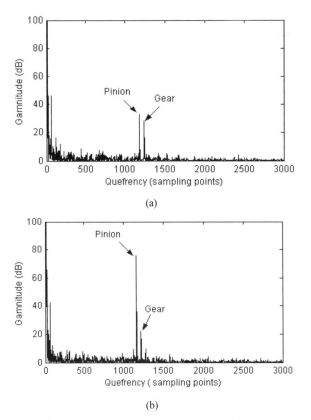

(a)

(b)

**Figure 5.6.** (a) Cepstrum of the signal as shown in Figure 5.4(a); (b) Cepstrum of the signal as shown in Figure 5.4(b)

value if the defect is in the pinion and a negative value if the defect is in the gear. An increase in the defect indicator value is an indication of the severity of the defect. The mean of the defect indicator is independent of signal amplitude, signal to noise ratio and sensor location [5.18].

The cepstrum can also be used to separate one defect from multiple defects. Location of surface defects can be obtained by using a moving cepstrum integral (MCI). A normalized rectangular window function is shifted in the time domain along the whole of the acceleration signal, and the power cepstrum integral is evaluated without considering the first sample. For localization of a gear defect, the window size should be less than the mesh period and more than the defect width. Faults are indicated by MCI minima values. A dominant minimum value in the MCI curve is associated with a larger fault [5.19].

### Automated Fault Detection

All the techniques presented in this section for fault diagnosis need human interpretation. Fault detection using statistical and spectral analysis techniques is more or less easy to interpret. Interpretation of time–frequency distributions for small changes becomes difficult and unreliable. Regardless of the signal processing technique used, automatic interpretation can improve the efficiency and reliability of fault diagnosis. Artificial neural networks (ANN) can identify and classify real data such as vibration signals. ANN consists of a number of richly interconnected artificial processing neurons called nodes, collected together in layers forming a network. A two-layered ANN is shown in Figure 5.7.

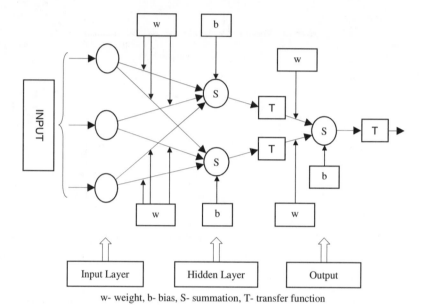

w- weight, b- bias, S- summation, T- transfer function

**Figure 5.7.** A two-layered neural network

The number of nodes within the input and output layers depends on the number of independent variables required to define the problem. The number of hidden layers and their nodes are selected by trial and error methods. All nodes, except the input nodes, sum their weighted input values. A bias value is added in the summed value to get a net value. This net value is the argument of the transfer function. The transfer function is generally a step function. The network can be trained by adjusting weights or bias parameters to solve a particular problem. The ANN is trained using back propagation as the most commonly used algorithm. Once the ANN is trained successfully, it can be used for unknown data. An ANN has been used for the fault detection of bearings [5.20].

## 5.4 Shock Pulse Analysis (SPA)

The SPA t0065chnique has been specifically developed for the condition monitoring of rolling element bearings. The technique is based on the fact that any damage in rolling element bearings will cause mechanical impacts that will generate ultrasonic shock waves. The magnitude of these impacts is a measure of the condition of the bearings. The magnitude of impacts depends on impact velocity, which depends on defect size and bearings speed and size. The transducer of the shock pulse meter (SPM) is a piezoelectric accelerometer tuned mechanically and electronically to a resonant frequency around 32 kHz. The shock wave is propagated through the bearing housing, and when the shock pulse hits the transducer, damped oscillations are initiated at the resonant frequency of the transducer. The amplitude increase of the damped resonant oscillation gives an indication of the condition of the rolling element bearings. The transducer signal is processed electronically to filter out low frequency vibrations such as inbalance, misalignment and other structure-related vibrations. The decibel (dB) unit is used to measure the shock value to accommodate a large range of shock values of good and damaged bearings.

The bearing race surfaces will always have a certain degree of roughness. So, when a bearing rotates, this surface roughness causes mechanical impacts with rolling elements. The shock pulse value generated by good bearings due to surface roughness has been found empirically to be dependent upon the bearing bore diameter and speed. This value, called initial value ($dB_i$), is subtracted from the shock value of the test bearing to obtain a 'normalized shock pulse value' ($dB_N$). The digital shock pulse meter gives the reading directly in $dB_N$. The shock pulse meter gives two values namely the 'maximum shock value' ($dB_M$) and the 'carpet value' ($dB_C$), as shown in Figure 5.8.

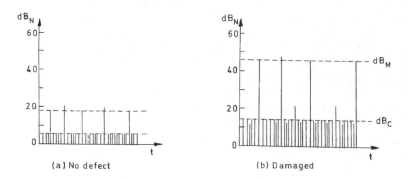

**Figure 5.8.** Normalized shock pulse values of a bearing

The maximum shock value is a measure of low rate (LR) impacts, and the carpet value is a measure of high rate (HR) impacts. HR impacts may exceed 1000 impacts per second and LR impacts may exceed 25 impacts per second. An increase in $dB_M$ value without an increase in $dB_C$ value is an indication of damaged bearings. Increase in both $dB_M$ and $dB_C$ value is an indication of lubrication problems. Manufacturers of SPM instruments supply a diagnostic table based on $dB_M$ and $dB_C$.

## 5.5 Current Monitoring

In recent years, stator current monitoring for the detection of broken rotor bars, rotor asymmetry and inbalance, and bearing defects in induction motors has attracted the attention of researchers [5.14][5.21][5.22]. A current sensor based on the Hall effect can be used in series with the motor supply line. The Hall element located in the air gap of the magnetic circuit converts the magnetic field generated by the primary current into a proportional Hall voltage. The majority of induction motor failures are either due to defects in rotor or bearings. A broken rotor bar generates frequencies in the current spectrum as given in Table 5.1. The air gap field produced by a slip frequency current flowing in a stator bar will consist of a fundamental component rotating at slip speed in the forward as well as backward direction with respect to the rotor. The backward components will sum to zero in a symmetrical rotor. The resultant is non-zero for a broken bar rotor. The field, which rotates at the slip frequency backwards with respect to the rotor, will induce electromotive forces (EMFs) in the stator side, which modulate the mains supply frequency at twice the slip frequency. Air gap eccentricity can be detected by monitoring the behavior of the current at the fundamental sidebands of the supply frequency, given by [5.14]:

$$f_{ecc} = f_e \left[ 1 \pm m \left( \frac{1-s}{p} \right) \right] \tag{5.6}$$

where $f_e$ is the electric supply frequency, $m = 1,2,3,\ldots$, $s$ is per unit slip and $p$ is number of pole pairs. The predicted frequencies for both air gap eccentricity and broken rotor bars are the same, but the frequency corresponding to a particular harmonic number is different, making it possible to distinguish the two faults [5.14]. The amplitude of the left sideband frequency component, $f_e(1-2s)$, is proportional to the amount of broken rotor bars. It is generally believed that when the amplitude of these components is more than 50 dB smaller than the stator current fundamental frequency component amplitude, the rotor can be considered healthy.

Ball bearings support the rotor of motors, so bearing defects result in radial motion between the rotor and stator. The bearing vibration can be related to the stator current because any air gap eccentricity will produce anomalies in the air gap flux density. These variations generate stator currents at the following frequencies [5.14]:

$$f_{brg} = \left| f_e \pm m f_{vd} \right| \tag{5.7}$$

where $m = 1,2,3,\ldots$ and $f_{vd}$ is the bearing element vibration defect frequencies as given in Table 5.1. Figure 5.9 shows the current spectrum of an induction motor ball bearing with a defect (pit) in its outer race. The bearing is of 25 mm bore with vibration outer race defect frequency of 84.1 Hz. Peaks at 34.1 and 134.1 Hz are clearly seen. Electric supply frequency $f_e$ was 50 Hz.

## 5.6 Acoustic Emission Monitoring

Acoustic emission (AE) is the phenomenon of high frequency elastic wave generation in materials under stress. Plastic deformation and growth of cracks are among the main sources of AE in metals. Hence, AE measurement is an important

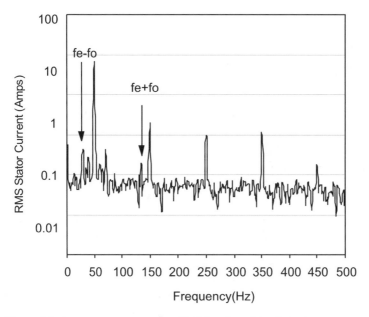

**Figure 5.9.** Stator current spectrum of ball bearing with a defective outer race

condition monitoring technique for the detection of crack propagation and failure detection in machinery and its components. It is possible to obtain information about the severity of changes occurring at defects in materials under load and to locate the position of the deformation within the component. AE monitoring has an added advantage that it can detect the growth of subsurface cracks. The signal is generated and measured in the 50 kHz to 2 MHz frequency range. Acoustic emission has been used for the condition monitoring of a wide range of machines for the last three decades.

Most commonly measured AE parameters include peak amplitude, counts, events/hits and energy of the signal. Counts are the number of times the signal rises above a threshold voltage level that has been set in the measuring system. An event or hit consists of a group of counts. When a hit occurs in all the sensors, it is defined as an event in some literature. Energy is measured as the area under the rectified signal envelope. In practice, signal voltage is first squared and then area under the curve of voltage squared against time is measured. This area is proportional to the signal energy. The threshold level is chosen in such a way that it is just higher than the background noise, and the desired signal counts or events are measured. The peak amplitude is directly related to the magnitude of the source, and like energy, is not dependent on the threshold setting. In AE measurements, the distribution plots of these parameters against time or against each other are also obtained, as shown in Figure 5.10 [5.23]. The distributions of AE events against peak amplitude for a pair of spur gears in good condition and when one tooth of a gear had an approximately 1 mm diameter pit, are shown. It is seen that, in this case, the distribution becomes broader because of the presence of a defect in the gear, and the peak amplitude is centered around 50 dB for the good gear and around 75 dB for the defective gear.

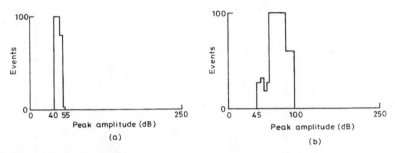

**Figure 5.10.** Distribution of AE events and peak amplitude of a spur gear, when (a) in good condition (b) with a defective tooth

The instrumentation system consists of AE transducer, preamplifier, bandpass filter, signal processing unit and display. Currently, AE measurement computer cards with software are available commercially. The threshold level can be set using the software, and the desired parameters can be measured and displayed on the computer monitor. Two types of transducers are used – resonant and broadband types. The most commonly used are resonant piezoelectric transducers whose sensitivity varies with frequency and are usually greatest in the range 100–400 kHz. Only frequency components around the resonant frequency are measured, using a bandpass filter. Broadband transducers have an almost flat frequency response within the 100 kHz to 2 MHz range. These can be used when frequency analysis of the AE signal is important.

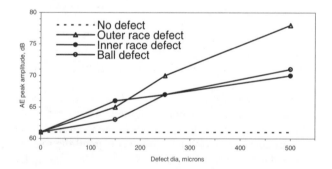

**Figure 5.11.** Acoustic emission peak amplitude for ball bearings

Acoustic emission monitoring has been successfully used for the detection of defects in rolling bearings and gears [5.23]–[5.26]. Figure 5.11 shows the increase in AE peak amplitude for a 6002 ball bearing due to the presence of different sizes of pits in its elements [5.25]. The measurements were performed with the bearings rotating at 1500 rpm. It is seen that there is a substantial increase in peak amplitude values with increase in defect diameter in inner and outer races and in the ball of the bearings. This shows that the AE technique can be successfully used for the detection of defects in bearings. Acoustic emission parameters have been shown to detect bearing defects before they appear in the vibration acceleration [5.24] and in very slowly rotating bearings [5.26]. The measurements can also detect contamination in rolling bearing grease [5.27]. The technique can also be used for detecting seal and blade rubbing in turbines, diagnosis of reciprocating machinery faults, and detection of partial discharge in the insulation and windings of power transformers [5.28][5.29]. Acoustic emission sources can be located by using three or more transducers around the source. The difference in AE wave arrival time at different sensors is measured and the location of the source is calculated using the velocity of waves in the material.

## 5.7 Wear Debris and Lubricating Oil Analysis

Lubricating oil is used in machines to minimize friction and for cooling purposes. This lubricating oil comes in direct contact with bearings, gears, piston–cylinder assembly and many other tribosystems. These components wear out gradually and wear particles are carried away by the lubricating oil. Analysis of these wear particles, known as wear debris, can give useful information about machine condition. The shape, size, quantity and composition of the wear particles need to be analyzed to assess machine condition. Wear debris analysis can determine wear from different wear mechanisms such as sliding, rolling, rubbing, combined sliding and rolling, abrasion *etc.*, which each produce a different kind of wear particle. An oil sample is taken from the machine to be analyzed. Sampling plays an important role in interpreting the results. The sample should be taken from the system where wear particles are mixed homogeneously in the oil (*e. g.* in a recirculating system, samples should be collected from the return line before the oil filter). Wear particles may vary from a few microns to 1000 microns. Various techniques like magnetic plugs, ferrography and spectroscopic oil analysis (SOA) are available for the analysis of wear debris. All these techniques are sensitive to particles size. Magnetic plugs are sensitive to particles of size larger than 50 microns; ferrography is sensitive to particles of size between 10 and 300 microns; and SOA is sensitive to particles of size below 10 microns.

### 5.7.1 Magnetic Plugs and Chip Detectors

A magnetic drain plug is basically used to remove wear debris in oil sumps to prevent secondary damage. Magnetic plugs are removed at regular intervals and inspected for signs of heavy wear particles. The positioning of the plugs should be such that it can gather the maximum amount of wear particles. The lubricating oil is first cleaned by solvent and then inspected under microscope. This can give a rough idea of the degree and type of wear.

Chip detectors are similar to magnetic plugs except that they have an in-built warning system whenever wear debris is detected. The debris bridges the gap between two electrodes. This bridging acts as a switch closer for an alarm.

### 5.7.2 Ferrography

To study the wear particles in the lubricating oil, the wear particles are arranged according to their size by applying a varying magnetic field. The instrument used to prepare such slides is known as a ferrograph and the slide is known as a ferrogram.

To prepare a ferrogram the oil sample is pumped and discharged over the transparent slide at a slow and steady rate. The oil sample is diluted with a solvent to improve particle precipitation. The glass slide is kept tapered and a magnet is kept below the slide resulting in a strong magnetic field at lower end and weak magnetic field at upper end. This diverging magnetic field collects larger particles at the upper end and smaller particles at the lower end. Particles other than ferrous, like dust, carbon, aluminum *etc.* are collected at the other end. Clean solvent then washes the oil, and when the slide is dried, the ferrogram is ready for microscopic analysis. The above-mentioned ferrograph is known as analytical ferrograph. In more advanced analytical ferrographs particles are precipitated in three circular rings using a rotary particle depositor (RPD). The RPD ferrograph offers a rapid and simple method of debris separation. A measured volume of sample is applied, by pipette, to a glass substrate located on a rotating magnet assembly. Particles of debris are deposited radially as three concentric rings by the combined effects of rotational, magnetic and gravitational forces. Removal of the lubricant by solvent washing and drying gives a stable well-separated deposit pattern ready for examination by optical or electron microscope.

Another type of ferrograph is known as a direct reader (DR) ferrograph. In a DR ferrograph the oil sample is passed through a glass tube under a diverging magnetic field, as shown in Figure 5.12. Light is passed across the tube near the inlet and outlet point. The light is sensed by photocells at the other end of the light source. This instrument works online and gives a quick reading of the difference between largest and smallest sized wear particles, which is a good indicator of severity of wear.

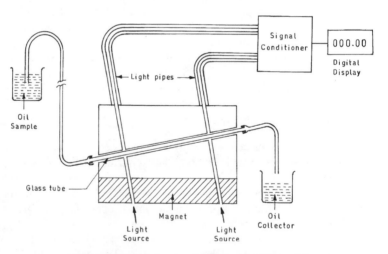

**Figure 5.12.** Schematic of direct reader ferrograph

### Hall Effect Analyzer

The principle of the Hall effect analyzer device is that when wear debris is captured in the air gap between the poles of the magnet, the magnetic flux density varies with the amount of wear debris. This change of magnetic flux can be transformed into an output voltage difference through the Hall sensor. The relationship between Hall voltage difference and the amount of wear debris follows a perfect power law. Therefore, this voltage difference can be used to detect the amount of wear debris. An online hall effect device for monitoring wear particle in oil has been developed [5.30].

*Optical Ferroanalyzer*

The optical ferroanalyzer is an improvement of the DR ferrograph. The analyzer's operating principle is based on analysis of the concentration of ferrous particles in the lubricant. The analysis is made by reading changes in the optical radiation reflected by the substrate in the zones of accumulation of large and fine particles under the effect of the magnetic field, and by rating the lubricity of the oil when registering its optical density changes [5.31].

*Wear Assesment*

To relate wear rate, severity, mode and source of wear, determination of wear debris quantity, size, morphology and composition is required. Wear rate can be assessed by measuring the quantity of wear debris. Severity of wear can be found by measuring the quantity, size and shape of wear debris. Wear mode can be determined by analyzing shape and size of wear debris. Shape and composition analysis of wear debris can be used to identify which component is wearing.

As explained earlier, ferrographs separate large ($L$) particles (near inlet) from small ($S$) particles (near outlet). For normal wear processes the values of $L$ and $S$ are almost the same (usually $L$ is slightly greater than $S$). The $L$ value is significantly larger than the $S$ value for severe wear. Therefore, $L$–$S$ can be used as an indication of severity. The inception of abnormal wear mode results in a significant increase in the quantity of wear debris. The $L$+$S$ value can be used as an indication of quantity of wear. By multiplying $L$+$S$ and $L$–$S$, a single index known as the wear severity index ($I_s = L^2 - S^2$) can be obtained. Ranges of $I_s$ and their associated wear modes can be determined by experimentation. For the complete life cycle of any tribosystem an initially large $I_s$ value indicates the running-in state, after some time low $I_s$ value indicates steady normal running and at the end very large $I_s$ indicates severe wear mechanism.

The wear debris particle composition can be obtained by examining it in a microscope under different light arrangements. When the particle is viewed under reflected white light, copper-based alloy appears yellow or reddish brown while most remaining metals appear silver white. By viewing wear particles under transmitted white light, the attenuation of light ascertains whether the particle is transparent, translucent or opaque. Free metal particles appear black, most of the other elements and all compounds appear translucent or transparent, the color displayed being characteristic of the material. In a bichromatic microscope wear particle can be viewed under reflected red light and transmitted green light. Free metal particles appear red while non-metals and all compounds appear green or yellow depending upon the degree of light attenuation. Further analysis in polarized light yields information on the crystal structure of various compounds.

*Automated Image Analysis*

To automate the analysis of wear debris, characterization of particle morphology is necessary. Perimeter/area ratio, aspect ratio (length/width), curl (length/fiber length), roundness $(4(area)/\pi(length)^2)$, form factor $(4\pi(area)/(perimeter)^2)$ *etc.* are some of the parameters which are used to assess the overall particle shape. Characterization of surface morphology is also essential. Laser scanning confocal microscopy (LSCM) allows the recording of three-dimensional image sets, thus allowing analysis of the true surface topography of wear debris. Computer images of wear particles are analyzed using software to calculate the above-mentioned parameters. Once the wear particle is defined properly an expert system can be used to determine the machine condition automatically [5.32].

### 5.7.3 Particle Counter

A particle counter gives the number of particles greater than a specified particle size per milliliter of oil in a bottle. A laser beam permits observation of particles by light scatter and a scanning and detection system quantifies the number of small particles above the set "threshold". It is possible to study the particle size distribution by making a series of measurements with different threshold settings. In the particle counter, a laser beam is focussed by lens assemblies to form a small illuminated volume within the oil as shown in Figure 5.13 [5.33]. Lateral displacement of this illuminated volume is provided by a mechanism scanning at constant speed. The beam travels through the oil to a photodetector assembly. As the illuminated volume moves across a particle suspended in the oil, some light from the laser beam is scattered and collected by the optical system of the photodetector assembly [5.33].

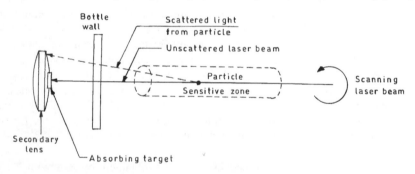

**Figure 5.13.** Laser particle counter

The flash of light reaching the photodetector causes an electrical pulse in the preamplifier connected to the photodetector. The magnitude and width of this pulse are related to the size of the particle. The walls of the bottle containing the oil are in the out-of-focus zone, so pulses caused by out-of-focus particles will be broader than those caused by an in-focus particle. Electronic circuits have been designed to count only the narrow pulses generated by in-focus particles. The illuminated beam sweeps through the oil at a constant speed, so a defined time period corresponds to a defined scanned volume. An electronic timer is provided in the system to give a count of particles per milliliter of oil.

### 5.7.4 Spectrographic Oil Analysis (SOA)

This technique is different from previously mentioned techniques as it does not look in to the morphology of wear particles. SOA looks for the type of metal element present in the oil sample quantitatively by either atomic emission or atomic absorption, to give an indication of which part is wearing and the rate of wear. Atomic absorption spectrometry is more accurate but time consuming when a number of metals have to be analyzed. The basic principle behind these techniques is that different metals radiate different wavelengths when their molecules are raised to an excited state by applying energy.

In emission spectroscopy, the oil sample is 'ashed' and put into the arc to vaporize. The radiation emitted from the vaporized metals is passed through an optical prism, which gives the spectrum. A photoelectric system measures the spectral radiation. Separate photodetectors are used for each element.

In atomic absorption spectrometry, the oil sample is diluted and sprayed into a flame. An air–acetylene flame is used to atomize the metallic elements in the sample. The atomized sample is passed through an energy source, which emits light characteristic of the element to be determined, *e. g.* a hollow cathode discharge lamp. A prism and slit arrangement is used to obtain a monochromatic wavelength. A photodetector is used to measure the energy.

### 5.7.5 Lubricating Oil Analysis

The basic purpose of lubricating oil analysis is to determine the most cost-effective interval between oil changes, but by examining various oil properties some indications of fault development in a mechanical system can be made.

Oil viscosity increases with time and use, therefore reduction in viscosity is abnormal. An increase in viscosity of 20% or a reduction of 10% from nominal grade is alarming. An abnormal increase in viscosity may be due to the presence of solid particles or the mixing of some leaking fluid of high viscosity. A drop in viscosity may be due to mixing of low viscosity fluids such as fuel or coolant. An increase in density may be due to the mixing of fluids.

Water is undesirable in oil. An increase in water content may be due to condensation of combustion products or leakage from the cooling system. High levels of water content cause corrosion and reductions in viscosity.

Fuel dilution of oil may be due to improper operation, fuel system leakage, ignition problems or improper timing. Fuel dilution reduces oil film strength and its ability to seal. A level of more than 2 to 5% is considered excessive. An increase in fuel soot content in the oil is due to inefficient burning of the fuel in a diesel engine.

Oxidation, total acid number (TAN), total base number (TBN), nitration, pour point, flash point are some of the other oil properties whose analysis can be useful in determining machine condition.

Additives are mixed in oil to increase certain oil properties. Monitoring of additives in the lubricating oil can give useful information regarding machine condition. A rapid reduction in anti-corrosive additive is an indication of increased wear rate. Rust inhibitor, corrosion inhibitor, anti-oxidant, anti-foaming agent, anti-wear agent, pour point depressant are some of the additives whose monitoring can give an indication of machine health.

## 5.8 Thermography

Temperature is one of the signatures that can give information regarding machine condition. All machines run at optimum efficiency only over a specific operating temperature range. An abnormal increase or decrease in temperature is an indication of malfunction. Temperature monitoring of critical machine components with other condition monitoring techniques may be helpful in preventing machine breakdown.

All objects emit heat that is constantly being absorbed and re-emitted by everything around it. Thermal radiation is electromagnetic radiation. The electro-magnetic spectrum is divided arbitrarily into a number of wavelength regions called wavebands. Thermal radiation consists of ultraviolet (0.1 μm to 0.4 μm), visible (0.4 μm to 0.8 μm) and infrared (0.8 μm to 100 μm) wavebands. In most mechanical systems surface temperatures give rise to radiation in the infrared region.

Thermography is the process of making infrared (IR) radiation visible and capable of interpretation. Thermography is a non-contact type measurement that gives the temperature distribution (thermal image) of a measured surface, in gray scale or colored scale, in real time. Thermography basically consists of an IR-camera unit and display unit. In the IR camera the most important part is the IR detector, which is a transducer that converts thermal energy into electrical energy. Cadmium mercury telluride (CdHgTe), platinum silicide (PtSi) and indium antimonide (InSb) are the detector materials used most commonly. These kinds of detectors need cooling to cryogenic temperatures: for this cooling, liquid nitrogen or argon gas was used in older IR cameras.

In modern IR-cameras, semiconductor material is used, which converts photon energy into electrical energy. The photon energy is inversely proportional to its wavelength. These types of detectors do not require cooling, therefore reducing size, weight, power and cost of IR-camera. High quality thermal image is possible by use of detector arrays.

Thermography is widely used in industry as a diagnostic tool. All rotating and reciprocating machines have some friction that generates heat, which is minimized by lubricants. In normal conditions the temperature is maintained within operating limits. The temperature of some machine components may rise if there is a lack of lubrication, excessive wear, increased load, *etc.* Thermography can detect the increase in particular machine components such as bearings, gears, couplings, flywheel, chain, clutches, *etc.* Thermography can also be used to monitor blast furnace lining, steel mixing vessels, electric transmission lines and switch boards.

In general, the thermographic image is captured manually and analyzed by experts. This is a tedious process, but it has low initial cost. Online thermography of plant and machines is becoming popular with the advancement of computer and IR camera output interfacing, optical fibers and the availability of software for artificial intelligence. Automatic thermographic condition monitoring is fast and more reliable, but its initial set-up cost is more.

## 5.9 Conclusions

Condition monitoring of machines is very useful for detecting the development of faults in them. The problem can be diagnosed and corrective action taken at an appropriate time. This avoids sudden breakdown of machines and the associated economic loss and safety problems. Maintenance actions are taken only when required, dependent on the condition of the machine. Performance-related parameters and parameters such as vibration, acoustic emission, thermograph, wear debris in oil, *etc.* are useful indicators of machinery condition. Their analysis is helpful in diagnosing the problem and estimating failures.

## References

[5.1]   Kim, S.-M., Joo, Y.-J., 2005, "Implementation of On-line Performance Monitoring System at Seoincheon and Sinicheon Combined Cycle Power Plant", *Energy*, **30**, pp. 2383–2401.

[5.2]   Wyatt, C., 2004, "Monitoring Pumps", *World Pumps*, December, pp. 17–21.

[5.3]   Vafaei, S., Rahnejat, H., Aini, R., 2002, "Vibration Monitoring of High Speed Spindle Using Spectral Analysis Technique", *Machine Tools & Manufacture*, **42**, pp. 1223–1234.

[5.4]   Kolbasseff, A., Sunder, R., 2003, "Lessons Learned with Vibration Monitoring Systems in German Nuclear Power Plants", *Progress in Nuclear Energy*, **43**(1–4), pp. 159–165.

[5.5]   Garcia, B., Burgos, J.C., Alonso, A., 2005, "Winding Deformations Detection in Power Transformers by Tank Vibration Monitoring", *Electric Power Systems Research*, **74**, pp. 129–138.

[5.6]   Tandon, N., Choudhury, A., 1999, "A Review of Vibration and Acoustic Measurement Methods for the Detection of Defects in Rolling Element Bearings", *Tribology International*, **32**, pp. 469–480.

[5.7]   Khemili, I., Chouchane, M., 2005, "Detection of Rolling Element Bearing Defects by Adaptive Filtering", *European Journal of Mechanics A/Solids*, **24**, pp. 293–303.

[5.8]   Tandon, N., Mata, S., 1999, "Detection of Defects in Gears by Vibration Monitoring", In *Proceedings of Asia-Pacific Vibration Conference*, Singapore, 13–15 December, pp. 161–164.

[5.9]   Walter, P.L., "Trends in Accelerometer Design for Military and Aerospace Applications", http://www.sensorsmag.com/articles/0399/0399_44/main.shtml.

[5.10]  Chen, Y.D., Du, R., Qu, L.S., 1995, "Fault Features of Large Rotating Machinery and Diagnosis Using Sensor Fusion," *Journal of Sound and Vibration*, **188**(2), 227–242.

[5.11]  ISO 2372, 1974, *Mechanical Vibration of Machines with Operating Speeds from 10 to 200 Rev/s – Basis for Specifying Evaluation Standards*.

[5.12]  Sahdev, M., "Centrifugal Pumps: Basic Concepts of Operation Maintenance and Troubleshooting, Part II", http://www.cheresources.com/centrifugalpumps3b.shtml.

[5.13]  Starr, A., Wynne, R., 1996, "A Review of Condition Based Maintenance for Electrical Machines", Chapter 12, In *Handbook of Condition Monitoring*, Rao, B.K.N., ed., Elsevier Science Ltd, pp. 267–284.

[5.14]  Benbouzid, M.E.H., 2000, "A Review of Induction Motors Signature Analysis as a Medium for Fault Detection", *IEEE Transactions on Industrial Electronics*, **47**(5), pp. 984–993.

[5.15] Chu, F., Lu, W., 2005, "Experimental Observation of Nonlinear Vibrations in A Rub-Impact Rotor System", *Journal of Sound and Vibration*, **283**, pp. 621–643.

[5.16] Lyon, R. H., 1987, *Machinery Noise and Diagnostics*, Butterworths, p. 13.

[5.17] Pachaud, C., Salvetat, R. and Fray, C., 1997, "Crest Factor and Kurtosis Contributions to Identify Defects Including Periodical Impulsive Forces," *Mechanical Systems and Signal Processing*, **11**(6), pp. 903–916.

[5.18] Badaoui, M. El, Guillet, F., Daniere, J., 2004, "New Applications of the Real Cepstrum to Gear Signals, Including Definition of a Robust Fault Indicator", *Mechanical Systems and Signal Processing*, **18**(2004), pp. 1031–1046.

[5.19] Badaoui, M. El, Antoni, J., Guillet, F., Daniere, J., 2001, "Use of the Moving Cepstrum Integral to Detect and Localise Tooth Spalls in Gears", *Mechanical Systems and Signal Processing*, **15**(5), pp. 873–885.

[5.20] Samanta, B., Al-Balushi, K. R., 2003, "Artificial Neural Network Based Fault Diagnostics of Rolling Element Bearings Using Time-Domain Features," *Mechanical Systems and Signal Processing*, **17**(2), pp. 317–328.

[5.21] Yang, D.-M., Penman, J., 2000, "Intelligent Detection of Induction Motor Bearing Faults Using Current and Vibration Monitoring", In *Proceedings of COMADEM 2000*, Texas, 3–8 December, pp. 461–470.

[5.22] Ramakrishna, K. M., Yadava, G. S., Tandon, N., 2001, "Fault Diagnosis in Electrical Motors Using Current and Vibration Monitoring – A Review", In *Proceedings International Seminar on Electrical Systems & Reliability*, New Delhi, 17-19 October, pp. 105–124.

[5.23] Tandon, N., Mata, S., 1999, "Detection of Defects in Gears by Acoustic Emission Measurements", *Journal of Acoustic Emission*, **17**(1–2), pp. 23–27.

[5.24] Yoshioka, T., Fujiwara, T., 1984, "Application of Acoustic Emission Technique to Detection of Rolling Bearing Failure", In *Acoustic Emission Monitoring and Analysis in Manufacturing*, Dornfield, D. A., ed., ASME, New York, pp. 55–75.

[5.25] Tandon, N., Nakra, B. C., 1990, "Defect Detection in Rolling Element Bearings by Acoustic Emission Method", *Journal of Acoustic Emission*, **9**(1), pp. 25–28.

[5.26] Jamaludin, N., Mba, D., 2002, "Monitoring Extremely Slow Rolling Element Bearings: Part I", *NDT&E International*, **35**, pp. 349–358.

[5.27] Miettinen, J. andersson, P., 2000, "Acoustic Emission of Rolling Bearings Lubricated with Contaminated Grease", *Tribology International*, **33**, pp. 777–787.

[5.28] Mba, D., Hall, L. D., 2002, "The Transmission of Acoustic Emission across Large-scale Turbine Rotors", *NDT&E International*, **35**, pp. 529–539.

[5.29] El-Ghamry, M. H., Reuben, R. L., Steel, J. A., 2003, "The Development of Automated Pattern Recognition and Statistical Feature Isolation Techniques for the Diagnosis of Reciprocating Machinery Faults Using Acoustic Emission", *Mechanical Systems and Signal Processing*, **17**(4), pp. 805–823.

[5.30] Chiou, Y. C., Lee, R. T. and Tsai, C. Y., 1998, "An On-line Hall-effect Device for Monitoring Wear Particle in Oils," *Wear*, **223**, pp. 44–49.

[5.31] Myshkin, N. K., Markova, L. V., Semenyuk, M. S., Kong, H., Han, H.-G. and Yoon, E.-S., 2003, "Wear Monitoring Based on the Analysis of Lubricant Contamination by Optical Ferroanalyzer", *Wear*, **255**, pp. 1270–1275.

[5.32] Kirk, T.B, Panzera, D., Anamalay, R. V. and. Xu, Z.L, 1995, "Computer Image Analysis of Wear Debris for Machine Condition Monitoring and Fault Diagnosis," *Wear*, **181–183**, pp. 717–722.

[5.33] Instruction Manual – *Spectrex Laser Particle Counter*, Spectrex Corporation, California, 1988.

# 6

## Advanced Diagnostic and Prognostic Techniques for Rolling Element Bearings

Thomas R. Kurfess[1], Scott Billington[2] and Steven Y. Liang[1]

[1] George W. Woodruff School of Mechanical Engineering, Georgia Institute of Technology
Atlanta, GA 30332-0405, USA
Email: tom.kurfess@me.gatech.edu

[2] Radatec, Inc., Atlanta, GA 30308, USA
Email: scottb@radatec.com

**Abstract**
Bearing failure is one of the foremost causes of breakdown in rotating machine, resulting in costly systems downtime. This chapter presents an overview of current state-of-the-art monitoring approaches for rolling element bearings. Issues related to sensors, signal processing as well as diagnostics and prognostics are discussed. This chapter also presents a brief discussion related to the typical failure modes of bearings. Such failures are more and more common on advanced, high speed, ultra precision production systems as higher spindle speeds are employed for increased accuracy, productivity and machining satiability. Models for rolling element bearing behavior are presented as well as mechanistic models for damage propagation. Examples are also presented from test systems to demonstrate the various approaches discussed in this chapter.

## 6.1 Introduction

Bearing failure is one of the foremost causes of breakdown in rotating machinery [6.1]. Such failures can be catastropic and can result in costly downtime [6.2][6.3]. Therefore, bearing monitoring is critical to machine maintenance and automation processes [6.3]. Past maintenance methods have included preventative maintenance based on predetermined time intervals, and scheduled maintenance based on historical performance. These methods, though still in use, are not entirely effective. When servicing bearings based on a set schedule, a nominal bearing life is assumed. However, many bearing failures arise from unpredictable factors, such as incorrect fitting, improper installation, the introduction of foreign matter and inappropriate lubrication. Additionally, even the principle process of fatigue failure – cyclic stresses leading to crack propagation and the subsequent flaking of material from the rolling surfaces – is random in nature. This leads to a broad spread in bearing lifetimes, with some units lasting far longer than expected but many failing earlier than intended.

Preventative and scheduled maintenance techniques waste useful operating life when bearings are replaced prematurely and, more importantly, provide no protection against early failures. Condition monitoring and condition-based maintenance (CBM) approaches have been investigated for over thirty years. The aim of these approaches is to determine the actual condition of the bearings and service them based on that information. Interest in the area increased markedly following a report in the early 1970s by a government working party that drew attention to the high cost of maintenance to industry in the UK [6.4].

Diagnostics and prognostics for rolling element bearings target the determination of the current state of the bearing and prediction of the useful remaining life of the bearing. Such capabilities are critical given modern ultra-precision and high speed machine tools where spindle speeds of 100,000 rpm and greater are employed. This chapter briefly discusses state-of-the-art diagnostic and prognostic techniques that are being employed for bearing CBM. The chapter is divided into four major sections, discussing measurement basics, models, diagnostics and prognostics. While the target application for this chapter is ultra-high precision spindles, most of the ideas and concepts presented in this chapter are applicable to a wide variety of systems employing rolling element bearings such as turbines, compressors, vehicles, *etc.*

## 6.2 Measurement Basics

While there are many sensors and techniques employed in rolling element bearing CBM, there are several quantities that are most often used. This section discusses several critical quantities that are used in diagnostic and prognostic applications, and then focuses on vibration type measurements for bearing monitoring.

### *Temperature*

Monitoring the temperature on the surface of the bearing housings with thermocouples has been used for many years. Under steady-state operating conditions, the temperature of a machine is monitored for changes that can indicate a problem. Experience has shown that temperature is a better indicator of load, speed or lubrication than of bearing condition [6.5]. An increase in temperature can warn of fault-producing conditions, such as overloading and faulty lubrication. The risk of failure can be decreased if these conditions are recognized and corrected. However, bearing defects have not been found to cause an appreciable increase in temperature until the damage has reached a severe state.

### *Debris*

Wear debris is always produced when there is contact between moving parts. When a circulated lubricant is used, wear and fatigue damage can be monitored by determining the amount and type of particles in the flow. One of the simplest particle collection methods is to place a magnetic plug in the flow stream, where ferrous material can be captured. Removable plugs exist that allow for extraction without disrupting the lubrication flow. Such plugs tend to collect particles greater than 50 μm in size. The samples can be weighed and examined for characteristic features. For example, spherical particles are associated with metal fatigue and the incidence of pitting on rolling elements [6.6][6.7]. A typical wear debris trend is shown in Figure 6.1.

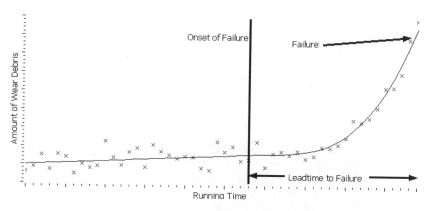

**Figure 6.1.** Typical wear debris trend

## Vibration

Vibryation techniques are widely used for health monitoring of rotating machinery. Even new bearings produce somey low energy vibration, due to varying compliance of their rotating assembly and manufacturing imperfections [6.8]. These imperfections, referred to as distributed defects, include surface roughness, waviness, misaligned races and off-size rolling elements. Once localized defects start to form – such as fatigue cracks, pits and spalls – the vibration energy is much greater. Each time a defect on one surface strikes another, it imparts a high-energy impact to the system.

The range or frequency of the oscillations usually determines the type of sensing and analysis chosen to sample and interpret them. The main frequency bands are [6.7]:

1. vibration: <1 Hz to 25 kHz
2. ultrasonics: 20 kHz to 100 kHz
3. shock pulse: 32 kHz
4. acoustic emission (stress wave): 100 kHz to 1 MHz.

Higher frequency vibration (>1 kHz) is best sensed by acceleration. The piezo-electric accelerometer, in particular, is widely used for this purpose. Because it has an extended frequency range of approximately 1 Hz to 10 kHz, it has become a standard device for machine vibration measurement.

Acoustic emission can be defined as transient elastic stress waves, generated at a source, by the rapid release of strain energy within a material. These waves are detected at the surface and recorded by a sensor. The source mechanism can arise from several phenomena, including asperity contact, micro-crack initiation and growth, and plastic deformation [6.9]. This potential to detect subsurface defects is an advantage of the acoustic emission sensing technique [6.10].

## Position, Acceleration and Vibration Sensors Example

To demonstrate the use of various sensors to measure vibration, a Kamen KD-2300-2S1 eddy current probe (non-contact position sensor), Radatec microwave sensor (non-contact position), and piezo-electric accelerometer were run in comparison tests to measure shaft vibration for the purpose of damage detection in bearings (Figure 6.2). The correlation between the eddy current probe and microwave sensor was investigated as both of these sensors are used to measure the shaft radial position deviation as it rotates. A 306K bearing with known natural spall damage on its inner race, as shown in Figure 6.3, is used in the test housing. Data are acquired from the eddy current probe, microwave sensor, and accelerometer when the shaft is rotating at a speed of 4,000 rpm.

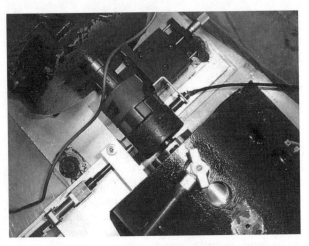

**Figure 6.2.** Picture shows microwave sensor and eddy current probe installation

**Figure 6.3.** Pictures of the spall damage on the inner race surface

The displacements from the microwave sensor and eddy current probe are plotted in Figure 6.4. The signals from both sensors are similar in terms of main features such as periodic peak amplitude and frequency. The differences in detail are expected because both sensors do not measure identical surfaces. In addition to the fact that the two sensors are located on opposite sides of the coupler, the spot size of the eddy current probe is significantly larger than that of the microwave sensor.

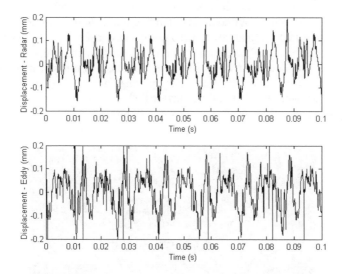

**Figure 6.4.** Displacements obtained from the microwave sensor and eddy current probe

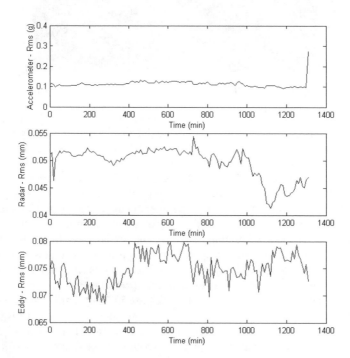

**Figure 6.5.** RMS plot over time from accelerometer, microwave and eddy current sensors

The accelerated bearing-life test was terminated when the RMS level from the accelerometer signal reached a predefined threshold value. The graphs in Figure 6.5 show RMS values of the accelerometer signal and the displacements computed from the microwave sensor and eddy current probe over the test period. The first graph in Figure 6.5 shows that, at the last acquisition, the accelerometer clearly indicates bearing damage from the obvious spikes standing out from the base RMS. However, the displacement RMS signals from the microwave sensor and eddy current probe do not detect this change, as shown in the two lower graphs. This result indicates that shaft vibration detection via position measurement is not sensitive enough for early bearing damage identification. The bearings were then disassembled and inspected for damage on their rolling surfaces. One of the bearings showed visual damage on its inner rolling surface as shown in Figure 6.6.

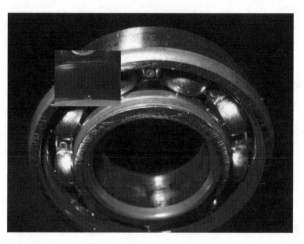

**Figure 6.6.** Small visual damage on the inner rolling surface

Frequency spectra were computed from the displacement signals in order to investigate changes in amplitude at the bearing defect frequency. Bearing 205K has ball frequency = 314 Hz, inner race frequency = 361 Hz, and outer race frequency = 239 Hz (computed from the SKF online calculator). Since the bearing was damaged at the inner rolling surface, the peak amplitude of displacements from the microwave sensor and eddy current probe at frequencies between 355 and 365 Hz are plotted with respect to time in order to monitor the amplitude at the inner race frequency. The displacement amplitudes at the inner race frequency from the microwave sensor and eddy current probe are shown in the second and third graphs of Figure 6.7. The spectral technique reduces fluctuation in the vibration signal that is a result of wideband noise. But even with the spectral technique, the amplitude at the defect frequency does not indicate the increase in shaft displacement amplitude at the last acquisition, before the test is terminated. This confirms that the measurement of shaft vibration using displacement sensors is not sensitive enough to indicate early bearing damage even by observing at the defect frequency. Thus, accelerometers are recommended for this type of detection.

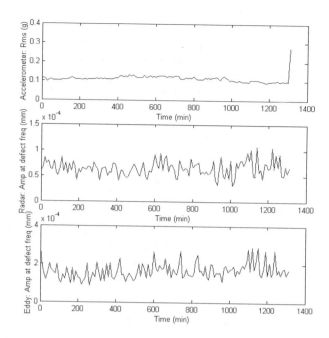

**Figure 6.7.** Microwave and eddy current sensors amplitude data at inner race frequency

The bearing damage was not detectable by the shaft sensors, because shaft vibration is not large enough at such small damage levels. Therefore, the test was continued with the same bearing using a lighter load for the purpose of retrieving more measurements as the damage increased. The test was performed with reduced axial and radial loads. The axial load for 1.13 N·m (10 in·lb$_f$) torque over 6 bolts yielded an axial load of 3,080 N. The radial load was applied by a hydraulic hand pump to 1,379 kPa (200 psi), corresponding to 4,910 N of load on each bearing. During this extension test the bearing damage quickly grew and became clearly audible. The data from all sensors were acquired more frequently at 2.5 minute intervals. The test was terminated when the vibration level reached eight times the base RMS (0.8 g). Figure 6.8 shows plots of data obtained from the accelerometer, microwave sensor, and eddy current probe with respect to time. The first graph shows RMS values of the acceleration from the accelerometer. The second and third graphs show displacement amplitudes from the microwave sensor and eddy current probe at the defect frequency, respectively.

The graphs from the extension test demonstrate that the bearing damage level is detectable through the shaft vibration as the level of damage increases. All the graphs in Figure 6.8 from the accelerometer, microwave sensor, and eddy current probe show heavily correlated signal increases resulting from progressing bearing damage. The signals from the microwave sensor and eddy current probe do not clearly identify the damage step increase at a time of 1,350 minutes, as seen by the accelerometer. However, after 1,400 minutes the microwave sensor and eddy current probe clearly identify the damage step increase corresponding to that from the accelerometer. It is also observed that the measurements of shaft and housing vibration are not linearly correlated. With this evidence, it may be concluded that both the microwave sensor and eddy current probe are capable of measuring shaft vibration produced by the bearing damage. However, in order to use shaft vibration as a means to identify bearing damage, the damage level of bearings must be considerable relative to the early damage that is detectable by accelerometer.

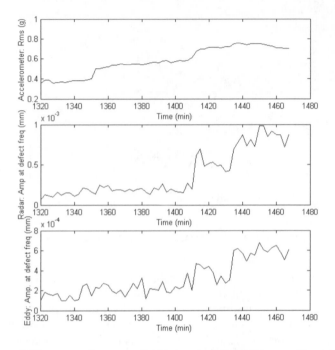

**Figure 6.8.** Inner race frequency displacement amplitude (damage detectable)

**Figure 6.9.** Damage found on the inner rolling surface after an extended test

After the test is stopped, the bearings are again inspected for damage on their rolling surfaces. Spall damage is found to have a larger size approximately 2 mm wide, and 2 mm long on the inner race of a bearing as shown in Figure 6.9.

In conclusion, this example demonstrates the capability of the shaft sensors in a vibration measurement application. Although, the test results show that the shaft sensor thresholds for early damage detection were higher, it is effective in detecting the larger shaft vibration as the damage size increases. In this particular application, the frequency spectrum of the shaft displacement is found to be useful in identifying the development of the bearing damage. The displacement amplitudes from both the microwave sensor and eddy current probe at the defect frequency are found to have good correlation with the damage level identified by the RMS level of the accelerometer. One difficulty when using accelerometers is that the location of these sensors on the system being monitored is critical. Slight changes in an accelerometer's position on the test unit can result in substantially different signal amplitudes. However, position sensors such as eddy current and microwave probes, directly target rotating components such as shafts. Such components are usually more accessible and the measurement consistency much more robust to sensor location.

## 6.3 Bearing Models

### *Bearing Deterministic Frequencies*

Bearing deterministic frequencies depend on the location of the defect, the bearing geometry, and the speed of rotation. However, due to slippage and axial loading, these critical bearing frequencies do not always precisely coincide with spectral frequencies. Derivation of the deterministic frequencies, shown in Equations (6.1)–(6.4), is beyond the scope of this section. However, three important explanations clarify these expressions: (1) the factor of ½ is the rate of rotation relative to the shaft center. The coefficient results from the contact of the rolling elements with the shaft-speed inner race and the stationary speed outer race; (2) the expressions neglect centrifugal forces or gyroscopic coupling; and (3) the equations assume a stationary outer race, rotating inner race and no skidding of the rolling element.

$$f_{bpor} = Z\frac{f}{2}\left(1-\frac{D}{d_m}\cos\alpha\right) \tag{6.1}$$

$$f_{bpir} = Z\frac{f}{2}\left(1+\frac{D}{d_m}\cos\alpha\right) \tag{6.2}$$

$$f_r = Z\frac{fd_m}{D}\left(1-\left(\frac{D}{d_m}\cos\alpha\right)^2\right) \tag{6.3}$$

$$f_{bpor} = \frac{f}{2}\left(1-\frac{D}{d_m}\cos\alpha\right) \tag{6.4}$$

White [6.3] and Harris [6.11] present Equations (6.1)–(6.4) in terms of the following bearing geometric parameters: α – rolling element contact angle, $D$ – mean rolling element diameter, $d_m$ – pitch diameter, $Z$ – number of rolling elements, and $f$ – fundamental rotational frequency. The defect frequencies stand for ball-pass outer race (bpor), ball-pass inner race (bpir), rolling element frequency (r), and fundamental train frequency (ftf). When bearing geometry is not known precisely, Equations (6.5)–(6.7) can approximate deterministic frequencies [6.3].

$$f_{bpor} = Zf(0.4) \tag{6.5}$$

$$f_{bpir} = Zf(0.6) \tag{6.6}$$

$$f_{ftf} = 0.4f \tag{6.7}$$

### Signal Interactions

Complex machines have multiple vibration excitations and often these excitations interact, resulting in somewhat different frequencies. Common interaction between signals can be linear as in signal summation or non-linear as in amplitude modulation or frequency modulation. Research has shown that other frequencies result from the interaction of deterministic defect frequencies [6.12]. For example, a single defect on a roller element can contact both the inner and outer races, hence the defect frequency is $2*f_r$. Shaft misalignment, inbalance, or excessive clearance can modulate bearing defect frequencies. In addition, axial pre-load can change the contact angle affecting deterministic frequencies [6.13].

The fundamental rotational frequency can amplitude modulate (AM) defects located in the inner race. This interaction produces sidebands around the carrier frequency; in this case, the deterministic defect frequency. Amplitude modulated signals are signals on machinery that change the amplitude of one signal due to the amplitude of a second signal, known as the modulating signal. A true AM signal clearly shows the carrier frequency and modulating frequency, plus sidebands. The sidebands are equal to the sum and difference of the two fundamental frequencies (Figure 6.10). As a bearing defect propagates, the bandwidth of the frequency spectrum may narrow and may become modulated by the shaft speed [6.1]. Furthermore, at advanced stages, ball pass frequencies may disappear resulting in a random spectrum.

Frequency modulation (FM), originates with a high frequency and constant amplitude carrier signal. A lower frequency signal modulates this carrier frequency. The following features characterize an FM signal: (1) a centered carrier frequency, (2) multiple sidebands at the carrier frequency plus or minus the orders of the modulator, and (3) constant amplitude in the time domain (Figure 6.11). Frequency modulation commonly occurs in gearboxes where the gear mesh frequency is modulated by the rotational frequency of the gear.

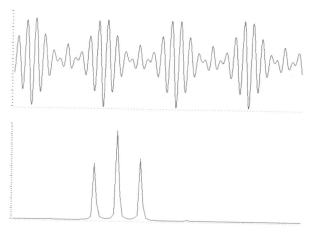

**Figure 6.10.** Amplitude modulation of a signal

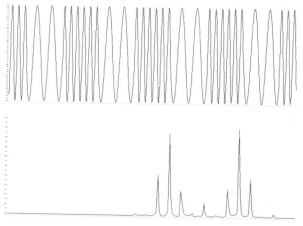

**Figure 6.11.** Frequency modulation of a signal

## 6.4 Diagnostics

This section presents basic concepts related to identifying and isolating signals indicative of an incipient bearing failure.

### 6.4.1 Signal Analysis

Most difficulties posed in bearing condition monitoring arise from the presence of mechanical noise in the wide spectrum of bearing defect signals. Therefore, signal processing methods and signal metrics attempt to emphasize defect signals over background noise. The current state of condition monitoring can be classified into time and frequency domain analysis. This section presents signal processing techniques and signal metrics commonly used to detect bearing defects.

### Time Domain Techniques

Traditionally, signal analysis in the time domain has been used to monitor simple machine conditions. However, complex signals, frequently encountered in industrial equipment, are difficult to analyze. Some of the time domain techniques applied to condition monitoring are: root mean square (RMS), mean, peak value, crest factor, kurtosis [6.2][6.14][6.15], and shock pulse counting [6.16].

### Root Mean Square

As described by Equation (6.8), root mean square (RMS), measures the overall level of a discrete signal.

$$RMS(x) = \sqrt{\frac{\sum x_i^2}{N}} \tag{6.8}$$

$N$ is the number of discrete points and $x_i$ represents the signal from each sampled point. The RMS is a powerful tool to estimate the average power in system vibrations. Often, condition monitoring establishes a baseline of a good machine and monitors the changes throughout bearing life. A substantial amount of research has employed RMS to successfully identify bearing defects using accelerometer and AE sensors. For example, Tandon [6.18] saw that the RMS for inner race damage is significantly lower than the RMS for outer race defects. While European industry uses RMS to quantify machinery vibrations, US industry seldom reports vibrations using RMS [6.1].

### Mean

The mean acceleration signal is the standard statistical mean value. Unlike RMS, the mean is reported only for rectified signals since for raw time signals the mean remains close to zero. As the mean increases, the condition of the bearing appears to deteriorate.

$$\bar{x} = \frac{1}{N} * \sum_{i=1}^{N} x_i \tag{6.9}$$

### Peak Value

In the time domain, peak value is the maximum acceleration in the signal amplitude, regardless of sing. In 1994, Tandon showed increasing peak values as defect diameter increases [6.17].

### Crest Factor

Crest factor is the ratio of peak acceleration over RMS. This metric detects acceleration bursts even if signal RMS has not changed. However, crest factor can be counterintuitive. At advance stages of material wear, bearing damage propagates, RMS increases, and crest factor decreases. Alfredson and Mathew discuss the limitations of crest factor as an indicator of bearing condition [6.2]; while Tandon shows that the level crest factor does not increase for outer race defects when compared to inner race defects [6.17]. For these reasons crest factor is unreliable to locate defects in rolling elements.

$$Crest\ Factor = \frac{Peak\ Acceleration}{RMS(x)}$$  (6.10)

*Statistical Moments: Skew and Kurtosis*

Machined or ground surfaces in bearings show a random distribution of asperities that are commonly described with the normal distribution function. For this reason, various statistical moments can describe the shape of distribution curves; therefore, assessing bearing surface damage level. Equation (6.11) defines the third moment or skewness as

$$Skew = \frac{1}{N-1}\sum_{i=1}^{N}(x_i - \bar{x})^3$$  (6.11)

where $\bar{x}$ is the mean value. For normally distributed data sets the odd moments are zero, unless the time domain signal is rectified. Hence, skew can easily track bearing conditions [6.15].

   The normalized fourth moment, kurtosis, is the ratio of the fourth moment to the square of the second moment (commonly known as variance). A good surface finish has a theoretical kurtosis of 3, and as the surface finish deteriorates, kurtosis increases [6.15]. Martin and Honarvar concluded that skew and kurtosis are insensitive to loads and speeds [6.15]. However, the level of noise between individual readings hampered the detection of bearing damage.

$$Kurtosis = \frac{(N-1)\sum_{i=1}^{N}(x_i - \bar{x})^4}{\left(\sum_{i=1}^{N}(x_i - \bar{x})^2\right)}$$  (6.12)

*Shock Pulse Counting*

Shock pulse counting records the number of pulses larger than a threshold value. This approach works because bearing defects generate sharply rising impulses with amplitudes larger than noise. Gustafsson and Tallian [6.16] successfully identified bearing conditions counting pulses. However, this technique does not localize the damage.

### Frequency Domain Methods

Frequency domain or spectrum analysis is applied most often to monitor machine condition. However, a simple Fourier transformation at low frequency has limitations in bearing defect detection. Fortunately, researchers have developed signal processing techniques to enhance defect frequencies over mechanical noise. The following discussion presents three standard techniques: the signal average method [6.18], cepstrum analysis [6.17], and bicoherence analysis [6.19]. Then, the high-frequency resonance technique [6.12][6.20][6.21] and the adaptive line enhancer, ALE are explained.

*Signal Average*

Signal averaging improves statistical accuracy, but it does not improve signal-to-noise ratio. In 1979, Braun and Datner [6.18] exploited this fact to develop an average scheme. Their work decomposes the vibration signature into generalized periodic functions, then spectrum analysis detects peaks at characteristic defect frequencies.

*Cepstrum Analysis*

Cepstrum, an anagram of spectrum, is the spectrum of the spectrum. In other words, a quefrency (unit of cepstrum) results from taking the Fourier transform of the logarithm of the power spectrum. This technique groups patterns of sidebands and harmonics from the spectrum into one signature. Cepstrum detects periodicities in the spectrum while being insensitive to transmission path [6.22]. The technique works well if the defect frequency signal has not been convoluted with other signals. Using this technique, Tandon [6.17] identified outer race defects, but was unable to identify inner race defects.

*Bispectrum/Bicoherence*

Bispectrum/bicoherence analysis shows the statistical dependence between harmonics. Equation (6.13) defines bispectrum, where $X(f)$ denotes Fourier transform of the discrete signal $x(t)$.

$$B(f_1, f_2) = E\left[X(f_1)X(f_2)X^*(f_1 + f_2)\right] \tag{6.13}$$

Bicoherence, the normalized bispectrum, is defined by:

$$b(f_1, f_2) = \sqrt{\frac{|B(f_1, f_2)|^2}{E\left[|X(f_1)X(f_1)|^2\right]E\left[|X^*(f_1 + f_1)|^2\right]}} \tag{6.14}$$

Li, Ma and Hwang [6.20] applied bicoherence to locate defects of tapered rolling bearings. Their observations showed that the method works independently from *a priori* knowledge of bearing structure resonance.

### High-frequency Resonance Technique (HFRT)

The high-frequency resonance technique, also known as demodulated resonance analysis or envelope power spectrum, exploits the large amplitude of defect signals around a resonance. HFRT, while being highly sensitive to initial bearing damage, provides an enveloped signal with a high signal-to-noise ratio free of mechanical noise. The following steps, as illustrated in Figure 6.12, obtain this envelope signal.

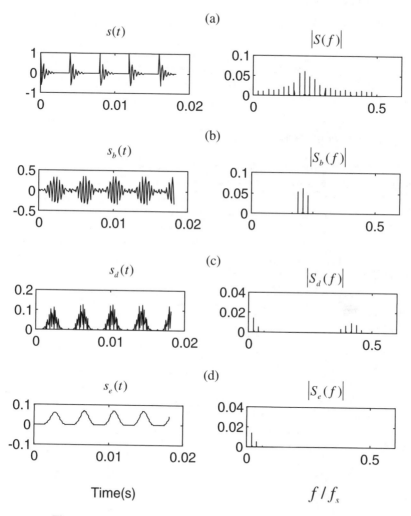

**Figure 6.12.** Illustration of the high-frequency resonance technique

1.  **Bandpass filtering around the system resonant frequency.** The signal in Figure 6.12(a) is bandpassed around a system resonant frequency resulting in Figure 6.12(b). The center frequency coincides with the largest amplitude resonant frequency. McFadden and Smith suggest a bandwidth of at least

twice (preferable four times) the highest deterministic defect frequency. However, the wider this window the higher the risk of interference.

2. **Non-linear rectification of the bandpassed signal.** A non-linear rectifier demodulates the bandpassed signal resulting in the signal of Figure 6.12(c).

3. **Cancel high frequency components with a lowpass filter.** The output signal of the lowpass filter, as shown in Figure 6.12(d), is denoted enveloped signal and will have nonzero values at the harmonics of the deterministic defect frequencies.

The HFRT is a powerful and widely used tool that separates defect frequencies from vibrations generated by other mechanical elements. Multiple bearing damages or severely damaged bearings may partially cancel various components in the envelope spectrum [6.20][6.24]. In practice, broadband noise corrupts the envelope signal of a damaged bearing. Hence, the Adaptive Line Enhancer (ALE) can further enhance this signal.

### Adaptive Line Enhancer (ALE)

The ALE enhances the envelope spectrum, which results from the HFRT, by reducing broadband noise. Widrow *et al.* introduced ALE [6.25] and Treichler studied its performance [6.26]. Figure 6.13 shows how the ALE cancels the broad band noise. The envelope spectrum, $s_e(k)$, plus the broadband noise, $n(k)$, enter the adaptive filter as $x(k)$. The reference input, obtained by a fixed number of samples, $\Delta$, enters the adaptive filter. The filter endeavors to predict a signal that then is subtracted from the primary input signal to form a discrete error $e(k)$. The loop is closed by feeding back the sequence error to adjust the filter weights.

$$x(k) = s_e(k) + n(k)$$

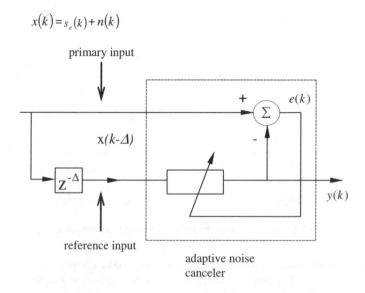

**Figure 6.13.** Block diagram of the adaptive line enhancer (ALE)

As explained by Y. Li [6.24], the delay of the input signal, $x(k\text{-}\Delta)$, decorrelates the broadband noise of $x(k)$, while the narrowband signal remains highly correlated. The output of the filter contains components that are unpredictable over the delay. Therefore, the adaptive filter rejects broad band noise while enhancing the narrowband signal.

### 6.4.2 Effects of Operating Conditions

This section presents the effects of typical operating conditions on bearing life, monitoring and functionality. For these critical operational parameters, example data from an experimental bearing test rig are provided. The test system is shown in Figure 6.14. The housing is located on the right of the picture, with the end of the housing located at the far right of the image. Figure 6.15 is a schematic of the bearing test housing depicting the locations of the specific test bearing.

**Figure 6.14.** Bearing test housing, load and drive mechanisms

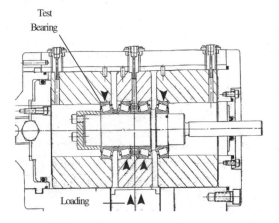

**Figure 6.15.** Life test housing schematic

## *Effects from Loading*

The applied load on a bearing is one of the fundamental design parameters of a bearing system. Bearing lifetime is directly dependent on load and speed applied to it. Though speed is easily measured, actual bearing loading conditions are difficult and sometimes impossible to measure in service. This is not an issue in systems functioning as they are designed. However, machine problems such as misalignment can increase effective bearing loads, and consequently reduce lifetime. Therefore, it is important for diagnostics to be robust to changes in load level. Figure 6.16 demonstrates that various load levels can change accelerometers. In this particular example (Figure 6.16), increasing the loads decreases the vibration. This does not seem intuitively correct. This issue is discussed in the next subsection.

## *Accelerometer Location Results*

As previously stated, accelerometers are often used to detect vibration levels in rotating systems. In many cases, these vibration levels can be used for diagnostic and prognostic purposes. However, the location of the accelerometer on the system being tested can greatly affect the signals generated from the sensor. To demonstrate this, several tests were conducted to test the effects of radial loading, with constant axial preloading. These tests were conducted under varying speeds An outer race groove defect 120 μm and 300 μm wide was placed in the top-center of the loaded zone for these tests. Sensor location was shown to be the dominant factor in load relationships. Two sensor locations, at the housing top and at the front end of the housing are used to demonstrate the differences in the signals generated by the same system, under the same conditions, but with the sensor in these two different locations.

   Figure 6.16 shows a plot of accelerometer RMS versus rotational speed and radial load for the accelerometer located at the housing top. A stud-mounted triaxial accelerometer was used to obtain these results. Tangential direction accelerations were found to be the most insensitive in earlier tests. Both axial and radial acceleration measurements show good agreement in these tests.

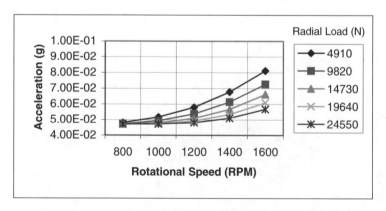

**Figure 6.16.** Z-direction RMS acceleration, housing above bearing

Figure 6.16 shows a counterintuitive relationship between load and speed. For all speeds tested, higher radial loads produced lower acceleration levels. Similar results were obtained for bandpassed RMS levels as well as the peak indicators for defect frequency and the maximum FFT at 4–6 kHz resonance. When the accelerometer was located at the front end of the housing, the same speed–load tests yielded different results. Figure 6.17 shows an opposite effect from load for the otherwise same conditions shown in Figure 6.16.

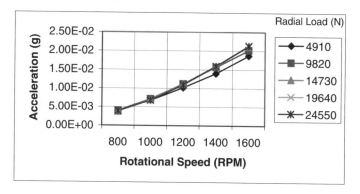

**Figure 6.17.** Axial direction, 4–6 kHz bandpassed RMS, front end housing

The key point from these examples is that the accelerometer must be located on the part of the system that increases its vibration if the roller impact increases. This is difficult to assess, because loading on bearings can cause changes to the vibratory response of the structure to which sensors are mounted. In this case, depending on sensor placement, a bearing with increased load can dramatically increase its wear rate without any indications to sensors.

### *Speed Effects*

#### *Peak Indicators*

Peak indicators (*e. g.* peak in envelope spectrum and the maximum FFT amplitude) also increase with bearing damage. These phenomena may not occur from the same causes of RMS signal increase. The amplitude spectrum FFT used in this chapter assigns signal amplitude to a frequency. Therefore, the race impact magnitude also increases with increasing speed. Figure 6.18 shows the defect frequency peak in the enveloped spectrum with a bandpass filter of 4,000–6,000 Hz. Note that the load relationships for the peak indicators are as discussed.

#### *Speed Effects on Defect Impulses*

In a periodic waveform of pulses, all the energy from the pulse is used to calculate the amplitude of the FFT values at their respective frequencies. Research suggests that the drop-off in signal peak indicators from 1,400 to 1,600 rpm is a result of defect impulses decaying vibration (as shown by the time constant) overlapping with the onset of the next defect impulse [6.10]. The second harmonic of the defect frequency also modulates the housing resonant frequency. This second harmonic cancels part of the signal from the primary defect frequency via destructive interference. At 1,600 rpm, this cancellation is the last part of the pulse signal that occurs before the next impulse. As a result, the impulse decay appears faster than the decay time constant of the hammer test. Figure 6.19 shows the impulse shape versus rotational speed.

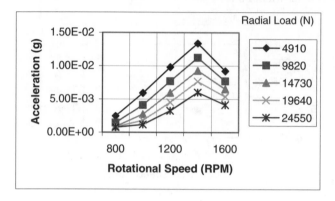

**Figure 6.18.** 4–6 kHz envelope spectrum peak, Z-direction accelerometer, housing top

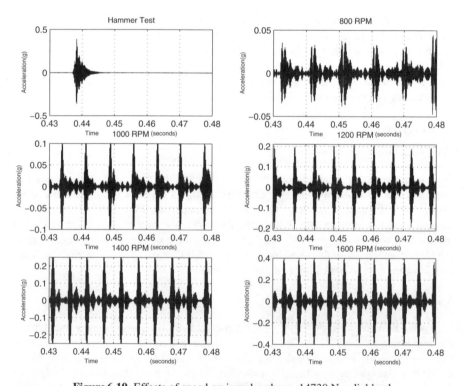

**Figure 6.19.** Effects of speed on impulse shape, 14730 N radial load

According to the hammer test, the time constant of decay is 0.0012 seconds. From Figure 6.19, the rise time to peak amplitude is roughly the same. The total duration for signal rise and three time constants of signal decay is approximately 0.0048. This "critical time" corresponds to a defect frequency of 208 Hz. 1,600 rpm corresponds to a defect frequency of 219 Hz.

The bump-test time-constant method does not provide a rigorous means for determining the speed sensitivity of peak indicators. However, by comparing the time-constant with the deterministic frequency of interest, one can quickly evaluate when some enveloping techniques begin to have a nonlinear effect on bearing damage state measurement.

### 6.4.3 Appropriate Use of Fast Fourier Transforms (FFTs)

There are a few helpful reminders to consider when looking at the multitude of signal processing options that are available with modern software processing. It is helpful to go over a couple of these, because they are heavily utilized in condition monitoring:

1. Zero padding does not add more information. Rather it only refines the underlying "analog" spectrum. It is important to know that zero padding before taking an FFT is the equivalent of interpolating points on the FFT spectra. This may have the appearance of adding resolution.
2. Increasing sample rate does not add information to frequency peaks of interest. The number of data points per FFT peak stays the same when the sampling rate increases since the Nyquist rate (bandwidth) goes up proportionally with increasing sample rate. The only direct benefit is that increased sample rate increases the signal to noise ratio, because it enables better filtering.
3. Better frequency resolution and accuracy can only be obtained by observing a given frequency for an extended period of time. However, this accuracy is eventually compromised by small changes in an observed frequency over the extended time period. This effect is called frequency smearing.

### 6.4.4 Trending

As can be seen in many of the examples presented, there is little hope to set a fixed absolute sensor signal value for all machines that can be directly related to an absolute bearing damage level. In practice, a baseline for a sensor signal is established, and the changes in the properties of that particular signal are used to determine a change in the overall system. This procedure is called "trending," and is employed by most diagnostic systems. Trending is a simple, but effective technique. However, it is still prone to variation in results based on sensor location, systems operating parameter variations (e. g. load, speed, etc.) and specific system configuration. For trending to work properly, the appropriate sensor location must be determined, a baseline must be developed, and the sensor signals must be monitored for changes. It should be noted that continuous monitoring is not typically done when trending is used. Rather, periodic sensor readings are made on the time scale of hours, days or even weeks, depending on the expected change and failure rate of any particular system.

## 6.5 Prognostics

Currently detectable bearing defects [6.27][6.28] are much smaller than 6.25 mm$^2$ (0.01 in$^2$), which is commonly considered to be a fatal failure size by industry standard [6.29]. However, to simply determine the existence of a bearing defect is insufficient for the purposes of condition-based maintenance and catastrophic failure avoidance. When a fatal defect is diagnosed, the machinery is often forced to shut down at inconvenient times with tremendous loss of time, productivity and capital. Therefore, it is important to predict the growth rate of defects and the remaining life of bearings in a prognostic mode in addition to diagnostic mode. With reliable prognostic capabilities, bearing maintenance and replacement can be scheduled at optimal times in the interest of overall system productivity and safety.

The primary difficulty for effective implementation of bearing prognostics is the highly stochastic nature of defect growth. For example, the propagation of a rolling contact fatigue spall beyond its initial appearance is a highly variable process [6.29]. The variation to reach the final failure size from the point where a defect can be detected may be greater than its $L_{10}$ life ($L_{10}$ life is the fatigue life defined as the number of revolution, or hours at some given constant speed that 90% of the bearing population will endure [6.11]). Deterministic models based on fracture or damage mechanics [6.30][6.31] do not accurately describe the variable process of bearing defect propagation. Condition symptoms such as vibration and acoustic emission are closely related to wear processes in machinery [6.32]. Reliability analysis and prognostics could be achieved by using Weibull and Frechet symptom models [6.33]. Therefore it is possible to predict breakdown time of machinery at the running stage based on vibration condition monitoring techniques. However, these techniques have not shown to have the capability to deal with time-varying nature of bearing defect propagation.

This section presents the formulation of a bearing prognostic methodology based on the in-process adaptation of defect propagation rate with vibration signal analysis. It utilizes a deterministic defect propagation model and an adaptive algorithm to fine tune the predicted rate of defect propagation in a real-time manner. The variable nature of defect propagation is addressed by a mechanistic model with time-varying parameters. The adaptive alteration of the model parameters offers the best prediction, in the least square error sense, of the bearing future state for any given diagnostic system. The ensuing text discusses the theoretical basis of the prognostic system. It is followed by an experimental study to verify its effectiveness [6.34].

### Damage Model

Results of laboratory experiments and observations of in-service structures clearly indicate that fatigue crack growth is affected by a variety of factors (*e. g.* stress states, material properties, temperature, lubrication and other environmental effects). The following deterministic fatigue crack propagation model based on Paris' formula has been widely accepted for many years:

$$-\frac{da}{dN} = C_0 (\Delta K)^n \qquad (6.15)$$

where $a$ is the instantaneous length of a dominant crack and $N$ represents running cycles. The parameters $C_0$ and $n$ are regarded as material-dependent constants. They are related to factors such as material properties, environment, *etc.* The term $\Delta K$ represents the range of stress intensity factor over one loading cycle. Equation (6.15) states that crack growth rate in terms of length per running cycle is an exponential function of stress intensity factor range $\Delta K$. Based upon experimental observations, other factors such as stress ratio and threshold stress intensity factors have been proposed and added to Equation (6.15) [6.35].

In the area of bearing defect analysis in industry, the defect severity of a bearing is represented by the surface area size, rather than the length, of the defect [6.30]. Thus, a deterministic bearing defect growth model can be given in a manner similar to Paris's formula as follows:

$$\dot{D} = \frac{dD}{dt} = C_0 (D)^n \qquad (6.16)$$

which states that the rate of defect growth is related to the instantaneous defect area $D$ under constant operating conditions. Again, the parameters $C_0$ and $n$ are material constants that need to be determined experimentally, and often vary with factors other than the instantaneous defect size.

### *Adaptive Scheme for Bearing Remaining Life Prediction*

A variety of adaptive schemes have been developed to estimate the size of defect and the rate of defect growth. One such successful scheme is presented in this chapter and is depicted in Figure 6.20. The system includes the deterministic defect propagation model, as specified by Equation (6.16), that calculates the future bearing defect size at time, $t+\Delta$, based on a given bearing running condition and the defect size at the current time $t$. Note that the defect area size at the current time is often unavailable from direct measurements without interrupting the machinery operation; therefore, indirect and non-intrusive measurements (such as the monitoring of vibration, temperature, or acoustic emission) are commonly exercised to infer the defect area through signal processing and diagnostic models.

The future bearing defect size as forecasted by the prognostic model is compared to the measurement-inferred bearing condition at time $t+\Delta$. The comparison shows a certain amount of prediction error due to the fact that the stochastic nature of defect growth cannot be accurately described by constant parameters $C_0$ and $n$. Therefore, an adaptive algorithm is employed to take advantage of the prediction error for the purpose of fine-tuning the model parameters. In this manner the propagation model is expected to continuously improve its accuracy in following the time-varying defect growth behavior. An adaptive algorithm based on the recursive least square principle, is used for this task and is discussed in the following section.

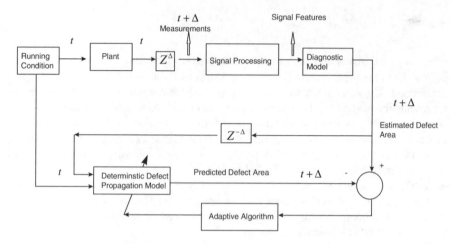

**Figure 6.20.** Adaptive prognostic methodology

*Nonlinear Recursive Least Square (RLS) Scheme*

Through time domain integration, Equation (6.16) can be written in the form

$$\ln(D) = \alpha + \beta \ln(t + t_0) \tag{6.17}$$

where $t_0 = \left(C_0/(1-n)\right) D_0^{n+1}$ is the time when the smallest defect area $D_0$ occurs, $D_0$ is the smallest defect area that can be detected by a given diagnostic system, $\alpha = \left(1/(n-1)\right) \ln\left(C_0/(1-n)\right)$, $\beta = \left(1/(n-1)\right)$, and $t_0$ is bearing running time relative to $t_0$.

Three parameters, $\alpha$, $\beta$ and $t_0$, need to be estimated in the model. Since these parameters are time-varying in a defect propagation process, a recursive least square (RLS) algorithm with a forgetting factor is used to adaptively update the values of $\alpha$, $\beta$ and $t_0$ [6.37]. The RLS algorithm is given as the following:

$$\mathbf{e}(t) = \mathbf{Y}(t) - \hat{\mathbf{Y}}\left(t, \hat{\theta}(t-1)\right) \tag{6.18}$$

$$\psi(t) = \left.\frac{d\hat{\mathbf{Y}}(t,\theta)}{d\theta}\right|_{\theta=\hat{\theta}(t-1)} \tag{6.19}$$

$$\mathbf{P}(t) = \lambda^{-1}\left(\mathbf{P}(t-1) - \frac{\mathbf{P}(t-1)\psi(t)\psi^T(t)\mathbf{P}(t-1)}{\lambda + \psi^T(t)\mathbf{P}(t-1)\psi(t)}\right) \tag{6.20}$$

$$\hat{\theta}(t) = \hat{\theta}(t-1) + \mathbf{P}(t)\psi(t)\mathbf{e}(t) \tag{6.21}$$

where the vector $\theta(t)$ of unknown parameters is $\theta(t)=[\alpha\ \beta\ t_0]^{\mathrm{T}}$, $e(t)$ is the prediction error, $\mathbf{Y}(t)=\ln(D)$, $\hat{\mathbf{Y}}(t)$ is the estimated value of $\mathbf{Y}(t)$ and $\mathbf{P}(t)$ is covariance matrix.

The initial covariance matrix is chosen as unit matrix scaled by a positive scalar that is typically in the region of 1 to 1,000. The scalar reflects uncertainty of a system. Without prior knowledge of a system, large scalar should be selected. The forgetting factor $\lambda$ falls within the range of $0<\lambda\leq1$.

*Experimental Validation*

The experimental system shown in Figure 6.14 and Figure 6.15 is employed to test and demonstrate the prognostic algorithms described in this section. In order to accelerate a defect propagation process, an initial defect is artificially generated on the cup raceway by an electrical discharge machine. The defect is a crack oriented along the bearing axial direction to simulate a real-time fatigue crack with a width of 300 μm. In order to simulate the propagation of a natural defect, the prognostic scheme is not exercised until the bearing has been run for 20 million cycles. At that point the maximum width of defect increased to 1,000 μm and a natural spall defect shape was generated. Experiments were performed at a shaft speed of 1,600 rpm, a preload of 1,300 lb, and a radial load of 5,522 lb. which is about 167% of the rated radial load. The defect is positioned in the loading zone. The system is lubricated by thin spindle oil with viscosity of 54–60 SSU at 100°F.

The experimental procedure began by recording the accelerometer measurements during the running of a defected bearing. The run was interrupted approximately every 10 hours, the defective bearing removed from the test set-up, and the defect size physically measured with a Hommelwerke T8000 profilometer. Then the bearing was then re-assembled into the set-up to repeat the experimental procedure. The growth of bearing defect as measured by the profilometer is shown by the cross-marks in Figure 6.21.

In the study, the diagnostic model estimated the defect size based on the RMS level of the accelerometer signal in the radial direction over the frequency band of 3,000–5,000 Hz where vibrations generated by other mechanical systems are absent. This is the frequency range where most of vibratory energy from the bearing defect is concentrated. It was found that RMS of bandpassed signals around the band were highly correlated to defect area size. The defect areas can be linearly related to the RMS of acceleration signal by

$$D = 3.28 + 4.56\left(RMS\right)\left(\mathrm{mm}^2\right) \tag{6.22}$$

with a standard deviation is 0.52 mm$^2$.

With various initial values of the model parameters, the bearing defect areas as forecasted by the adaptive prognostic model are shown in Figure 6.21. Note that the prognostic model utilized only the diagnostic model, Equation (6.22), for adaptation purpose while remaining ignorant of the profilometer measurements. This situation emulates practical applications in which bearing operations cannot be interrupted for physical defect inspection. The results in Figure 6.21 imply that the adaptive prognostic system can effectively predict the bearing defect propagation process. The fact that the prediction accuracy is not strongly affected by the choice of initial parameter values further suggests that the prognostic system can perform well without *a priori* knowledge of the model parameters. This aspect is particularly important for real life applications since *a priori* and precise knowledge of the fracture mechanics model is usually unavailable.

The major prediction error sources are attributed to the uncertainty in the defect propagation process as well as in the diagnostic model based on accelerometer signals. Since the prognostic system cannot rely upon accurate measurement of the defect size in practical application cases, the reliability of the diagnostic model can be a critical factor to the overall performance of the prognostic system.

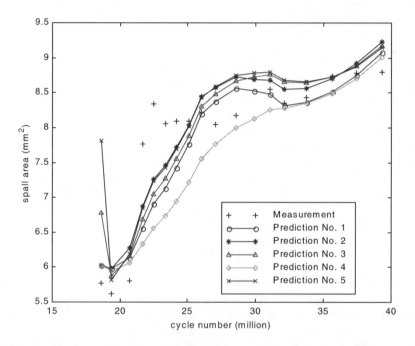

**Figure 6.21.** Predicted defect areas by the adaptive algorithm and measured defect areas by the profilometer with respect to running cycle numbers

This methodology incorporates the time-variant nature of defect growth while providing the best prediction possible, in the least square error sense, for any given diagnostic system. The defect size as predicted by a fatigue crack propagation model is compared to the estimation from a diagnostic model in the future to fine-tune the propagation model parameters. The adaptive prognostics effectively predict the bearing defect propagation process without *a priori* knowledge of the prognostic model parameters. In addition, the reliability of the diagnostic model is suggested to be a critical factor to the overall performance of the adaptive prognostic system.

## 6.6 Conclusions

The ability to effectively and accurately monitor the health of rolling element bearings is a key factor in providing optimal use in a variety of systems. In this chapter, basic rolling element bearing models have been presented along with functional and successfully implemented diagnostic and prognostic techniques. For both diagnostics and prognostics, models of the system are critical in understanding the current state of the bearing, as well as predicting its future life expectancy. Furthermore, the location and proper choice of a sensor determines the sensitivity of the diagnostic and prognostic systems. In general, some sort of trending is employed in both diagnostic and prognostic techniques, as every system has a different base line of satisfactory operation.

Finally, when attempting health prognostics, or predicting the remaining life of a rolling element bearing, uncertainties exist. Typically, accurate life predictions cannot be generated until incipient bearing failure occurs. This is due to the fact that bearing failure can occur for a variety of reasons, and in many cases these reasons cannot be accurately predicted. For example, a sub-surface crack or void may be the nucleation point for a fatal bearing defect. This condition provides the initial condition for the damage propagation model. However, unless detected during inspection, this initial condition is unknown. It is not until the initial condition develops into a detectable condition, that either diagnostic or prognostic approaches can be used. Once some type of incipient failure condition is identified, diagnostic algorithms may be used to determine the extent of the defect condition, and the prognostic algorithms may be used to determine the remaining life of the bearing.

## References

[6.1]    Eisenmann, R. C., Eisenmann, R. C., Jr., 1998, *Machinery Malfunction Diagnosis and Correction*, Hewlett Packard. London, UK.

[6.2]    Alfredson, R. J. and Mathew, J., 1985, "Frequency Domain Methods for Monitoring the Condition of Rolling Element Bearings," The Institution of Engineers, Australia, *Mechanical Engineering Transactions*, **10**(2), pp. 108–112.

[6.3]    White, G., 1997, *Introduction to Machine Vibration*, DLI Engineering Corp.

[6.4]    Thompson, G., 1991, "Condition Monitoring in the Process Industries," *Condition Monitoring and Diagnostic Engineering Management,* IOP Publishing Ltd, Bristol, England.

[6.5]    Henry, T., 1985, "Advances in Monitoring Bearings," *Bearings: searching for a longer life*, Mechanical Engineering Publications Ltd, Suffolk, England.

[6.6]    Bartwell, F. T., 1979, *Bearing Systems*, Oxford University Press, Oxford, UK.

[6.7]    Hunt, T. M., 1996, *Condition Monitoring of Mechanical and Hydraulic Plant*, Chapman and Hall, London, England.

[6.8]    Reif, Z. and M. S. Lai, 1989, "Detection of Developing Bearing Failures by Means of Vibration," *Rotating Machinery Dynamics*, The American Society of Mechanical Engineers, New York, New York.

[6.9]    Boness and S. L. McBride, 1991 "Condition monitoring of adhesive and abrasive wear processes using acoustic emission techniques," *Condition Monitoring and Diagnostic Engineering Management*, IOP Publishing Ltd, Bristol, England.

[6.10]  Williams, T., Ribadeneira, X, S. Billington, S. A., T. Kurfess, T. R., 2001, "Rolling Element Bearing Diagnostics in Run-To-Failure Lifetime Testing," *Mechanical Systems and Signal Processing*, Academic Press, London, **15**(5), pp. 979–993.

[6.11]  Harris, Tedric A., 1991, *Rolling Bearing Analysis*, Third Edition, John Wiley & Sons, Inc. pp. 672.

[6.12]  Su, Y.-T. and Lin, S.-J., 1992, "On Initial Fault Detection of a Tapered Rolling Bearing: Frequency Domain Analysis," *Journal of Sound and Vibration*, **155**, pp. 75–84.

[6.13]  Hewlett-Packard, *Effective Machine Measurements Using Dynamic Signal Analyzers*, Application Note 243–1.

[6.14]  Dyer, D. and R. M. Stewart, 1978, "Detection of Rolling Element Bearing Damage by Statistical Vibration Analysis," *Trans. of the ASME, J. of Mechanical Design*, **100**, pp. 229–235.

[6.15]  Martin, H. R. and F. Honarvar, 1995, "Application of Statistical Moments to Bearing Failure Detection," *Applied Acoustics*, **44**, pp. 67–77.

[6.16]  Gustaffson, Olof G. and T. Tallian, 1962, "Detection of Damage of Assembled Rolling Element Bearings," *ASLE Transactions*, **5**, pp. 197–209.

[6.17]  Tandon, N., 1994, "A Comparison of Some Vibration Parameters for the Condition Monitoring of Rolling Element Bearings," *Measurement*, **12**, pp. 285–289.

[6.18]  Braun, S. and B. Datner, 1979, "Analysis of Roller/Ball Bearing Vibrations," *Trans. of the ASME, J. of Mechanical Design*, **101**, pp. 118–125.

[6.19]  Li, James C., J. Ma, and B. Hwang, 1995, "Bearing Localized Defect Detection by Bicoherence Analysis of Vibrations," *Trans. of ASME, Journal of Engineering for Industry*, **117**, pp. 625–629.

[6.20]  McFadden, P. D., Smith, J. D., 1984, "Vibration Monitoring of rolling Element Bearings by the High Frequency Resonance Technique-A Review," *International J. of Tribology*, **17**, pp. 1–18.

[6.21]  Martin, K. F. and P. Thorpe, 1992, "Normalised Spectra in Monitoring of Rolling Bearing Elements," *Wear*, **159**, pp. 153–160.

[6.22]  Ray, A. and S. Tangirala, 1996, "Stochastic Modeling of Fatigue Crack Dyanmics for On-Line Failure Prognostics," *IEEE Trans. on Control Systems Technology*, **4**(4), pp. 443–451.

[6.23]  Li, Y., 1998, "Dynamic Prognostics of Rolling Element Bearing Condition," *Ph.D. Dissertation*, George W. Woodruff School of Mechanical Engineering, Georgia Institute of Technology, Atlanta, GA.

[6.24]  McFadden, P. D. and Smith, J. D., 1985, "Vibration Produced by Multiple Point Defects in Rolling Element Bearings," *Journal of Sound and Vibration*, **98**(2), pp. 263–273.

[6.25]  Widrow, B., et al., 1975, "Adaptive Noise Canceling: Principles and Applications," *Proceedings of the IEEE*, **63**(12), pp. 1692–1716.

[6.26]  Treichler, John R., 1979, "Transient and Convergent Behavior of the Adaptive Line Enhancer," *IEEE Trans. on Acoustics, Speech, and Signal Processing*, **27**(1), pp. 53–62.

[6.27]  Shiroishi, J. Y. Li, S. Liang, T. Kurfess and S. Danyluk, 1997, "Bearing Condition Diagnostics via Multiple Sensors" *Mechanical Systems and Signal Processing*, **11**(5), pp. 693–705.

[6.28]  Li, Y., J. Shiroishi, S. Danyluk, T. Kurfess and S. Y. Liang, 1997, "Bearing Fault Detection via High Frequency Resonance Technique and Adaptive Line Enhancer," *21st Biennial Conference on Reliability, Stress Analysis and Failure Prevention (RSAFP)*, Virginia Beach, Virginia, April, pp. 763–772.

[6.29]  Hoeprich M. R., 1992, "Rolling Element Bearing Fatigue Damage Propagation," *Trans. of ASME, J. of Tribology*, **114**, pp. 328–333.

[6.30]  Murakami, Y., M. Kaneta and H. Yatsuzuka, 1985, "Analysis of Surface Crack Propagation in Lubricated Rolling Contact," *ASLE Trans.*, **28**, pp. 60–68.

[6.31]  Keer, L. M. and M. D. Bryant, 1983, "A Pitting Model for Rolling Contact Fatigue," *Trans. Of ASME, J. of Tribology*, **105**, pp. 198–205.

[6.32]  Cempel, C., 1985, "The Tribovibroacoustic Model of Machines," *Wear*, **105**(4), pp. 297–305.

[6.33]  Cempel, C. H. G. Natke, and M. Tabaszewski, 1997, "A Passive Diagnostic Experiment with Ergodic Properties," *Mechanical Systems and Signal Processing*, **11**(1), pp. 107–117.

[6.34]  Li, Y., Billington, C., Zhang, C., Kurfess, T., Danyluk S. and Liang, S. Y., 1999, "Adaptive Prognostics for Rolling Element Bearing Condition," *Mechanical Systems and Signal Processing*, **13**(1), pp. 103–113.

[6.35]  Parton, V. Z. and E. M. Morozov, 1985, *Mechanics of Elastic-plastic Fracture*, Hemisphere Publishing Corporation, pp. 219–313.

[6.36]  Goodwin, G. C. and K. S. Sin, 1984, *Adaptive Filtering Prediction and Control*, Prentice-Hall, Inc., Englewood Cliffs, New Jersey 07632.

# 7

## Sensor Placement and Signal Processing for Bearing Condition Monitoring

Robert X. Gao, Ruqiang Yan, Shuangwen Sheng and Li Zhang

Department of Mechanical & Industrial Engineering
University of Massachusetts, Amherst, MA 01003, USA
Email: gao@ecs.umass.edu

**Abstract**
The effectiveness and reliability of measurement techniques for bearing condition monitoring are affected by both the locations of the sensors and the signal processing algorithms selected for defect feature extraction. This chapter describes a structural dynamics-based sensor placement strategy by investigating the mechanisms of signal propagation from the source of its generation to the sensor location. Numerical simulation of a group of sensors for measuring vibration measurement on two custom-designed bearing test beds is presented, and an approach to optimizing the sensor placement based on the Effective Independence (*EfI*) method is introduced. The chapter then comparatively investigates several commonly employed signal processing techniques for feature extraction, such as wavelet transform-based signal enveloping, the Wigner–Ville distribution, and the wavelet packet transform, and evaluates performance using vibration signals measured from the bearing test beds.

## 7.1 Introduction

Rolling bearings have been widely used as low friction joints in rotating machinery. Proper functioning of bearings over their designed life cycle is of critical importance to ensure product quality, prevent tool damage, and minimize costly machine downtime [7.1]–[7.4]. Due to faulty installation, inappropriate lubrication, overloading, and other unpredictable adverse operating conditions, premature and sudden bearing failures often occur in real-world applications. Even under ideal operating conditions, vibrations resulting from discontinuous load support in a bearing due to the discrete number of rolling elements and the resulting periodically-varying compliance would lead to bearing structural deterioration, in addition to material fatigue and failure to meet manufacturing tolerances [7.5]. The goal of bearing condition monitoring is to identify the inception and propagation of bearing structural defects at an early stage for timely remedy, and ultimately, enable condition-based instead of schedule-based "intelligent" maintenance. The specific tasks involed in achieving this goal include defect detection, diagnosis of defect causes, estimation of defect severity level, and prediction of bearing remaining service life [7.6]–[7.10]. Such tasks are typically accomplished using a bearing condition monitoring system as illustrated in Figure 7.1, which consists of components for sensing (conversion of physical symptom to electric signal), data acquisition (data transmission from sensors to processing unit), feature extraction (extraction of characteristic information from raw signals), pattern classification (diagnosis of defect types and severity level) and life prediction (prognosis of bearing remaining life).

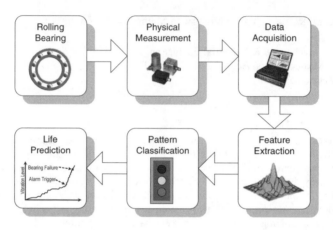

**Figure 7.1.** Configuration of a typical bearing condition monitoring system

During the past several decades, various sensors have been developed and employed for bearing condition monitoring. Representative parameters measured include displacement, vibration (acceleration), dynamic force, acoustic emission, temperature, *etc.* Because vibration signals are directly associated with the structural dynamics as well as working condition of a bearing or a machine, vibration measurement has been widely accepted as an effective tool for bearing condition monitoring. Due to noise and machine-borne vibrations that commonly exist in a manufacturing environment, characteristic components uniquely associated with the bearing to be monitored are often contaminated, resulting in poor signal-to-noise ratio in vibration measurement, thus making it difficult to effectively identify and diagnose defects that are indicative of potential bearing failure. Such difficulty is compounded if the placement of vibration sensors and/or the signal processing algorithms are not properly chosen. This chapter presents a systematic study of sensor placement strategy for high quality sensing, and compares the performance of several representative signal processing techniques for effective feature extraction and reliable bearing condition monitoring. After introducing the analytical framework, numerical simulations are performed, and the results are subsequently compared with experiments conducted on two realistic bearing test beds The study has demonstrated the need for carefully choosing the locations of vibration sensors and suitable signal processing algorithms to ensure measurement quality and effectiveness in defect diagnosis, which is the basis for reliable bearing prognosis.

## 7.2 Sensor Placement

Given the mechanical coupling between a bearing and the surrounding structures (*e. g.* spindles, gearboxes, motors), vibration sensors (accelerometers) mounted on machine housing, as commonly practiced, will pick up structural-borne vibrations in addition to bearing defect-induced vibrations, thus resulting in a generally low signal-to-noise ratio. Understanding the mechanism of structural attenuation of a vibration signal during its propagation throughout the machine structure is critical to devising a sensor placement strategy that optimizes its effectiveness.

### 7.2.1 Structural Attenuation

Given the generally small size of a defect at the incipient stage, it can be viewed as a point source where impulsive mechanical waves due to the rolling element–raceway interactions are generated. The amplitude ($H$) of the waves propagating in solid materials can be expressed as [7.11]:

$$H(r,t) = \frac{A}{r} e^{-j\phi(r,t)} \tag{7.1}$$

where $r$ is the distance between the source and the sensor location, $A$ is an amplitude constant depending on the source vibration strength, and $\phi$ represents the phase function of distance $r$ and time $t$. Equation (7.1) indicates that the strength of the vibration signal measured by a sensor is attenuated proportionally to the distance that the signal travels.

Structural damping due to energy dissipation in materials, as illustrated by the timely decreasing amplitude of the impulse response in Figure 7.2 (measured from bearing test bed II), is another source of signal attenuation.

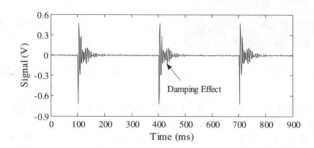

**Figure 7.2.** Effect of structural damping on the impulse response of a ball bearing

Figure 7.3 illustrates the scenario where a defect-induced vibration originated at the outer raceway of a test bearing propagates through several possible pathways (*e. g.* ① through ③), before reaching the vibration sensor placed on top of the bearing housing. When a defect-induced vibration signal propagates through such a long chain of components, the energy ultimately transmitted to the sensor can be estimated as:

$$E_o = E_p \prod_{i=1}^{j}(1-L_i)$$    (7.2)

where $E_p$ is the energy at the source of signal generation and $L_i$ is the energy loss ratio of the $i$th component along the signal propagation path. The energy loss ratio is defined as the ratio between the actual energy content available at the $i$th structural component and that under an ideal situation when no energy loss exists:

$$L_i = 4\pi\xi_i \frac{T_{n_i}}{T}$$    (7.3)

where the symbols $\xi_i$, $T_{n_i}$, and $T$ denote the damping ratio of the $i$th structural component, the natural period of the $i$th structural component, and the duration of the defect-induced vibration, respectively.

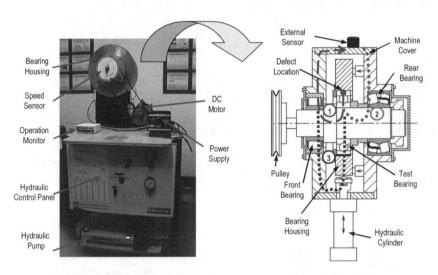

**Figure 7.3.** Bearing test bed I (left) and its cross-sectional view (right)

Assume a defect-induced vibration signal propagates through pathway ①, which consists of six components connected in series: rolling elements of the test bearing, its inner raceway, the shaft, the bearing housing, the support bearing, and the machine cover. Also assume a damping ratio of $\xi_i=0.02$ for each of the components and a vibration period $T_{ni}=T$. For each component, the energy loss ratio $L_i$ is calculated as 25%, and the energy content of the vibration signal reaching the sensor would be only 18% of the signal where it was generated. Given that the signal may propagate simultaneously along multiple paths (e. g. paths ① through ③) before reaching the sensor, and the energy contained within the signal would consequently be dissipated at multiple locations, the net level of signal energy could be too small for a sensor to measure and subsequently detect the existence of the defect. Such findings were also demonstrated in experimental studies reported in [7.12], where the average loss of signal strength between a shaft and a ball bearing was shown to be almost 20 dB, meaning that less than 10% of the original signal energy could reach the sensors.

In addition to structural damping, boundary conditions along the vibration signal propagation path also have a significant effect on signal attenuation. For example, signal reflection occurs at interfaces between dissimilar materials, even when sufficient pressure is applied through tight interference [7.13]. If a normal incident wave with amplitude $H_1$ propagates through a cohesive interface between two dissimilar materials, the energy reflection ratio between the source and transmitted waves $H_1$ and $H_2$ can be expressed as:

$$\gamma_r = 1 - \frac{H_2}{H_1} = 1 - \frac{4\rho_1 c_1 \rho_2 c_2}{(\rho_1 c_1 + \rho_2 c_2)^2} \tag{7.4}$$

where $\rho_1$ and $\rho_2$ are material densities and $c_1$ and $c_2$ are the wave propagation speeds of the two materials. Thus, if a vibration signal propagates through a steel–aluminum interface, the reflection ratio can be as high as 20%, the density for steel and aluminum being $\rho_1 = 8,030$ and $\rho_2 = 2,770$ kg/m$^3$, and the wave propagation speed being $c_1 = 5,790$ and $c_2 = 6,420$ m/s, respectively.

Because of the various structural attenuation effects, vibration sensors for bearing condition monitoring should be placed close to the bearing to be monitored to retain maximum energy content in the signal, which can be viewed as a measure of the signal's "detectability" under given operating conditions. Because of the complexity of real machine systems, parameters in Equations (7.2)–(7.4) may not be readily available, thus accurate evaluation of the sensor location effectiveness in terms of the energy loss ratio $L_i$ may be difficult. To establish a reference basis for sensor location comparison, numerical analysis using the finite element (FE) method is performed, based on the geometry of the bearing test bed I shown in Figure 7.3, as described below.

### 7.2.2 Simulation of Structural Effects

Finite element modeling has been applied to the analysis of rotating machine dynamics and defect detection [7.14][7.15]. To investigate the effect of sensor location on signal strength in response to an impulsive input and subsequently, a total of six representative sensor locations, noted as S1 through S6, were identified, as illustrated in Figure 7.4. Selection of location S2–S6 was based on common measurement practice and accessibility to the test bed, whereas S1 was selected based on the concept of "embedded" sensing [7.9][7.10][7.16]–[7.21]. When setting up the FE model for the bearing, contact elements were used to reflect the interactions between the rollers and the bearing raceways. To model the bearing housing structure, the degree-of-freedom along the z-axis was constrained such that only responses in the x and y planes were considered. The direction of the applied signal and noise input ($F_a$ and $F_b$) was chosen to match realistic experimental scenarios in the bearing test bed I system, where vibration input along the vertical direction has been shown to produce the strongest structural response.

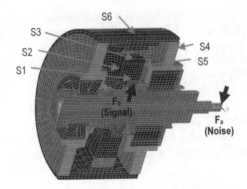

**Figure 7.4.** Finite element model of the bearing support structure as shown in Figure 7.3

To simulate impulsive impacts generated by interactions between a localized defect on the bearing raceway and the rolling elements, a dynamic force impulse of $F_b = 12,700$ N with 1.5 ms duration was applied to the bearing FE model. Such a dynamic force represents 10% of the bearing's dynamic load rating [7.22]. To simulate machine-borne vibrations from other structural sources, a noise signal $F_a$ of the same duration but 1/10 of the dynamic force in magnitude was applied to the driving end of the shaft on which the test and support bearings are installed. The nodal displacements responding to the signal and noise loads along the sensor measurement direction were simulated using the software package ANSYS, and the performance of each sensor location was evaluated by the *Nodal-Signal-to-Noise-Ratio (NSNR)* [7.23]:

$$NSNR(i) = 20\log(\frac{d_j^i}{d_j^n}) \tag{7.5}$$

where $d_j^i$ and $d_j^n$ are the FE nodal displacements at the $j$th sensor location as the result of the $i$th defect-induced and noise-induced vibrations, respectively. The *NSNR* values, along with the absolute and relative displacements responses at the six candidate sensor locations, are listed in Table 7.1. The relative response values are obtained by normalizing individual values against response at sensor location S1, which is assumed to be 100%. A higher *NSNR* value implies that the corresponding defect-induced vibration is easier to differentiate from the background noise. As an example, a vibration sensor placed at location S6 (bearing housing cover) would sense only 27.4% of the force impulse $F_b$, compared to a sensor placed at location S1. Accordingly, a sensor placed at S1 would have a higher sensitivity to the localized bearing defect than the same sensor placed at S6. On the other hand, under the noise load $F_a$, the noise strength picked up by the sensor at location S1 was about 0.4 to 13.8 times weaker than the same sensor placed at locations S2 to S6. This means that a sensor placed closer to the bearing would be less disturbed by structure-borne vibrations, and therefore have a higher immunity against background noises.

**Table 7.1.** Effectiveness of specified sensor locations

| Sensor | Responses from $F_b$ | | Responses from $F_a$ | | NSNR |
|--------|----------|--------------|------------|--------------|--------|
| Location | Absolute (m) | Relative (%) | Absolute(m) | Relative (%) | (dB) |
| S1 | -1.00E-05 | 100 | -7.69E-10 | 100 | 82.3 |
| S2 | -4.95E-06 | 49.5 | -1.07E-08 | 1,391.4 | 53.3 |
| S3 | -7.85E-06 | 78.5 | -1.08E-09 | 140.4 | 77.3 |
| S4 | -2.06E-06 | 20.6 | -8.92E-09 | 1159.9 | 47.3 |
| S5 | -5.79E-06 | 57.9 | -1.14E-08 | 1482.4 | 54.1 |
| S6 | -2.74E-06 | 27.4 | -2.23E-09 | 289.9 | 61.8 |

As illustrated by the results from finite element modeling, defect-induced vibrations in a bearing will generally reach a vibration sensor only after structural attenuations, which may render the sensing technique ineffectual. While the specific machine structure may vary in each application, the above simulation based on a custom-designed bearing test bed illustrates a generally applicable principle: to achieve high quality sensing, the monitoring sensors must be placed as close as possible to the machine component or structure to be monitored.

### 7.2.3 Experimental Evaluation

The effect of signal attenuation due to structural dynamics was evaluated on bearing test bed I, as illustrated in Figure 7.3. A ball bearing of type 6220C3S1 with a seeded defect (0.25 mm diameter on the outer raceway) was used in the experiment. The bearing was operated at 1,000 rpm under 11 kN radial load. Main specifications of the sensors used for the experiments are listed in Table 7.2.

**Table 7.2.** Experimental sensor specifications

| Sensor Notation | Sensor Type | Physical Quantity Measured | Sensor Location |
|-----------------|-------------|----------------------------|-----------------|
| S1 | Piezoceramic plate | Dynamic force | Within bearing outer raceway |
| S2 | Piezoceramic plate | Shock wave/strain | Bearing housing front |
| S3 | 8636B50 | Acceleration | Bearing housing top |
| S4 | Piezoceramic plate | Shock wave/strain | Bearing housing back |
| S5 | 8636B50 | Acceleration | Bearing housing top |
| S6 | CMSS787A | Acceleration | Top of bearing test bed I cover |

The three sensors at locations S1, S2 and S4 consist of identical rectangular piezoceramic plate (PKI502 material) with a piezoelectric constant ($g_{33}$) of 400 pC/N and dimensions 10×4.5×2.5mm. The sensor at S1 was configured as a dynamic force sensor and embedded into the cavity of a phenolic module, which was sandwiched between the bearing outer raceway and the bearing housing. The signal conditioning circuitry had a sensitivity of 2.5 mV/N. The two sensors at S2 and S4 were configured as shock wave sensors [7.21]. They were glued to the bearing housing without a proof mass and reacted to compressive shock waves originated from the ball–defect interactions in the form of an electrical charge proportional to the magnitude of the compressive wave. The sensitivity of the shock wave sensors was calculated to be 80 mV/ppm strain. To enable a meaningful comparison among these customized sensors as well as with commercial accelerometers, the sensor responses were normalized against the respective maximum peak value. The results of signal strengths detected by the sensors at locations S1 to S6 are presented in Figure 7.5. Blanks at a specific frequency indicate that either the magnitude of the peak was too small to be differentiable, or no signal was detected at that frequency. Depending on the sensing mechanism of the individual sensors, (dynamic force, shock wave-induced strain, or acceleration), the units used were kN, ppm, or G, respectively.

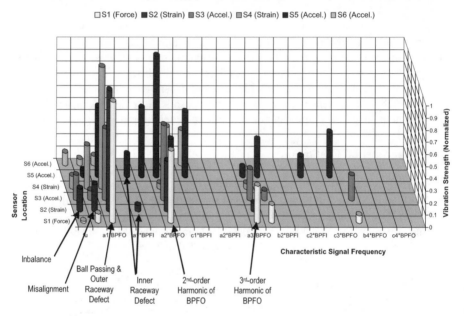

**Figure 7.5.** Comparison of measurement results at different sensor locations

From Figure 7.5, the general trend can be identified that the closer a sensor is placed relative to the bearing being monitored, the better the defect detection result. For example, sensor locations S1–S5 are within the bearing test bed I structure, whereas S6 is outside. As a result, locations S1–S5 have returned an overall better measurement quality than that of S6, where significant signal contamination from other machine components is seen. Of the six sensors, the custom-designed shock wave-based strain sensor at S2 was the most effective in detecting defects in the test bearing (*e. g.* an inner raceway defect developed during the experiment), in addition to detecting components related to bearing inbalance and misalignment at $f_u$ and $f_m$, respectively. The embedded dynamic force sensor at S1 was effective in detecting load variations, as illustrated by the high amplitude of the 2nd and 3rd BPFO (ball passing frequency in outer race), but was less effective than the S2 sensor in detecting the inner raceway defect. The accelerometer at S5, while able to detect the inner raceway defect of the test bearing, shows stronger signal contamination from the two support bearings and thus had a lower signal-to-noise ratio for the test bearing than the custom-designed sensor at S2.

### 7.2.4 Sensor Location Ranking

In real-world applications where rotating machines and machine components are monitored, multiple instead of single defects may be present, and their specific locations are generally unknown *a priori* and can be time-variant. To enable comprehensive coverage of multiple defect features given a limited number of sensors but multiple candidate sensor locations, where to place the sensors becomes a critical issue. Two aspects arise when devising a sensor placement strategy for multiple defects detection: (1) how to evaluate the relative effectiveness of various locations in terms of a performance index, and (2) how to optimize the location determination process [7.24]. Various performance indexes can be applied to sensor location evaluation, *e. g.* error covariance matrix [7.25][7.26], information entropy [7.27], controllability and observability gramian [7.28], Hankel singular values [7.29], kinetic energy [7.30], and Fisher information matrix (FIM) [7.31]–[7.35]. Since the Fisher information matrix can be built from either a finite element model of the machine structure being monitored or results from an experimental modal analysis, it can incorporate more structural information than a simplified analytical model, as required by other indexes. Thus it has been extensively used.

Effective Independence (EfI) [7.31] is a measure for selecting sensor locations, based on the FIM. Assuming $N$ possible defect positions within a bearing and $S$ sensor locations needed to monitor the defects, vibration signals generated at each of the $N$ defect positions will propagate to the $S$ locations and be measured by the sensors. These vibration data can be formulated by a column vector $\mathbf{d}^i$ with a dimension of $S \times 1$:

$$\mathbf{d}^i = \{d_1^i, d_2^i, \cdots d_j^i, \cdots, d_S^i\}^T, \ i = 1, 2, \cdots, N \tag{7.6}$$

where $d_j^i$ is the vibration generated by the $i$th defect at the $j$th sensor location. The measured data obtained at the $S$ locations can be presented by a vector as [7.36]:

$$\mathbf{y} = [\mathbf{d}^1 \, \mathbf{d}^2 \cdots \mathbf{d}^N]\mathbf{q} + \mathbf{w} = \mathbf{D}\mathbf{q} + \mathbf{w} \tag{7.7}$$

where $\mathbf{q} = \{q_1, q_2, \cdots, q_N\}^T$, and $q_i$ ($i=1, 2, \ldots, N$) represents the individual contribution of the $\mathbf{d}^i$ to the measurement data matrix $\mathbf{y}$. The symbol $\mathbf{w}$ represents the measurement noise, and $\mathbf{D}$ is a matrix consisting of the vibrations $\mathbf{d}^i$ ($i=1, 2, \ldots, N$) obtained at the $S$ sensor locations. The measurement data matrix $\mathbf{y}$ is optimized if the best estimation of vector $\mathbf{q}$ is obtained. This is achieved by minimizing the error covariance matrix defined as follows [7.32]:

$$X = E[(\mathbf{q} - \hat{\mathbf{q}})(\mathbf{q} - \hat{\mathbf{q}})^T] = [\mathbf{D}^T R^{-1} \mathbf{D}]^{-1} = Q^{-1} \tag{7.8}$$

where $\hat{\mathbf{q}}$ is the estimation of $\mathbf{q}$, $E(\cdot)$ denotes the expectation operator, $R$ is the covariance matrix of the noise, and $Q$ is the Fisher Information Matrix. If the noise w is uncorrelated among the various sensors, and each sensor is subject to the same noise statistics, then the noise will have no impact on the sensor locations, and the Fisher Information Matrix in Equation (7.8) can be simplified to [7.31]:

$$Q = \mathbf{D}^T \mathbf{D} \tag{7.9}$$

Equations (7.8) and (7.9) indicate that minimizing the error covariance matrix is the equivalent of maximizing $Q$. In the *EfI* method, this is achieved by maximizing the determinant $|Q|$, which means physically maximizing the signal strength in the sensor output. To determine which sensor locations lead to high signal strength, the contribution from each sensor location to the measurement data matrix y is evaluated. A sensor location that does not produce high signal output will be deleted from the candidate location set, and its contribution will be correspondingly deleted from $Q$. As a result, a modified Fisher Information Matrix can be re-written as:

$$Q_T^j = Q - \mathbf{D}_j^T \mathbf{D}_j \tag{7.10}$$

where $\mathbf{D}_j$ is the *j*th row of the vibration data matrix associated with the *j*th candidate sensor location. The determinant of the new FIM $Q_T^j$ is given as [7.32]:

$$\det(Q_T^j) = \det(Q) \det(1 - E_j) \tag{7.11}$$

where $E_j$ is the *EfI* value corresponding to the *j*th sensor location, defined as:

$$E_j = \mathbf{D}_j^T (Q)^{-1} \mathbf{D}_j \tag{7.12}$$

It is evident from Equation (7.12) that $E_j$ represents the contribution of the *j*th sensor location to the Fisher Information Matrix and ultimately, to the measurement data matrix. Therefore, a candidate sensor location can be ranked based on its *EfI* value. As an iteration process, sensor locations with the relatively lowest *EfI* values will be removed at each step, and a modified *EfI* is calculated for the resulting new set of candidate locations. Through such an iterative process, sensor locations with the relatively largest *EfI* values will form the largest possible FIM determinant [7.32], and the candidate sensor location set will contain only a relatively small number of sensor locations, compared with the original set. To guarantee that a critical sensor location is not incorrectly eliminated, only one sensor location can be removed at each time during the iteration process.

### 7.2.4.1 Ranking Simulation

The performance of an *EfI*-based sensor location selection approach was evaluated on bearing test bed II, as shown on the left-hand side in Figure 7.6, whereas a geometry-true finite element model of the bearing housing is shown on the right-hand side. The nodal displacements of the bearing were obtained through a transient solution of the finite element model. To reduce computational load, coupling between the DC motor and the shaft through a universal joint, as well as between the hydraulic cylinder and the bearing housing, were not directly included in the model, but modeled as a *noise* load ($F_a$) and a *static* force load ($P_s$), respectively. Input to the test bed, a bearing defect-induced vibration, was modeled as a transient dynamic force ($F_b$) of 2,860 N (10% of the bearing's dynamic load rating) with 1 ms duration, and is applied radially to the test bearing. The *noise* load was chosen to be 1/10 in magnitude of the excitation input (transient dynamic force). The *static* force load represents the preload applied to the bearing, when the hydraulic cylinder pulls the bearing housing.

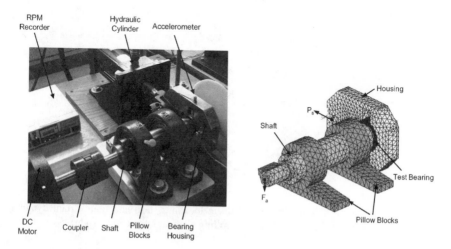

**Figure 7.6.** Bearing test bed II (left) and an FE model of the bearing structure (right)

To model the time-variant defect position within the bearing as it rotates with the shaft, eight equal-distance positions (Pos 1 through Pos 8 in Figure 7.7) around the bearing periphery were selected. The arrangement was made in consideration of the trade-off between modeling accuracy and computational complexity. A transient dynamic force $F_b$ was applied successively at each of the eight representative defect positions, to simulate the time-varying, impulsive interactions between the rolling element and the defect. Given the presence of background noise, selection of the optimal sensor locations need to consider nodal responses to both the signal input $F_b$. and the noise input $F_a$. Certain locations, while showing large nodal displacement response to defect-induced signal input in the simulation, may also be sensitive to noise input, resulting in an overall low signal-to-noise ratio that renders the specific location unfavorable. To address this issue, the *NSNR* as defined in Equation (7.5) was used to form the FIM, instead of using nodal displacements excited only by signal input.

A total of 68 nodes on the bearing housing, as marked by the dotted lines in Figure 7.7, were selected as the initial candidate sensor locations. The housing plate contains three structural recesses machined for sensor placement comparison. Each two adjacent nodes were separated from each other by a 5–10 mm space. These initial nodal locations and separations were chosen based on the geometrical dimension of commercially available accelerometers. For each of the 68 candidate locations, nodal displacement along the Y directions was computed. The resulting *nodal-signal-to-noise-ratio (NSNR)* for the 68 nodes were compiled as a column vector $\mathbf{d}^i$. Given there are eight possible defect positions (Pos 1 through Pos 8), $i = 1$, 2, ... 8, resulting in a total of eight 68×1 vectors. These eight column vectors were then integrated into the nodal displacement matrix $\mathbf{D}$ with a dimension of 68×8. On the MATLAB® platform, the 68 candidate sensor locations were ranked using the *EfI* method. The eight nodal locations with the highest eight *EfI* values are listed in Table 7.3.

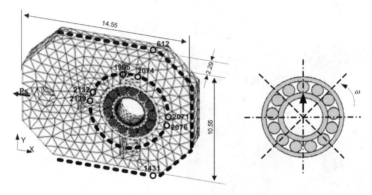

**Figure 7.7.** Illustration of initial sensor candidate locations (left) and time-variant dynamic impact on the bearing assembly due to defect–rolling element interactions (right)

**Table 7.3.** *EfI* values at selected nodal locations (Y-direction)

| Node Number | EfI Value | Node Number | EfI Value |
|---|---|---|---|
| 2071 | 1.0000 | 1995 | 0.9596 |
| 2076 | 1.0000 | 2132 | 0.8212 |
| 2074 | 0.9971 | 612 | 0.8069 |
| 1431 | 0.9842 | 2139 | 0.7155 |

As shown in Figure 7.7, the eight locations can be grouped as two individual (nodes 612 and 1431) and three cluster locations (represented by nodes 2071/2076, with 5 mm separation, nodes 1995/2074, 10 mm separation, and nodes 2132/2139, 5 mm separation). Given the geometrical closeness of these nodes to each other, it was considered sufficient to place one sensor at each of the three clusters. Thus, each cluster can be viewed as a candidate sensor location.

Based on the *EfI* values in Table 7.3, the five candidate sensor locations are ranked as follows: (1) node group 2071/2076, within the bearing static load zone, (2) node group 1995/2074, located within the recess on top of the bearing, 90⁰ counterclockwise from the load zone, (3) node 1431, along the bottom edge of the housing plate, (4) node 612, top edge of the housing plate, and (5) node group 2132/2139, within the recess opposite the bearing load zone.

### 7.2.4.2  Ranking Experiments

The *EfI* method provides a structural effect-based numerical approach to comparatively selecting candidate sensor locations. To evaluate the validity of such an approach, an experimental study was conducted in which accelerometers were placed at four sensor candidate locations selected as above, marked as ① through ④ in Figure 7.8. For comparison with locations not recommended by the *EfI* method, accelerometers were also placed at locations ⑤ and ⑥. The test bearing, containing a 0.1 mm hole as a seeded defect in the outer raceway, was set to rotate at 900 rpm, under a preload of $P_s = 1,833$ N, applied opposite to the X-direction. The characteristic frequency related to the outer raceway defect (BPFO) was calculated to be 79 Hz, based on bearing geometry and rotational speed.

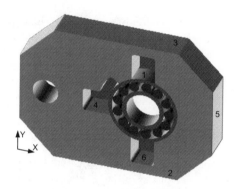

**Figure 7.8.** Sensor locations selected for the experimental ranking evaluation

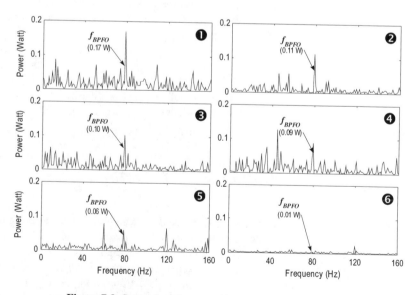

**Figure 7.9.** Sensor locations used in experimental evaluation

The sensor outputs at the six selected locations were evaluated with a wavelet-based enveloping technique, as described in Section 7.3.2.3. As shown in Figure 7.9, sensor location ① has returned the strongest BPFO amplitude among the six locations, followed by locations ② through ④. This result is consistent with that of the simulation described above. Sensor output at location ⑤ is over 30% weaker than that at the first four locations, and location ⑥ almost failed to identify the existence of the BPFO frequency. These experimental results demonstrate that placing sensors at locations recommended by the *EfI* method is beneficial to ensuring effectiveness in bearing defect detection and consequently, improving the performance of bearing condition monitoring.

## 7.3 Signal Processing Techniques

In addition to sensor location, selection of appropriate signal processing techniques to extract characteristic features from bearing vibration signals and consequently, diagnose its status, is also of significance to the quality of bearing condition monitoring. This section introduces several commonly used signal processing techniques and comparatively evaluates their performance using realistic bearing vibration signals.

### 7.3.1 Frequency Domain Techniques

As a bearing wears out, localized structural defects will be initiated on the surface of its raceways (inner or outer) or of the rolling elements, in the form of spalling, due to periodic interactions between the rolling elements and the raceway. As the rolling elements interact with the structural defect, dynamic impacts will be generated, in the form of impulsive vibrations that repeat multiple times per shaft revolution. Such dependency on the shaft frequency lends itself naturally to frequency domain techniques for bearing defects detection and identification.

#### 7.3.1.1 Power Spectrum

The power spectrum descibes how the power of a signal is distributed among its frequency components. The power spectrum density of a signal $x(t)$ is defined as the square of the magnitude of the Fourier transform of the signal:

$$PSD(f) = \left| \frac{1}{\sqrt{2\pi}} \int_{-\infty}^{\infty} x(t)e^{-i2\pi ft} \right|^2 = X(f)\overline{X}(f) \qquad (7.13)$$

where $X(f)$ is the Fourier transform of the signal $x(t)$ and $\overline{X}(f)$ is the complex conjugate of $X(f)$. Given that the Fourier transform utilizes a series of sine and cosine functions to express the average frequency information during the entire period of the signal, the power spectrum is insensitive to subtle, local changes in the signal, which are often indicative of structural defect initiation. Therefore, the effectiveness of the power spectrum is limited for bearing defect diagnosis.

### 7.3.1.2 Enveloping from Bandpass Filtering

Impacts caused by defect–rolling element interactions may excite one or more resonance modes of the machine structure in the high frequency range, which are amplitude modulated by the defect characteristic frequency. Through bandpass filtering in the high frequency range and subsequent lowpass filtering, the envelope of the related vibration signals can be extracted. By means of subsequent spectral analysis, the occurrence of repetitive frequencies indicative of localized structural defects can be detected in the envelope spectrum. In Figure 7.10, major computational steps for bandpass based enveloping are illustrated.

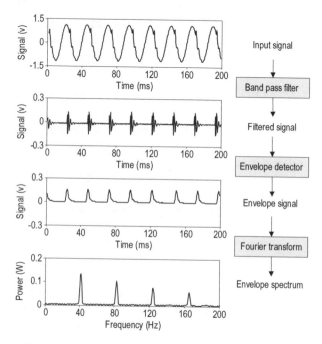

**Figure 7.10.** Procedure for signal enveloping based on bandpass filtering

The Hilbert transform is a viable alternative to rectification and lowpass filtering for envelope extraction. Performing the Hilbert transform on a signal leads to the formulation of a corresponding *analytic* signal, with its real and imaginary parts being the original signal itself and its Hilbert transform, respectively. The modulus of the *analytic* signal represents the signal's envelope.

The effectiveness and reliability of the bandpass enveloping technique depend on the selection of an appropriate filtering band to ensure consistent performance under varying operating conditions and increasing defect size, as the resulting impacts may excite different resonance modes. In practice, impact testing using hammer strikes is often applied to determining possible resonant modes and helping choose matching bandwidth of the bandpass filters. The accuracy of the determined frequency range, however, is subject to the experience of the operators.

## 7.3.2 Time–frequency Techniques

Defect inception and propagation in a rolling bearing throughout its life span follow different time scales, and vibration signals measured are generally non-stationary in nature, with the constituent frequency components changing over time. Time–frequency techniques translate a one-dimensional time domain signal into a two-dimensional representation (time versus frequency), and provide an enabling tool for non-stationary signal processing.

### 7.3.2.1 Short-time Fourier Transform

By employing a sliding window function, the short-time Fourier transform (STFT) performs a *time-localized* Fourier transform to reveal how the signal's frequency contents evolve over time. The STFT is defined as

$$STFT(\tau, f) = \int x(t)g(t-\tau)e^{-2j\pi ft}\,dt \tag{7.14}$$

where $x(t)$ is the signal being analyzed, and $g(t)$ is the window function centered at time $\tau$, which determines the time and frequency resolutions. While high resolution translates into better separation of the characteristic components of the signal from other components, the time and frequency resolutions ($\Delta t$ and $\Delta f$) of the STFT cannot be chosen arbitrarily small simultaneously, according to the uncertainty principle [7.37]. The product of the time–frequency resolution is lower bounded by

$$\Delta t \cdot \Delta f \geq \frac{1}{4\pi} \tag{7.15}$$

This means that once the window function is chosen, the time–frequency resolution is fixed. Since the signal is generally not known a *priori*, selection of a suitable window size for effective signal analysis is not always guaranteed.

### 7.3.2.2 Wigner–Ville Distribution

The Wigner–Ville distribution (WVD) can be viewed as the Fourier transform of the auto-correlation function of a signal $x(t)$, defined as [7.37]

$$WVD(t, f) = \int x(t+\frac{\tau}{2})x^*(t-\frac{\tau}{2})e^{-2j\pi f\tau}\,d\tau \tag{7.16}$$

where the product $x(t+\tau/2)x^*(t-\tau/2)$ is the auto-correlation function of the signal at time $t$. In general, the Wigner–Ville distribution has better time and frequency resolution than the STFT technique, as it does not require any template for its calculation. However, the existence of cross-term interference [7.37] limits its application. This is seen in the WVD of a signal consisting of $n$ elements:

$$WVD(t,f) = \sum_{i=1}^{n} WVD_{x_{ii}}(t,f) + \sum_{i=1}^{n} \sum_{\substack{j=1 \\ j \neq i}}^{n} WVD_{x_{ij}}(t,f) \tag{7.17}$$

which indicates that the WVD of the sum of the $n$ components is not equal to the sum of their corresponding WVDs. In addition to the $n$ auto-correlation sub-terms $WVD_{x_{ii}}(t,f)$, there also exist $n \cdot (n-1)/2$ cross-correlation sub-terms $WVD_{x_{ij}}(t,f)$ between the $n$ components. These cross-correlation sub-terms make it difficult to achieve a clear distinction of the various constituent components in the time–frequency spectrum, and thus diminish the effectiveness and reliability of the WVD technique.

### 7.3.2.3 Wavelet Transform

The wavelet transform employs the concept of *scale* instead of *frequency* to decompose a signal onto a two-dimensional, time-scale plane, in which each scale corresponds to a specific frequency band. The wavelet transform of a signal $x(t)$ can be calculated through the convolution of $x(t)$ with the complex conjugate $\bar{\psi}(\bullet)$ of a scaled and shifted mother wavelet $\psi(\bullet)$:

$$a(s,\tau) = |s|^{-1/2} \int_{-\infty}^{\infty} x(t)\bar{\psi}\left(\frac{t-\tau}{s}\right) dt \tag{7.18}$$

where $s$ is the scaling parameter that dilates or contracts the mother wavelet, and $\tau$ is the shifting parameter, which enables translation of the wavelets along the time axis.

Two methods exist to compute the wavelet transform of a signal: the discrete wavelet transform (DWT), for which the scaling parameter $s$ must be chosen as a power of two, and the continuous wavelet transform (CWT), for which the scaling parameter $s$ can be chosen arbitrarily. The DWT is generally implemented through a pair of lowpass and highpass wavelet filters [7.38], whereas the CWT is computed either from Equation (7.18) directly or by means of the convolution theorem, which states that the Fourier transform of the convolution operation is the product of the respective Fourier transforms in the frequency domain [7.39].

The Fourier transform of Equation (7.18) can be written as

$$A(s,f) = F\{a(s,\tau)\} = \frac{|s|^{-1/2}}{2\pi} \int_{-\infty}^{\infty} \left[ \int_{-\infty}^{\infty} x(t)\bar{\psi}\left(\frac{t-\tau}{s}\right) dt \right] e^{-j2\pi f\tau} d\tau \tag{7.19}$$

where the symbol $F$ denotes the Fourier transform. Based on the scaling property of the Fourier transform and the convolution theorem, Equation (7.19) can be expressed as

$$A(s,f) = |s|^{1/2} X(f)\bar{\Psi}(sf) \tag{7.20}$$

where $X(f)$ denotes the Fourier transform of $x(t)$ and $\bar{\Psi}(\bullet)$ denotes the Fourier transform of $\bar{\psi}(\bullet)$. By taking the inverse Fourier transform, Equation (7.20) is converted back into the time domain as

$$a(s,\tau) = F^{-1}\{A(s,f)\} = |s|^{1/2} F^{-1}\{X(f)\bar{\Psi}(sf)\} \tag{7.21}$$

where the symbol $F^{-1}$ denotes the inverse Fourier transform. Equation (7.21) implies that the wavelet transform can act as a bandpass filter. Furthermore, since the imaginary part of a complex wavelet is the Hilbert transform of its real part, the wavelet coefficients of a transformed signal, in which the complex wavelet is used as the mother wavelet, are analytic and their corresponding modulus forms the signal's envelope [7.40]. As a result, the complex wavelet-based signal transform combines bandpass filtering and enveloping into one single step, which can be utilized to enable wavelet-based enveloping. As illustrated in Figure 7.11, the input signal is first filtered using the wavelet transform, and the envelope signal is then calculated from the wavelet coefficient modulus of the signal. By performing the Fourier transform on the envelope signal, the enveloping spectrum of the signal can be obtained.

**Figure 7.11.** Procedure for wavelet-based signal enveloping

### 7.3.2.4 Wavelet Packet Transform

The wavelet packet transform (WPT) is an extension of the discrete wavelet transform, which applies a recursive algorithm to further decompose the *detailed* information of the signal, as shown in the shaded boxes in Figure 7.12, where a four-level wavelet packet transform is illustrated. The transform results in sixteen groups containing the *detailed* information, with each group corresponding to one sixteenth of the signal frequency region.

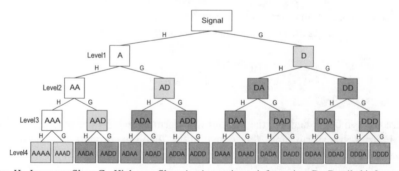

Note: H - Low pass filter; G - High pass filter; A - Approximate information; D - Detailed information

**Figure 7.12.** Illustration of signal decomposition using the wavelet packet transform

To obtain the WPT of a specific level (*e. g.* level 4), the results for all of its upper decomposition levels (*i. e.* levels 1, 2, and 3) need to be obtained first, which increases the computational cost. The harmonic wavelet packet transform (HWPT) [7.41][7.42] allows all of the groups at a specific decomposition level to be calculated directly, as illustrated in Figure 7.13, based on a compact FFT-based algorithm, thus is computationally more efficient than conventional wavelet packet transforms implemented through a pair of low and highpass filters. Experiment has shown that the HWPT is over four times as fast as the conventional WPT to perform on a 2.0 GHz computer with 512 MB RAMs, when a signal is decomposed into 4-level, 16 groups [7.42].

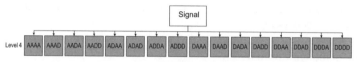

Note:  A: Approximate information; D: Detailed information

**Figure 7.13.** Illustration of signal decomposition using HWPT

The HWPT uses the harmonic wavelet as the mother wavelet, which is expressed in the frequency domain as [7.43]

$$\Psi_{m,n}(f) = \begin{cases} 1/(n-m) & m \le f \le n \\ 0 & elsewhere \end{cases} \qquad (7.22)$$

where $m$ and $n$ are the scale parameters. These parameters are real but not necessarily integers. Similar to Equation (7.21), the harmonic wavelet packet transform can be expressed as:

$$hwpt(m,n,t) = (m-n)F^{-1}\{X(f)\overline{\Psi}((m-n)f)\} \qquad (7.23)$$

Figure 7.14 illustrates the algorithm for the harmonic wavelet packet transform. After taking the Fourier transform of the signal $x(t)$ to obtain its frequency expression $X(f)$, the inner product between the $X(f)$ and $\overline{\Psi}((n-m)f)$ at the given parameters $m$ and $n$ is calculated. Finally, the harmonic wavelet packet transform of the signal $x(t)$, denoted as $hwpt(m,n,t)$, is obtained by taking the inverse Fourier transform on the inner product of $HWPT(m,n,f)$.

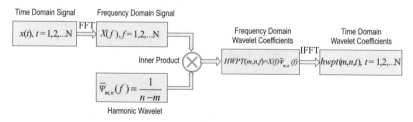

**Figure 7.14.** Procedure for implementing the harmonic wavelet packet transform

As defined in Equation (7.22), the parameter pair $(m, n)$ specifies the lower and upper cut-off frequencies of the frequency band defined by the harmonic wavelet:

$$\begin{cases} m = i \times f_h / 2^s \\ n = (i+1) \times f_h / 2^s \end{cases} \quad i = 0, 1, ..., 2^s - 1 \qquad (7.24)$$

where $f_h$ is the highest frequency component contained in the signal being analyzed, and $s$ is the signal decomposition level.

With the harmonic wavelet packet transform, the signal is decomposed into a number of sub-frequency bands. From the time–frequency domain distribution of the energy content of the signal, information on the working condition of the machine component being monitored can be obtained.

### 7.3.3 Performance Comparison

To evaluate the effectiveness of the various signal processing techniques introduced above, two case studies were conducted where bearing vibration signals measured from experiments were analyzed comparatively. In the first case study, sensor output at location ① as specified in Figure 7.8 was analyzed using the traditional power spectrum, bandpass filtering-based signal enveloping, and wavelet transform-based enveloping.

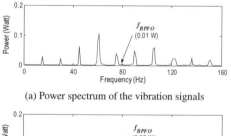

(a) Power spectrum of the vibration signals

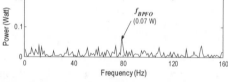

(b) Traditional enveloping spectrum of the vibration signals

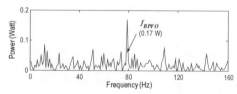

(c) Wavelet-based enveloping spectrum of the vibration signals

**Figure 7.15.** Performance comparison of three signal processing techniques for sensor location ① as specified in Figure 7.8

As shown in Figure 7.15, the power spectrum technique was not able to detect the existence of the BPFO component, as it was submerged in the spectrum of other components. The wavelet-based enveloping spectrum is significantly more effective in displaying the BPFO component than the traditional enveloping technique, showing a ratio of 2.6 between BPFO and the next strongest component in the spectrum, as compared to a ratio of 1.7 for the traditional enveloping technique.

The second case study investigated the effect of bearing defect propagation. As different intrinsic modes of the bearing system are excited due to the defect growth, frequency changing in the defect-induced vibrations occur. Such changes in the time–frequency distribution chart are indicative of changes in the bearing health condition. Figure 7.16 illustrates a representative data segment from a run-to-failure experiment conducted on a ball bearing, together with its power spectrum. The signal segment was analyzed using the short-time Fourier transform, Wigner–Ville distribution, and harmonic wavelet packet transform, respectively, and the results are shown in Figure 7.17.

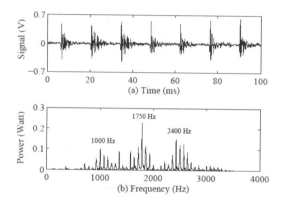

**Figure 7.16.** Vibration signals measured on a ball bearing

For the short-time Fourier transform, a time window of 12.8 ms was chosen, after a series of trial-and-error runs to identify the best-suited window size. From part (a) in Figure 7.17, two major frequency components at around 1,750 and 2,400 Hz are seen to be dominant in the time–frequency plane. After the time point 45 ms, another major frequency component of 1,000 Hz became evident, indicating a change in the bearing's working condition. However, due to the resolution limitation typically associated with the STFT, the signal components within the 1,000–1,750 Hz range cannot be clearly differentiated. The Wigner–Ville distribution, while able to identify the two major frequency components at around 1,750 and 2,400 Hz, was not able to detect the frequency change caused by defect propagation, as shown in Figure 7.17(b). In comparison, the harmonic wavelet packet transform, shown in Figure 7.17(c), was able to clearly identify all the major frequency components at 1,000 Hz, 1,750 Hz, and 2400 Hz before and after the 45 ms time point, as well as the frequency change at the 45 ms time point. Also, it demonstrated a higher frequency resolution in differentiating the various components than the other two techniques, thus is more effective for bearing signal decomposition.

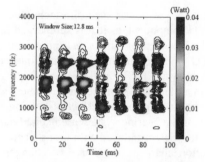

(a) Short-time Fourier transform of the data segment

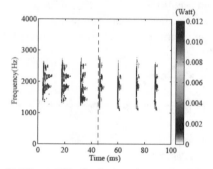

(b) Wigner–Ville distribution of the data segment

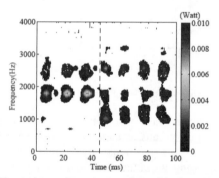

(c) Harmonic wavelet packet transform of the data segment

**Figure 7.17.** Time–frequency distribution of the ball bearing (type 1100KR) signal

## 7.4 Conclusions

To ensure effective and reliable bearing condition monitoring, both the specific locations of the vibration sensors and the signal processing algorithm for data analysis need to be carefully chosen. While placing sensors close to the component of interest will improve the signal-to-noise ratio in general, the specific locations of the sensors should be determined, based on the dynamics of the structure to be monitored. The effective independence-based sensor location ranking method provides a viable approach to devising a systematic sensor placement strategy, although the performance of individual locations is affected by the accuracy of the structural model, based on which the iterative location ranking process is conducted. Experimental evaluation is needed to ultimately verify the appropriateness of the recommended sensor locations. Because of the transient nature of the vibration signals associated with bearing defect propagation, multi-domain, time–frequency techniques are generally more effective than the traditional spectral analysis technique in extracting characteristic features hidden in the vibration signals. However, understanding the physical nature of the signals to be analyzed is key to selecting an appropriate signal processing technique for the specific monitoring task. In addition to rolling element bearings, the presented techniques are applicable to the condition monitoring of other machines commonly seen in manufacturing.

## Acknowledgement

The authors gratefully thank the National Science Foundation for funding provided to the research reported here, under grants DMI-9624353, 9988757, and 0218161, and for the financial and experimental support from the SKF and Timken companies.

## References

[7.1]    Anon, 1995, "Condition Monitoring Puts the Squeeze on Plant Downtime," *Noise and Vibration Worldwide*, **26**(9), pp. 12–14.

[7.2]    Mathew, J. and Alfredson, R. J., 1984, "Condition Monitoring of Rolling Element Bearings using Vibration Analysis," *Journal of Vibration, Acoustics, Stress, and Reliability in Design*, **106**(3), pp. 447–453.

[7.3]    Harris, T. A., 1991, *Rolling Bearing Analysis*, John Wiley & Sons, Inc., 3$^{rd}$ edition.

[7.4]    Tandon, N. and Nakra, B. C., 1992, "Comparison of Vibration and Acoustic Measurement Techniques for the Condition Monitoring of Rolling Element Bearings," *Tribology International*, **25**(3), pp. 205–212.

[7.5]    Sunnersjo, C., 1978, "Varying Compliance Vibrations of Rolling Bearings," *Journal of Sound and Vibration*, **58**(3), pp. 363–373.

[7.6]    Li, Y., Billington, S., Zhang, C., Kurfess, T. and Danyluk, S., 1999, "Dynamic Prognostic Prediction of Defect Propagation on Rolling Element Bearings," *Tribology Transactions*, **42**(2), pp. 385–392.

[7.7]    Barkov, A., Barkova, N. and Mitchell, J., 1995, "Condition Assessment and Life Prediction of Rolling Element Bearings - Part 1," *Sound and Vibration*, **29**(6), pp. 10–17.

[7.8]    Malhi, A. and Gao, R., 2004, "Recurrent Neural Networks for Long-term Prediction in Machine Condition Monitoring", In *IEEE Instrumentation and Measurement Technology Conference*, Como, Italy, **3**, pp. 2048–2053.

[7.9]    Gao, R., Jin, Y. and Warrington, R. O., 1994, "Microcomputer-based Real-time Bearing Monitor," *IEEE Transactions on Instrumentation and Measurement*, **43**(2), pp. 216–219.

[7.10] Holm-Hansen, B. T. and Gao, R., 2000, "Vibration Analysis of a Sensor-integrated Ball Bearing," *ASME Journal of Vibration and Acoustics*, **122**(4), pp. 384–392.

[7.11] Kolsky, H., 1963, *Stress Wave in Solids*. Dover Publications, Inc.

[7.12] Singh, A., Houser, D. R. and Vijayakar, S., 1999, "Detecting Gear Tooth Breakage Using Acoustic Emission: A Feasibility and Sensor Placement Study," *Journal of Mechanical Design, Transactions of the ASME*, **121**(4), pp. 587–593.

[7.13] Rinehart, J., 1975, *Stress Transients in Solids*. University of Colorado.

[7.14] Kiral, Z. and Karagulle, H., 2003, "Simulation and Analysis of Vibration Signals Generated by Rolling Element Bearing With Defects," *Tribology International*, **36**(9), pp. 667–678.

[7.15] Debray, K., Bogard, F. and Guo, Y., 2004, "Numerical Vibration Analysis on Defect Detection in Revolving Machines Using Two Bearing Models," *Archive of Applied Mechanics*, **74**(1–2), pp. 45–58.

[7.16] Gao, R. and Phalakshan, P., 1995, "Design Considerations for a Sensor-integrated Roller Bearing," In *ASME International Mechanical Engineering Congress and Exposition*, San Francisco, CA, USA.

[7.17] Holm-Hansen, B. and Gao, R., 1997, "Monitoring of Loading Status Inside Rolling Element Bearings Through Electromechanical Sensor Integration," In *ASME International Mechanical Engineering Congress and Exposition*, Dallas, TX, USA.

[7.18] Wang, C. and Gao R., 1999, "Sensor Module for Integrated Bearing Condition Monitoring," In *ASME International Mechanical Engineering Congress and Exposition*, Nashville, TN, USA.

[7.19] Varghese, B., Pathare, S., Gao, R., Guo, C. and Malkin, S., 2000, "Development of a Sensor-integrated 'Intelligent' Grinding Wheel for In-process Monitoring," *CIRP Annals - Manufacturing Technology*, **49**(1), pp. 231–234.

[7.20] Suryavanshi, A. and Gao, R., 2002, "Mechatronic Design and Vibration Analysis of a Sensor Module for Integrated Bearing Condition Monitoring," In *The 5th International Conference on Frontiers of Design and Manufacturing*, Dalian, China.

[7.21] Gao, R., Wang, C. and Sheng, S., 2004, "Optimal Sensor Placement Strategy and Sensor Design for High Quality System Monitoring," In *SPIE-Sensors and Smart Structures Technologies for Civil, Mechanical, and Aerospace Systems*, San Diego, CA, USA.

[7.22] Barkov, A. and Barkova, N., 1995, "Condition Assessment and Life Prediction of Rolling Element Bearings - Part 2," *S V Sound and Vibration*, **29**(9), pp. 27–31.

[7.23] Sheng, S. and Gao, R. X., 2004, "Structural Dynamics-based Sensor Placement Strategy for High Quality Sensing," In *Proceedings of the IEEE Sensors Conference*, Vienna, Austria, pp. 642–645.

[7.24] Guo, H., Zhang, L., Zhang, L. L. and Zhou, J. X., 2004, "Optimal Placement of Sensors for Structural Health Monitoring Using Improved Genetic Algorithms," *Smart Materials and Structures*, **13**(3), pp. 528–534.

[7.25] Arbel, A., 1982, "Sensor Placement in Optimal Filtering and Smoothing Problems," *IEEE Transactions on Automatic Control*, **AC-27**(1), pp. 94–98.

[7.26] Omatu, S., Koide, S. and Soeda, T., 1978, "Optimal Sensor Location for a Linear Distributed Parameter System," *IEEE Transactions on Automatic Control*, **AC-23**(4), pp. 665–673.

[7.27] Papadimitriou, C., Beck, J. L. and Au, S.-K., 2000, "Entropy-based Optimal Sensor Location for Structural Model Updating," *Journal of Vibration and Control*, **6**(5), pp. 781–800.

[7.28] Longman, R. W. and Alfriend, K. T., 1990, "Energy Optimal Degree of Contollability and Observability for Regulator and Maneuver Problems," *Journal of the Astronautical Sciences*, **38**(1), pp. 87–103.

[7.29]  Cherng, A.-P., 2003, "Optimal Sensor Placement for Modal Parameter Identification Using Signal Subspace Correlation Techniques," *Mechanical Systems and Signal Processing*, **17**(2), pp. 361–378.

[7.30]  Salama, M., Rose, T. and Garba, J., 1987, "Optimal Placement of Excitations and Sensors for Verification of Large Dynamical Systems," In *Proceedings of the 28th Structures, Structural Dynamics, and Materials Conference*, Monterey, CA, pp. 1024–1031.

[7.31]  Kammer, D.C., 1991, "Sensor Placement for On-orbit Modal Identification and Correlation of Large Space Structures," *Journal of Guidance, Control, and Dynamics*, **14**(2), pp. 251–259.

[7.32]  Kammer, D.C. and Yao, L., 1994, "Enhancement of On-orbit Modal Identification of Large Space Structures Through Sensor Placement," *Journal of Sound and Vibration*, **171**(1), pp. 119–140.

[7.33]  Yao, L., Sethares, W.A. and Kammer, D.C., 1993, "Sensor Placement for On-orbit Modal Identification via a Genetic Algorithm," *AIAA Journal*, **31**(10), pp. 1922–1928.

[7.34]  Kammer, D.C., 2004, "Optimal Placement of Triaxial Accelerometers for Modal Vibration Tests," *Mechanical Systems and Signal Processing*, **18**(1), pp. 29–41.

[7.35]  Kirkegaard, P.H. and Brincker, R., 1994, "On the Optimal Location of Sensors for Parametric Identification of Linear Structural Systems," *Mechanical Systems and Signal Processing*, **8**(6), pp. 639–647.

[7.36]  Camelio, J.A., Hu, S.J. and Yim, H., 2005, "Sensor Placement for Effective Diagnosis of Multiple Faults in Fixturing of Compliant Parts," *Journal of Manufacturing Science and Engineering, Transactions of the ASME*, **127**(1), pp. 68–74.

[7.37]  Qian, S., 2002, *Introduction to Time-Frequency and Wavelet Transforms*, Prentice Hall.

[7.38]  Mallat, S.G., 1989, "A Theory for Multiresolution Signal Decomposition: The Wavelet Representation", *IEEE Transactions on Pattern Analysis and Machine Intelligence*, **2**(7), pp. 674–693.

[7.39]  Bracewell, R., 1999, *The Fourier Transform and Its Applications*, New York: McGraw-Hill, 3rd edition.

[7.40]  Mallat, S.G., 1999, *A Wavelet Tour of Signal Processing*, Academic Press.

[7.41]  Newland, D.E., 1993, "Harmonic Wavelet Analysis", *Proceedings of the Royal Society of London*, A, **443**, pp. 203–225.

[7.42]  Yan, R. and Gao, R.X., 2005, "An Efficient Approach to Machine Health Diagnosis Based on Harmonic Wavelet Packet Transform", *Robotics and Computer-Integrated Manufacturing*, **21**, pp. 291–301.

[7.43]  Yan, R. and Gao, R.X., 2005, "Generalized Harmonic Wavelet as an Adaptive Filter for Machine Health Diagnosis", In *SPIE International Symposium on Sensors and Smart Structures Technologies for Civil, Mechanical and Aerospace Systems*, paper # SN5765_87, San Diego, CA.

# 8

# Monitoring and Diagnosis of Sheet Metal Stamping Processes

R. Du

The Chinese University of Hong Kong
Hong Kong, China
Email: rdu@cace.cuhk.edu.hk

**Abstract**
Sheet metal stamping is one of the most commonly used manufacturing processes. Every day, millions of parts are made by stamping, ranging from small battery caps to large automobile body panels. Yet, it is a difficult process involving the press, the dies (including the binder), the material (the blank) and the forming process with very large forces. Even with advanced technologies today, such as finite element modeling (FEM) and computer control, malfunctions occur from time to time. As a result, condition monitoring and fault diagnosis are important. This chapter presents research on monitoring and diagnosis of sheet metal stamping processes. It consists of five sections. Section 8.1 introduces some of the authors's research on the sheet metal stamping process. Section 8.2 is a brief description of the sheet metal stamping process. Understanding this section is essential to the rest of the chapter. Section 8.3 presents an effective online monitoring method based on support vector regression (SVR). Section 8.4 gives a new diagnosis method based on infrared thermal imaging. Finally, Section 8.5 contains conclusions.

## 8.1 Introduction

A number of years ago, the author was asked to write an article for the *Encyclopedia of Electrical and Electronics Engineering* [8.1] about fault diagnosis. After giving various signal processing and decision making methods, it was stressed that the key to the success is to understand the process and the signal.

In the years that followed, the author worked on the sheet metal stamping process. Sheet metal stamping is one of the most commonly used manufacturing processes. Every day, millions of parts are made by stamping, ranging from small battery caps to large automotive body panels. Hence, even a small technological gain may result in significant corporate benefits.

Under the support of the industry and the government, four different projects were undertaken:

(a) Develop an online monitoring system. In this project, an industry PC computer was used to collect strain signals from the anvil of the press, analyze the signals using various methods, such as the latent process model [8.2], snake skeleton graph [8.3][8.4], hidden Markov model [8.5], support vector machine (SVM) [8.6], marginal energy method [8.7], wavelet transform [8.8], and bispectrum [8.9].

(b) Develop a fault diagnosis system. In this project, an infrared camera was used to acquire the thermal image of a part right after it was made. The thermal image represents the energy distribution incurred during the stamping process and hence, helps to pinpoint problems that cause quality deterioration [8.10][8.11].

(c) Develop a finite element analysis (FEA) software system. In this project, a one-step FEA (also called an inverse FEA) software system was developed that predicts the original blank shape, the thickness strain distribution of the part and the quality variation. The system is especially useful for multi-step stamping die design [8.12]–[8.15].

(d) Design and build a new type of controllable mechanical metal forming press. The new press combines the advantages of a mechanical press and a hydraulic press. It is energy efficient and fast, like the mechanical press, and yet is controllable, like a hydraulic press [8.16][8.17].

Through these projects, a good understanding was acquired of the sheet metal stamping process, the monitoring signals (including the tonnage signal, the vibration signal, *etc.*), as well as monitoring and diagnosis methods. In this chapter, some of our ideas are shared with you in the hope these ideas can be helpful to your practice and research.

## 8.2 A Brief Description of Sheet Metal Stamping Processes

The sheet metal stamping process is one of the most commonly used manufacturing processes. As a result, it is taught as a basic component in manufacturing engineering courses in universities and is thoroughly investigated in industry. Although the rest of the section is designed to familiarize the reader with the sheet metal stamping process, you may wish to quickly review a manufacturing engineering textbook, such as [8.18]. Those familiar with the process can simply skip this section.

Like many other manufacturing processes, the sheet metal stamping process involves the machine (the press), the dies (including progressive dies and transfer dies), the material (the sheet metal blank) and the process itself (during which time the blank deforms to the shape of the die). Figure 8.1 shows a typical small C-shape stamping press (the other type is the large straight-sided press). The press is driven by an electrical motor through a crank mechanism, which converts the rotational motion of the motor to the linear motion of the punch. The punch provides the force for metal forming.

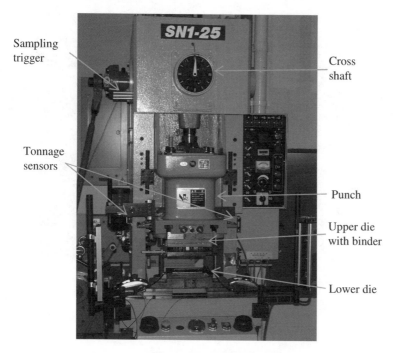

**Figure 8.1.** A typical small C-shape cross-shaft press

In general, there are three different types of sheet metal stamping processes: blanking, bending and drawing. We will focus on drawing, also called deep drawing, as it has the most problems. Figure 8.2 shows some sample parts made by progressive stamping dies. The part is the back of a small motor about 25 mm in diameter and 1 mm in thickness. It is made in five progressive steps. They are, from the right to the left: positioning, 1st drawing, 2nd drawing, 3rd drawing, and trimming. Note that between the 3rd drawing and trimming there is an idling step in which no stamping operation is taking place. In practice, as shown in the figure, a number of malfunctions may occur, such as mis-feed, slug (dirt left on the surface of the part), and material defects (the blank is too thin or too thick). Since the press runs at a high speed (ranging from 30–300 shots per minute), it is not possible for press operators to detect the defects manually. Therefore, condition monitoring is very important.

Condition monitoring relies on monitoring signals. Various signals can be used, such as power (motor current), vibration (acceleration), and strain. The most commonly used signal is the strain signal, also called the tonnage signal as it is calibrated to correlate the stamping force. The tonnage sensor is usually installed on the anvil of the press as shown in Figure 8.1. For a C-shape press, the best sensor location will be at the middle of the "C", where the deformation is the largest.

From a physical point of view, the tonnage signal reflects the deformation of the anvil under the stamping force. As shown in Figure 8.3, from the signal, one can

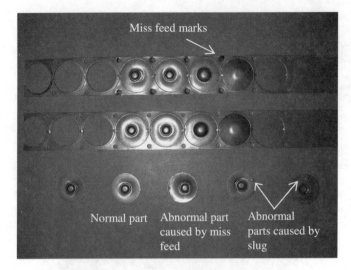

**Figure 8.2.** Sample parts made by progressive stamping

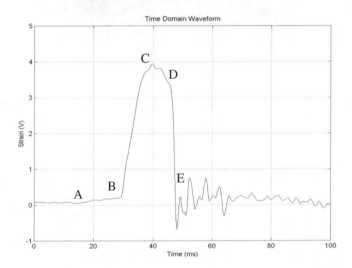

**Figure 8.3.** A typical tonnage signal

see how a stamping process proceeds First, as the punch runs down, the binder first touches the workpiece at Point A. As a result, the signal gradually increases as the stamping force increases. At Point B, the upper die touches the workpiece and metal forming starts. Consequently, the signal quickly increases to its maximum at Point C. At Point D, the metal forming process is completed. Hence, the signal quickly drops to zero. Finally, starting from Point E, the last portion of the signal corresponds to anvil vibration.

It should be pointed out that the stamping force is determined by the metal forming process. One can estimate the stamping force based on finite element modeling (FEM). Currently, there are a number of commercial FEM software systems available, such as DYNAFORM® and PAMSTAMP®; however, special training may be needed to run these software systems. In general, stamping forces are large and impulsive since the metal forming process is usually completed within milliseconds. Besides, it has a spatial distribution depending on the shape of the die.

Furthermore, it has been found that the signal also depends on the position of the crank [8.19]. For example, the two signals in the setup of Figure 8.1 will be different because of the crank motion. Figure 8.4 shows a typical example, in which "position 1" is the signal from the left tonnage sensor and "position 2" is the signal from the right tonnage sensor. The "1" signal peaks first because the crank rotates from left to right counter-clockwise.

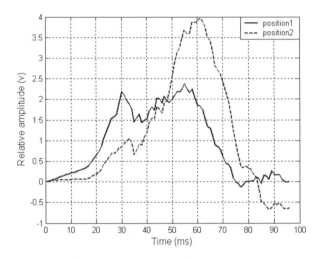

**Figure 8.4.** The difference in the two signals is attributed to the crank position

If four tonnage sensors are used to monitor a straight-sided hydraulic press, however, the difference among the four signals is only attributed to the distribution of the stamping force. Figure 8.5 shows a typical example, where $fa$, $fb$, $fc$ and $fd$ are four force signals. Note that the four sensors are installed on the four columns of the press at the same horizontal plane and hence, the difference in the signals is mainly caused by the shape of the die. In order to ensure product quality and to protect the press, it is usually required that the four forces shall not be different by more than 50%.

In order to better visualize the spatial distribution of the force in time, the so-called snake skeleton graph (SSG) [8.3][8.4] was developed. Figure 8.6 shows the SSG corresponding to the signals in Figure 8.5. SSG is essentially a special way of plotting the four signals in a 3D graph: the horizontal plane $(x, y)$ is the base, its four corners representing the locations of the four sensors. From the base plane, the readings of the four sensors at time $t = 1$ form the four corners of the next surface. Note that the four readings may not be on the same plane and hence, the surface is not a plane but a surface (we use a Bázier surface to model it). The surface can be

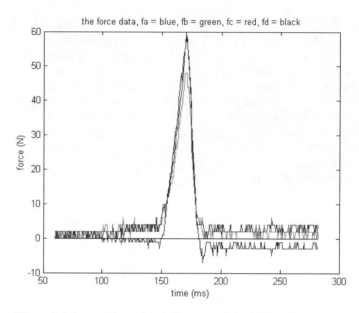

**Figure 8.5.** A set of force signals from a straight-sided hydraulic press

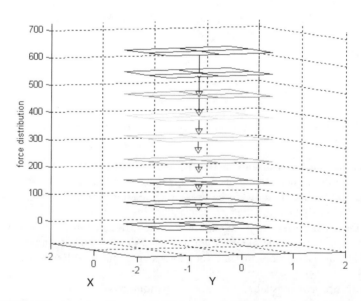

**Figure 8.6.** The corresponding snake skeleton graph (SSG) of the forces in Figure 8.5

characterized by its averaged normal vector: the amplitude is associated with the amplitude of the forces and the orientation represents the direction of the forces. The next surface is constructed in the same way with a shift up in the vertical direction representing the four forces at $t = 2$. This process continues until $t = N$, where $N$ is the number of data points in the force signals. Since the resulting normal vectors look like a wrinkling snake, it is called the snake skelton graph.

The other important factor in the stamping process is the blank holding force. For simple dies, blank holding is done by a binder with a set of springs (refer to Figure 8.1). The actual blank holding force is determined by the stamping force and the spring constant, and is not controllable. For large complex dies, blank holding is done by another set of punches. Hence, one can measure the blank holding force and control it to some extent. SSG is also useful [8.4]. In fact, SSG can be extended to analyze the spatial distribution of all kinds of multiple sensor signals, 1-dimensional or 3-dimensional, on the same installation plane or different planes.

It should be stressed that understanding the fundamentals of the process and the signals is the key to the success of condition monitoring and diagnosis. Without that, the best one can do is but trial-and-error. Luck usually does not prevail on shop floor.

## 8.3 Online Monitoring Based on the Tonnage Signal and Support Vector Regression

### 8.3.1 A Study of the Tonnage Signal

As mentioned in [8.1], condition monitoring is used primarily to classify whether a sensor signal corresponds to the normal system/process or otherwise. Since the signal is only a window to the system, it is usually biasd (the view from the window does not show the whole thing), unclear (the glass is not clean), and distorted (the glass is not perfect). Therefore, classification can only be done based on some sort of preassumption leading to different methods.

The condition monitoring methods can be divided into two types: a model-based method and a feature based method. The former uses a model to describe the signal and/or its latent process. Various models can be used, such as auto-regressive and moving average (ARMA) model, hidden Markov model (HMM), artificial neural network (ANN), etc. Note that these models may or may not have physical meaning. Based on the model, condition monitoring can be carried out by checking either the variations in model parameters or prediction errors in the model outputs.

The latter consists of two steps: first, extract the features from the signal, and then make decisions based on changes in the features. Various features can be used, including

- time-domain features (such as mean, standard deviation, range, root mean squares, skewness, kurtosis, crest factor, etc.);
- frequency-domain features (such as frequencies, damping ratios, energy in different frequency bands, etc.);
- spatial domain features;
- time–frequency domain features (i. e. time–frequency distribution);

- time–spatial domain features (the aforementioned SSG is aimed at capturing the time-spatial domain features);
- frequency–spatial domain features (*i. e.* holospectrum) [8.20].

Note that you can also model the features. In other words, one can use a combination of the model-based method and the feature-based method. Such a combination is nearly infinite and hence, many papers have been published. The successful application, however, depends on the understanding of the process, the signal, and then the method.

Examining the tonnage signals in Figures 8.3, 8.4, and 8.5, it is obvious that the signals are transient dynamic signals. Moreover, the signal is not periodic and hence, we cannot use the frequency-domain features. Of course, we can use the spatial domain features or the time–spatial domain features (the SSG method). However, this will be costly: a good tonnage sensor ranges from US$1,000 to US$10,000, not to mention the signal conditioner, the data acquisition system and the computer, while a small C-shape press costs only US$20,000! Therefore, in the subsequent discussion of this section, we focus on using just one sensor for monitoring the stamping operation.

In the previous section, we have seen that the tonnage signal has physical meaning (Figure 8.3). Armed with this understanding, we can try to model the signal. Since the signal is a transient signal, conventional methods, such as the ARMA model will not work. Therefore, we use the so-called support vector regression (SVR) model as shown below.

### 8.3.2 A Brief Introduction to Support Vector Regression (SVR)

Support vector regression (SVR) is a relatively new statistical classification method. Its theory is rather straightforward [8.21]. Briefly, given a set of independent and identically distributed (i.i.d.) samples $S = \{(\mathbf{x}_1, y_1), (\mathbf{x}_2, y_2), ..., (\mathbf{x}_n, y_n)\}$, where, $\mathbf{x} \in \mathbf{R}^N$ is the input vector and $y \in \mathbf{R}$ is the output value. We wish to find a function $f$ that correlates the input and the output:

$$y = f(\mathbf{x}, \mathbf{w}) = \langle \mathbf{w} \cdot \mathbf{x} \rangle + b = \sum_{i=1}^{n} \mathbf{w}_i \mathbf{x}_i + b \qquad (8.1)$$

where, $\mathbf{w} = (w_1, w_2, ..., w_n) \in \mathbf{R}^n$ and $b \in \mathbf{R}$ are the weighting factors, <•*•*> denotes the dot product, and $f: \mathbf{R}^n \to \mathbf{R}$. Introducing a nonlinear function $\varphi(\mathbf{x})$, a more generalized form is as follows:

$$y = f(\mathbf{x}, \mathbf{w}) = \sum_{i=1}^{n} \mathbf{w}_i \varphi(\mathbf{x}_i) + b = \langle \mathbf{w} \cdot \varphi(\mathbf{x}) \rangle + b \qquad (8.2)$$

From a mathematical point of view, Equation (8.2) represents a nonlinear regression. The nonlinear function $\varphi$ maps the input vector $\mathbf{x}$ onto a feature space

$\langle \mathbf{w} \cdot \varphi(\mathbf{x}) \rangle \in Z$. In the feature space, it finds a function in the family $F(\mathbf{x}, \mathbf{w}) = \{f \mid f : \mathbf{R}^N \rightarrow \mathbf{R}\}$ that minimizes the risk function:

$$R(\mathbf{w}) = \int c(\mathbf{x}, y, f(\mathbf{x}, \mathbf{w}))dp(\mathbf{x}, y) \tag{8.3}$$

where, $c$ denotes a loss function determining how we will penalize the estimation error based on the training samples. $p(\mathbf{x}, y)$ is the probability distribution of $(\mathbf{x}, y)$. Due to the lack of information on the $p(\mathbf{x}, y)$, one usually uses the empirical risk function estimated from the training samples $S_f$. This results in the following empirical risk function:

$$R_{emp}(\mathbf{w}) = \frac{1}{n}\sum_{i=1}^{n} c(\mathbf{x}, y, f(\mathbf{x}, \mathbf{w})) \tag{8.4}$$

In addition, to overcome the overfitting, a capacity control term is often added leading to the following equation [8.21],

$$R_{req}(\mathbf{w}) = R_{emp}(\mathbf{w}) + \frac{\lambda}{2}\|\mathbf{w}\|^2 = \frac{1}{n}\sum_{i=1}^{n} c(\mathbf{x}, y, f(\mathbf{x}, \mathbf{w})) + \frac{\lambda}{2}\|\mathbf{w}\|^2 \tag{8.5}$$

where, $\lambda > 0$ is a regularization constant. The last term controls the shapes of the function, which has a significant effect onto the generalization capability of the algorithm. Intuitively, SVR is to map the input vectors onto a high-dimensional feature space, and then find an optimal function $f$ there that cannot only minimize the fitness error, but also controls the generalization capability for the subsequent classification.

It has been shown that solving the SVR is equivalent to solving the following optimization problem:

$$\min_{\mathbf{w}, b, \xi} \quad \frac{1}{2}\langle \mathbf{w} \cdot \mathbf{w} \rangle + C\sum_{i=1}^{n}(\xi_i + \xi_i^*) = \frac{1}{2}\|\mathbf{w}\|^2 + C\sum_{i=1}^{n}(\xi_i + \xi_i^*) \tag{8.6}$$

subject to:    $y_i - \langle \mathbf{w}, \mathbf{x}_i \rangle - b \geq \varepsilon + \xi_i,$

$\langle \mathbf{w}, \mathbf{x}_i \rangle + b - y_i \geq \varepsilon + \xi_i^*,$

$\xi_i, \xi_i^* \geq 0,$

where, $i = 1, 2, \ldots, n$, $\xi_i$ and $\xi_i^*$ are slack variables used to deal with the noise, and $C = 1/(\lambda n)$ is a constant. In practice, $C$ balances the two terms: the shape of $f$ and the precision error $\varepsilon$. This is the so-called Vapnik's $\varepsilon$-SVR algorithm.

In order to solve Equation (8.6), a Lagrange function is constructed:

$$L = \frac{1}{2}\|\mathbf{w}\|^2 + C\sum_{i=1}^{n}(\xi_i + \xi_i^*) - \sum_{i,j=1}^{n}\alpha_i(\varepsilon + \xi_i - y_i + \langle\mathbf{w},\mathbf{x}_i\rangle + b)$$

$$- \sum_{i,j=1}^{n}\alpha_i(\varepsilon + \xi_i^* + y_i - \langle\mathbf{w},\mathbf{x}_i\rangle - b) - \sum_{i=1}^{n}(\eta_i\xi_i + \eta_i^*\xi_i^*)$$

(8.7)

where, $\alpha_i$, $\alpha_i^*$, $\eta_i$, $\eta_i^* \geq 0$ are the Lagrange factors. A more generalized form of SVR involves the use of a kernel function, $K$:

$$K(\mathbf{x},\mathbf{z}) = \langle\varphi(\mathbf{x})\cdot\varphi(\mathbf{z})\rangle$$

(8.8)

which satisfies the Mercer condition:

$$\iint K(\mathbf{x},\mathbf{z})\varphi(\mathbf{x})\varphi(\mathbf{z})d\mathbf{x}d\mathbf{z} > 0$$

(8.9)

where, $K$ is symmetric, and $\varphi(x)$ does not always equal to zero and $\int\varphi^2(x)dx < \infty$. As a result solving Equation (8.7) is equivalent to solving the quadratic optimization problem:

$$\min L = -\sum_{i=1}^{n}(\alpha_i - \alpha_i^*)y_i + \frac{1}{2}\sum_{i,j=1}^{n}(\alpha_i - \alpha_i^*)(\alpha_j - \alpha_j^*)K(\mathbf{x}_i,\mathbf{x}_j) + \varepsilon\sum_{i=1}^{n}(\alpha_i + \alpha_i^*) \quad (8.10)$$

Subject to     $\sum_{i=1}^{n}(\alpha_i - \alpha_i^*) = 0$

$$0 \leq \alpha_i, \alpha_i^* < C$$

$$i = 1, 2, ..., n$$

Note that this is a quadratic problem and can be solved using various methods, such as the conjugate gradient method and the quasi-Newton method, *etc*. The solution is guaranteed to be the global optimum. After the dual, $\alpha$ and $\alpha^*$, are found, it is easy to find the weighting factor:

$$\mathbf{w} = \sum_{i=1}^{n}(\alpha_i - \alpha_i^*)\varphi(\mathbf{x}_i)$$

(8.11)

Furthermore, according to the Karush-Kuhn-Tucker (KKT) complimentary condition, at the optimal solution the product between the dual and the constraints has to vanish. That is:

$$\alpha_i(y_i - \langle\mathbf{w}\cdot\varphi(\mathbf{x}_i)\rangle + b - \varepsilon - \xi_i) = 0$$

$$\alpha_i^*(\langle\mathbf{w}\cdot\varphi(\mathbf{x}_i)\rangle + b - \varepsilon - \xi_i^* - y_i) = 0$$

(8.12)

Thus, we can calculate the other weighting fact $b$:

$$b = y_i - \langle \mathbf{w} \cdot \varphi(\mathbf{x}_i) \rangle - \varepsilon \quad \text{for } \alpha_i \in (0, C)$$
$$b = y_i - \langle \mathbf{w} \cdot \varphi(\mathbf{x}_i) \rangle + \varepsilon \quad \text{for } \alpha_i^* \in (0, C)$$

(8.13)

Then, the SVR model is completed:

$$f(\mathbf{x}, \boldsymbol{\alpha}, \boldsymbol{\alpha}^*) = \sum_{j=1}^{m} (\alpha_j - \alpha_j^*) K(\mathbf{x}_j, \mathbf{x}) + b$$

(8.14)

Note that based on the KKT condition, $m$ input vectors would lie on the margin of the $\varepsilon$-tube. These vectors are referred to as the support vector as the others need not be considered. Furthermore, the method is referred to as the support vector regression (SVR).

Let us examine Equation (8.5) again. Given a set of training samples $S_f$, we wish to find a function $f$ that has at most $\varepsilon$ deviation from the targets for all the training samples, and at the same time, is as general as possible so that it can cover the unseen new samples. This relates to the selection of the cost function $c(\mathbf{x}, y, f(\mathbf{x}, \mathbf{w}))$. Under the assumption that the training sample set $S$ was generated by a true function $f_{\text{ture}}$ plus additive noise $\xi$ with density $p(\xi)$,

$$y_i = f_{\text{ture}}(\mathbf{x}_i) + \xi_i$$

(8.15)

The likelihood of an estimate $F_f = \{(\mathbf{x}_1, f(\mathbf{x}_1)), (\mathbf{x}_2, f(\mathbf{x}_2)), \ldots, (\mathbf{x}_n, f(\mathbf{x}_n))\}$ based on the training samples is

$$P(F_f \mid F) = \prod_{i=1}^{n} P(f(\mathbf{x}_i, \mathbf{w}) \mid (\mathbf{x}_i, y_i)) = \prod_{i=1}^{n} P(f(\mathbf{x}_i, \mathbf{w}) \mid y_i)$$
$$= \prod_{i=1}^{n} p(y_i - f(\mathbf{x}_i, \mathbf{w})) = \prod_{i=1}^{n} p(\xi_i)$$

(8.16)

To maximize the likelihood function $P(F_f | F)$ is equivalent to maximizing its logarithm, $\log P(F_f | F)$:

$$\log P(F_f \mid F) = \log \prod_{i=1}^{n} p(\xi_i) = \sum_{i=1}^{n} \log p(\xi_i) = \sum_{i=1}^{n} \log p(y_i - f(\mathbf{x}_i, \mathbf{w}))$$

(8.17)

Therefore, the cost function, or rather the loss function, is

$$c(\mathbf{x}, y, f(\mathbf{x}, \mathbf{w})) = -\log p(y - f(\mathbf{x}, \mathbf{w})) = -\log p(\xi)$$

(8.18)

The importance of this result is twofold: on one hand, it presents a way to construct a loss function. Once the noise density of the system is defined, we can obtain its related loss function. In the real-world application, the standard Gaussian density $N(0, 1)$ is a common model to describe the noise. Consequently, the loss function is defined as shown in Equation (8.19). This function is depicted in Figure 8.7 and is employed for monitoring the sheet metal stamping processes.

$$p(y - f(\mathbf{x}, \mathbf{w})) = p(\xi) = \frac{1}{\sqrt{2\pi}} \exp\left(-\frac{1}{2}\xi^2\right)$$

$$c(\mathbf{x}, y, f(\mathbf{x}, \mathbf{w})) = \frac{1}{2}(y - f(\mathbf{x}, \mathbf{w}))^2 = \frac{1}{2}\xi^2$$

(8.19)

On the other hand, it helps to determine whether a new signal is generated under the homologue condition as the training samples (we assume that the training samples are acquired when the monitored process/system is knowingly operated under normal conditions). As the process/system becomes defective or malfunctioning, the corresponding signal will carry deviations generating the so-called "noise". Through the noise density model, or the related loss function, as shown in Equation (8.19), their likelihood can then be worked out. Accordingly, the classification (*i. e.* the monitoring decision) can be done.

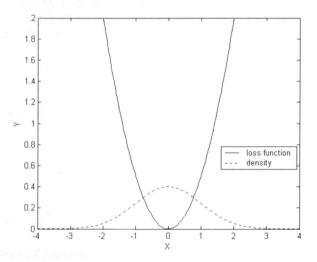

**Figure 8.7.** Gaussian loss function and its density model

It should be mentioned that the precision error $\varepsilon$ may play an important role in the SVR. In the original SVR paper [8.22], the $\varepsilon$-insensitive loss function is introduced:

$$c(\mathbf{x}, y, f(\mathbf{x}, \mathbf{w})) = \left|y - f(\mathbf{x}, \mathbf{w})\right|_\varepsilon = \max(0, \left|y - f(\mathbf{x}, \mathbf{w})\right| - \varepsilon)$$

(8.20)

We set the $\varepsilon = 0$ so that the condition **monitoring** has the highest level of alert.

Like other monitoring methods, SVR-based condition monitoring is carried out in two phases: training and classification. The implementation procedures are summarized in Table 8.1 and Table 8.2, respectively.

**Table 8.1.** The training procedure in SVR-based condition monitoring

| | |
|---|---|
| Step 1: | Acquire training sample sets under normal condition, $S = \{(\mathbf{x}_i, y_i), i = 1, 2, ..., n$; where $n$ is the number of samples in the training set. |
| Step 2: | (optional) Process the signals to eliminate / minimize the sampling noise. |
| Step 3: | Define a kernel function, $K(\mathbf{x}, \mathbf{x}_i)$. |
| Step 4: | Solve Equation (8.10) to obtain $\alpha_i$, $\alpha_i^*$. |
| Step 5: | Find those $\alpha_i$, $\alpha_i^* \neq 0$, they correspond to the support vectors, $\mathbf{x}_j$, $j = 1, 2, ..., m$, where $m$ is the number of support vectors. Also, find $b^*$ using Equation (8.13). |
| Step 6: | Construct the SVR model using Equation (8.14). |
| Step 9: | Find the residual error of the SVR using Equation (8.19) including its mean $\mu$ and standard deviation $\sigma$. |
| Step 10: | Construct the monitoring threshold $\mu - 3\sigma$. |

**Table 8.2.** The classification procedure in SVR-based condtion monitoring

| | |
|---|---|
| Step 1: | Acquire the new data $(\mathbf{x}, y)$. |
| Step 2: | (optional) Process the signal to eliminate / minimize the sampling noise. |
| Step 3: | Find the residual error of the SVR using Equation (8.14). |
| Step 4: | If residual > threshold, then alarm, else continue. |

In practice, three factors need to be considered. First, an appropriate kernel $K(\mathbf{x}, \mathbf{x}_j)$ can make the classification more effective. From a mathematical point of view, the kernel function reflects the geometric relationship between the new input sample and a support vector. Define the Gram matrix, $\mathbf{K} = [\mathrm{K}_{i,j}] = [K(\mathbf{x}_i, \mathbf{x}_j)]$, where, $i, j = 1, 2, ..., m$. It contains all the information acquired from the training. In the classification, the fused information in $\mathbf{K}$ is used to measure the similarity between the support vector and the new sample. Obviously, different kernel functions may suit different types of signals. Our experience indicates that the exponential radial based function (ERBF) kernel function in Equation (8.21), as illustrated in Figure 8.8, is rather effective. Its performance can be further improved by tuning the variance factor $\sigma$.

$$K(\mathbf{x}, \mathbf{z}) = \exp\left(-\sqrt{(\mathbf{x} - \mathbf{z})^2} / 2\sigma^2\right) \tag{8.21}$$

Second, the length of the sample, $n$, is important. Clearly, it should be large enough to cover all the information in the signal. However, too many data points do not help. This is because repeated information is not useful but repeated noise will reduce modeling accuracy. Moreover, it also prolongs the calculation time, affecting the monitoring decision speed. Therefore, we recommend using as few data points as possible.

Third, the quality of the signal is very important. The adage "garbage in garbage out" is well known. Therefore, monitoring signals must have high signal-to-noise ratio (SNR). In fact, good signal acquisition accounts for more than 70% of the success. We strongly encourage readers to study the signals as described in Section 8.1 of this chapter.

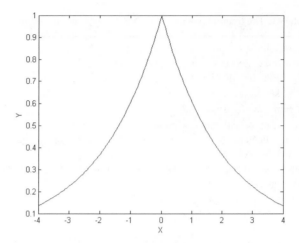

**Figure 8.8.** The exponential RBF kernel function

### 8.3.3 Experiment Results

Many experiments were undertaken on different presses with different workpieces. The press in Figure 8.1 is the one used most. It is a small C-frame cross-crank mechancial press (manufacturer: SEYI; model: SN1-25) typical in mass production of small domestic products. The maximum stamping force is 25 tons and the maximum speed is 110 shots per minute (SPM). The monitoring signals are acquired from a strain sensor (Kistler 9232A) mounted on the middle of the press frame. The signals were sampled using a data acquisition card (manufacturer: National Instruments; model: PCI 4452) installed in an Intel Pentium III 550 MHz PC. In order to capture the signal every time at the same instance, a sampling trigger sensor was installed on the crank of the press. It is a proximity sensor. Every time the keyway of the crank shaft rotates passing the sensor, a pulse is generated and sent to a computer to trigger the start of sampling. The stamping speed is set at 60 SPM. Accordingly, the sampling frequency is set at 4 kHz and every sample contains 1,024 data points covering 250 ms. Note that the stamping operation is completed in less than 100 ms. Figure 8.3 shows typical sampled data. Since the signal is very clean, filtering is not necessary. Note that the signal magnitude is in terms of voltage: since the sensor was not calibrated against the actual tonnage reading, the magnitude is just a relative term.

One of the first workpieces tested is the part shown in Figure 8.2. It is the back of a small motor. The workpiece material is SPCC steel. The diameter of the part is 35 mm and the height of the part 6.6 mm. The blank thickness is 1 mm. As shown in the figure, two different faults are considered: slug and mis-feed. The former occurs when small particles, such as dirt and chips from the positioning hole, are left on the surface of the workpiece. The latter is most likely caused by wear of the dies. They both are rather common on the shop floor. Figure 8.9 shows a set of typical signals corresponding to these conditions. It is seen that the variation is rather small. In fact, the variations caused by defects/malfunctions are less than 5% in terms of the root mean squares of the signal! Clearly, condition monitoring is not an easy job. The other major challenge is drift in the signal. Once production stops for a period of time, the signal may noticeably change. This may be attributed to various factors, such as change of the lubrication condition. Fortunately, the signal is very stable if production is continuous. Thus, every time a new production run starts, training is needed. Usually, training requires 30 sets of training samples and takes no more than a couple of minutes. Many runs were made, each run containing at least 90 samples. The average success rate was about 98%. In comparison, we also used an artificial neural network (ANN) and the success rate was about 93%.

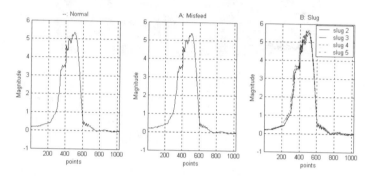

**Figure 8.9.** Comparison of signals acquired under different conditions

Another workpiece tested was the AAA size battery cap shown in Figure 8.10. The workpiece material was copper, the diameter of the part 12 mm, and the height 4.3 mm. The thickness of the blank was 0.7 mm. The part was formed in 11 progressive steps. Two kinds of malfunctions were encountered: blank jam (as shown in the figure) and part fracture. The latter, which usually occurred at the shoulder of the part, was the most common because of the small blank thickness. Even with a slight change of lubrication condition, for example, fracture may occur. This test was even more difficult than the previous one because the signal is rather weak (it does not take a lot of force to deform such a small part made of copper). A typical signal is shown in Figure 8.11. As in the previous test, we did a number of runs and the average success rate was 90%. In comparison, the success rate of the ANN was only 37%.

### 8.3.4 Remarks

Condition monitoring is essentially the classification of sensor signals. There are a large number of different classification methods available. Compared to other methods, SVR has a number of advantages. First, it is very general. A linear regression model, for instance, requires that the output is linearly related to the input. This is often impractical. Using the kernel function, $K(\mathbf{x}, \mathbf{x}_i)$, the SVR model maps the inputs into a high-dimensional feature space. Hence, it has more flexibility to depict correlation between input and output. In practice, various kernel functions can be employed as long as they comply with the Mercer condition. As a result, given a signal, it is almost always possible to fit an SVR model.

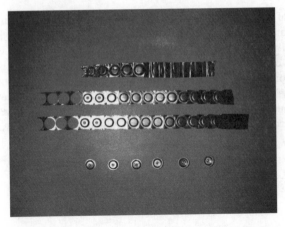

**Figure 8.10.** Battery caps made by progressive stamping

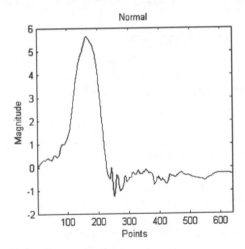

**Figure 8.11.** A typical signal corresponding to the workpiece in Figure 8.10

Second, the solution for SVR model fitting is globally optimal. Moreover, as pointed out in [8.23], the actual risk in training consists of two parts: the empirical risk of the training samples $R_{emp}(\mathbf{w})$ and the confidence interval $\Phi$:

$$R(\mathbf{w}) \leq R_{emp}(\mathbf{w}) + \Phi(n/h) \tag{8.22}$$

where, $h$ is the Vapnik–Chervonenkis (VC) dimension, which implies that $\Phi$ is a function of the complexity of the classification function structure. Note $\Phi(n/h)$ is inversely proportional to $h$. In other words, optimization will find a classification function that not only minimizes the model fitting error as defined by $R_{emp}(\mathbf{w})$ but also has the simplest structure as defined in $\Phi$. Therefore, SVR does not have the overfitting problem found with other methods.

Third, it does not require complicated calculation and hence, can be used for real-time applications. It is interesting to note that although SVR maps the data to a higher-dimension feature space, it does not significantly increase computation load. This is because of the introduction of the kernel function. By replacing the dot product with an appropriately chosen kernel function in Equation (8.8), non-linear mapping is completed without increasing the number of tunable parameters. As a result, the computational complexity is low.

Fourth, it only uses the training samples under normal conditions and considers all malfunctions/defects to be discrepancies. Moreover, its noise density model can work out the classification threshold automatically. All of these features make the SVR very attractive for practical applications. It is expected that in addition to the monitoring of sheet metal stamping processes, it will have many other applications.

Last but not least, the following is a useful source of information about SVR: www.support-vector.net, from which one can find details of the fundamental theory, updated research progresses, and shared computer software.

## 8.4 Diagnosis Based on Infrared Imaging

### 8.4.1 A Study of Diagnosis Methods

When a person is sick, everyone can tell. However, it takes a trained medical professional to diagnose the root cause and hence, find a remedy. Clearly, diagnosis is different from monitoring and usually is more difficult: the latter needs only to detect defects or malfunctions while the former needs to pinpoint the causes of the defects and the malfunctions. Though the latter must be done as quickly as possible to prevent catastrophic consequence, while the latter can afford a little delay as milliseconds are no longer crucial. As a result, diagnosis methods could be rather different from the monitoring methods.

For sheet metal forming processes, most product defects and/or machine malfunctions are caused by the dies: their design, setup and usage (lubrication, cleaning, worn out, *etc.*). To pinpoint on the dies that causes the problem is not always easy. That is why much time is spent on die tryout in the shop floor. Moreover, useful tools are limited.

The most commonly used method is finite element modeling (FEM). One may argue that it is not diagnosis but prediction. Well, proper prediction is even better than diagnosis. In general, FEM calculates the thickness strain distribution of the part and accordingly, predicts possible defects such as wrinkling, tearing and *etc.* Since this can be done even before dies are made, it is very cost effective. With ever improving computer technology and numerical methods, it has become an industry standard, especially for high-end products such as automobile body panels. However, it cannot include factors such as workpiece positioning error, improper lubrication, die manufacturing error, die worn out, *etc.* With the modeling error, its accuracy is about 90% at best. Moreover, it cannot identify the cause of the defects. In order to solve this problem, recently, new method was proposed [8.12]. First, it divides the restrain areas, called the flange, into a number of zones and converts the variation factors into equivalent restrain forces in these zones. Then, following the principle of Monte Carlo simulation, it runs a large number of FEMs to predict the part quality variations and the root causes. In order to reduce the computation time, instead of using conventional FEM, so-called inverse FEM (also called the one-step FEM, as it computes the thickness strain distribution in one step in inverse order from the part to the blank) is used. Presently, there are two different inversions of inverse FEM [8.24][8.25]; each has its advantages and limitations. As an example, Figure 8.12 shows the FEM of a rectangular box. From the figure, it is seen that there are four regions of interest (ROI), where thickness strains are the largest. These are the places that defects are most likely to occur. In order to study what may cause the defects, the flange areas of the part are divided into 12 segments as shown in Figure 8.13. Then, by Monte Carlo simulation with inverse FEM, we can find what affects the ROIs. For example, Figure 8.14 shows the sensitivity of the flange with

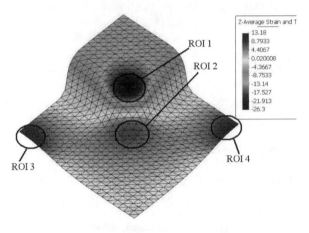

**Figure 8.12.** FEM of drawing a rectangle box

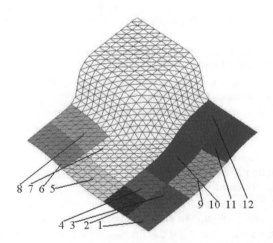

**Figure 8.13.** Definitions of the flange areas for the example in Figure 8.12

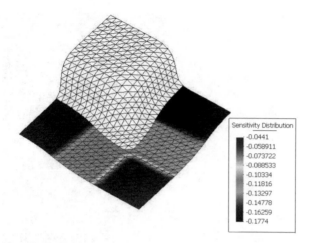

**Figure 8.14.** Sensitivity distribution for ROI1 for the example in Figure 8.12

respect to ROI1. From the figure, it is seen that the most sensitive areas (and hence, restrictions must be applied) are on the two sides but not the corner. This method can also be extended for design optimization. Currently, we are working on it.

The other method is direct measurement of the thickness strain distribution [8.25]. First it is required to mask the workpiece with square grids before the stamping operation and then measure the deformation of the grids. This method is effective but not convenient. The masking process is tedious and the imaging process is not very accurate. As a result, it is not very popular on the shop floor.

A new method has been developed based on infrared imaging as shown below.

### 8.4.2 Thermal Energy and Infrared Imaging

The basic idea of the proposed new method is straightforward [8.10][8.11]: During the stamping process, the workpiece deforms under the force. According to the principle of energy conservation, the force converts to heat giving a temperature rise. Therefore, by acquiring and analyzing the themal image of a part right after it is produced, it is possible to diagnose problems. As an example, Figure 8.15 shows the infrared image of the part in Figure 8.2 directly after it was made. From the figure, it is seen that there is a hot spot around the position pin. Since the part is symmetrical, the thermal energy distribution should be symmetrical. The hot spot is therefore an indication of a problem. In this case, by simple reasoning, the cause is traced to the bending of the position pin. Knowing the exact cause of the problem, repair can be carried out without much difficulty.

In order to use this idea, two problems must be solved. First, we need to find the "ideal" thermal energy distribution. Second, we need to acquire and analyze the actual thermal energy distribution. If a part is simple, as the one in Figure 8.15, we may rely on common sense, such as part symmetry, to deduce the thermal distribution. If a part is complex, however, FEM will be necessary. According to [8.26], during the stamping process, the temperature variation is governed by the following equation:

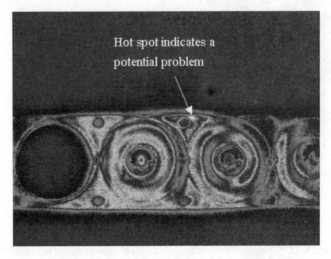

**Figure 8.15.** Thermal energy distribution of a part, and diagnosis

$$\rho c \dot{T} = \text{div}[k\,\text{grad}(T)] + \boldsymbol{\sigma}:\dot{\boldsymbol{\varepsilon}}^p + T\frac{\partial \boldsymbol{\sigma}}{\partial T}:\dot{\boldsymbol{\varepsilon}}^e \qquad (8.23)$$

where, $\rho$ is the density of the workpiece material, $c$ the specific heat, $k$ the thermal conductivity, $T$ the temperature, $\boldsymbol{\sigma}$ the stress, $\boldsymbol{\varepsilon}$ the strain, and the superscripts $p$ and $e$ represent plastic deformation and elastic deformation respectively. Note that only the plastic deformation energy is important and hence, the elastic strain-rate tensor can be neglected. This results in

$$\rho c \dot{T} = \text{div}[k\,\text{grad}(T)] + \eta\boldsymbol{\sigma}:\dot{\boldsymbol{\varepsilon}}^p \qquad (8.24)$$

where $\eta \approx 0.9$ is a constant representing the fraction of work converted to heat.

Assuming that the sheet metal stamping process is carried out at high speed (*e. g.* over 60 SPM), the heating will be locally adiabatic, *i. e.* no heat transfer takes place during the stamping. As a result, the temperature rise $\Delta T$ can be approximated by the following equation:

$$\Delta T = \frac{\eta}{\rho c}\int_0^{\Delta \bar{\varepsilon}} \bar{\sigma}\,d\bar{\varepsilon} \qquad (8.25)$$

where, $\bar{\sigma}$ and $\bar{\varepsilon}$ are the effective stress and the effective strain, respectively. Suppose that the material property follows the simple power law:

$$\bar{\sigma} = K\bar{\varepsilon}^n \qquad (8.26)$$

where, $K$ is the stress flow constant, and stress flow is a constant, then

$$\Delta T = \frac{\eta}{\rho c} \frac{K \Delta \bar{\varepsilon}^{(1+n)}}{(1+n)} \tag{8.27}$$

In other words, the temperature rise is proportional to the strain distribution. Therefore, to find the strain distribution one can use the temperature distribution. It should be pointed out, however, that the approximations in Equation (8.27) do not take bending and unbending effects into consideration (for example, in the drawing operation, some parts of the blank are first stretched and then compressed), and hence, might not be very accurate. In order to improve the accuracy, it is necessary to conduct an incremental FEM and then find the temperature rise using Equation (8.25) step-by-step. However, based on experience, the approximation in Equation (8.27) is usually sufficient.

To acquire the actual temperature distribution, infrared (IR) imaging is the clear choice. A sample IR image is shown in Figure 8.15. Note that when choosing an infrared camera, one needs to consider not only spatial resolution but also the temperature resolution as well as the temporal resolution. We had only a lower-end infrared camera; its imaging results are often less than desirable.

For relatively flat parts, such as the part shown in Figure 8.2, 2D IR imaging is usually good enough. For other parts, for instance the deep drawing of a cup, 2D IR imaging can no longer reveal the complete thermal distribution. Thus, it is necessary to take multiple IR images and reconstruct a 3D IR image. This is a somewhat tricky task because IR images are always fuzzy and it is difficult to find corresponding points on the IR images, as are required by traditional stereo matching methods. We use the octree carving technique to construct thermal distribution based on the silhouettes of the IR images and the camera orientations [8.11][8.27]. It consists of two steps: First, from the apparent contour and the extrinsic camera positions, a volumetric model is generated. The octree is initialized as a single large cube that encloses the reconstructed model. Second, the cube is projected onto the images and classified as either (1) completely outside the apparent contour, (2) completely inside the apparent contour, or (3) ambiguous (partially inside the contour). If the cube is classified as (3), as shown in Figure 8.16, it is subdivided into 8 sub-cubes, and each of them is projected onto the images and classified again. This process is repeated until a predefined maximum level is reached. If the cube is classified as (1), it is thrown away. So, only category (2) and (3) remain and are used to constitute the volumetric model of the object. The octree carving algorithm is summarized in Table 8.3.

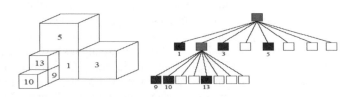

**Figure 8.16.** Illustration of the octree carving technique

**Table 8.3.** Implementation procedure of the octree algorithm

---

Initialize a cube that encloses the whole model
**While** max level not reached **do**
    **For** each cube in the current level **do**
        Project the cube onto each image and classify either as follows:
        (1) Completely outside the apparent contour
        (2) Completely inside the apparent contour
        (3) Ambiguous
        **if** the cube is classified as category (3) **then**
            subdivide the cube into 8 sub-cubes
            add the sub-cube to the next level
        **end if**
    **end for**
    increase the level number
**end while**

---

In order to display the reconstructed 3D thermal energy distribution model effectively and efficiently using conventional graphics-rendering hardware and software, a triangulated surface mesh is extracted from the octree using a standard marching cubes algorithm [8.28]. This uses the occupancy information of the 8 corners of a cube to determine how the surface intersects the edges of that cube, and eventually generates triangular patches that best approximate the surface. For the thermal information of the object, we use coloring of the vertex instead of texture mapping. This gives a smooth thermal signature rather than discrete and fragmentary information. The color of each vertex in the mesh is estimated based on the weighted average of all the pixel colors of the projections of that vertex. The weighting factor is the cosine of the angle (dot-product) between the viewing direction and the surface normal at that vertex [8.29]:

$$\omega_i = \begin{cases} -\mathbf{n} \cdot \mathbf{v}_i & \text{(if visible)} \\ 0 & \text{(otherwise)} \end{cases} \tag{8.28}$$

where $\mathbf{n}$ is the unit normal vector of the vertex (pointing outward) and $\mathbf{v}_i$ is the unit viewing direction vector of view $i$.

Figure 8.17 shows a set of four IR images taken from different angles immediately after the part was made. Note that the images were taken in sequence approximately 12 seconds apart. This introduced some errors because the part was quickly cooling down. A good IR camera can solve this problem. Also, in order to construct a good quality 3D thermal distribution, we usually need 16 or more images from different viewing angles. Figure 8.18 shows the 3D thermal distribution constructed from 16 2D IR images. From the figure, it is seen that higher temperatures occur around the shoulder, where deformation is largest. This agrees with the simulation results of the FEM. If disagreement is found, we suspect the presence of defects and/or malfunctions.

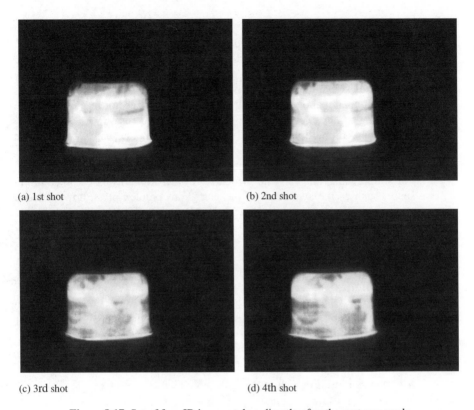

(a) 1st shot

(b) 2nd shot

(c) 3rd shot

(d) 4th shot

**Figure 8.17.** Set of four IR images taken directly after the part was made

**Figure 8.18.** The correpsonding 3D thermal distribution of the part in Figure 8.17

## 8.5 Conclusions

This chapter summarizes studies carried out over six years by the author and co-workers on monitoring and fault diagnosis of sheet metal stamping processes. From the discussions above, the following conclusions can be drawn:

(a) A good understanding of the process is essential for condition monitoring and fault diagnosis. One may need to study the process first using FEM.
(b) Condition monitoring is usually done using the tonnage signal. The tonnage signal reflects the stamping process and is highly nonstationary. One may use various signal classification methods, such as artificial neural networks (ANN), hidden markov model (HMM), and Support vector machine (SVM). According to our experience, support vector regression (SVR) is effective. In addition, if multiple tonnage sensors are used, one can use the snake skeleton graph (SKG).
(c) Fault diagnosis can be conveniently conducted using infrared (IR) imaging. For simple parts, one can find the causes of the problems based on common sense, such as the symmetry of the part. For complex parts, however, one may need finite element modeling (FEM) and/or 3D thermal distribution reconstruction. A good IR camera is always important.

## Acknowledgments

Studying the sheet metal stamping process is laborious and expensive. In the past seven years, many people have helped the author. A partial list of them includes:

- Dr Diane Xu, Ford Motor Company Vehicle Operations
- Dr Evangles Liasi, Ford Motor Company Vehicle Operations
- Prof. Y. S. Xu, Chinese University of Hong Kong
- Mr Xing Chu, Miko Manufacturing Ltd
- Mr Harry W. H. To, Mansfield Manufacturing Co.
- Mr Ivan Ho, Innovation and Technology Commission Hong Kong
- Mr Jack Ma, Kistler A. B.
- Dr M. Ge, Hong Kong Productivity Council
- Dr X. L. Li, Birmingham University
- Mr Y. Huang, Chinese University of Hong Kong
- Mr M. L. Yu, Chinese University of Hong Kong
- Mr Harry Y. M. Ng, Chinese University of Hong Kong

Without their help, this chapter would not have been possible. Also, I wish to acknowledge the financial support of the following organizations:

- Ford Motor Company Vehicle Operations under the P. O. No. 0020011598
- Hong Kong Innovation and Technology Fund (ITF) and Miko Manufacturing Ltd under the grant No. AF/79/99
- Mansfield Manufacturing Co. Ltd And ITF under the grant No. UIM/94
- Chinese University of Hong Kong under the grant No. 2050257
- Hong Kong Research Grant Council (RGC) under the grant No. 2150410
- ITF, Guangdong Metal Forming Works Ltd and Mansfield Manufacturing Co. Ltd under the grant No. GHS/008/04

# References

[8.1]   R. Du, "Fault Diagnosis," *Encyclopedia of Electrical and Electronics Engineering*, edited by John G. Webster, Article No. 2608, John Wiley, 2000.

[8.2]   X. Li and R. Du, "Condition Monitoring Using Latent Process Model with an Application to Sheet Metal Stamping Processes," *Trans. of ASME, J. Manufacturing Science and Engineering*, **127**(2), pp. 376–385, 2005.

[8.3]   D. Xu, E. Liasi, W. Z. Guo and R. Du, "Visual Comparison of Tonnage Signatures Using the Snake Skeleton Graph," *Mechanical Systems and Signal Processing*, **19**, pp. 311–328, 2005.

[8.4]   R. Du, D. Xu and E. Liasi, "Snake Skeleton Graph – A New Method for Analyzing Signals that Contain Spatial Information," *Trans. of ASME, J. of Dynamic Systems Measurement and Control*, **125**(3), pp. 294–302, 2003.

[8.5]   M. Ge, R. Du and Y. S. Xu, "Hidden Markov Model-Based Fault Diagnosis for Stamping Operations," *Mechanical Systems and Signal Processing*, **18**(2), pp. 391–408, 2004.

[8.6]   M. Ge, R. Du. G. C. Zhang and Y. S. Xu, "Fault Diagnosis Using Support Vector Machine with an Application in Sheet Metal Stamping Operations," *Mechanical Systems and Signal Processing*, **18**(1), pp. 143–159, 2004.

[8.7]   M. Ge, R. Du and Y. S. Xu, "Condition Monitoring Using Marginal Energy and Hidden Markov Model," *Int. J. of Control and Intelligent Systems*, **32**(1), pp. 1–9, 2003.

[8.8]   M. Ge, G. Zhang, R. Du, and Y. S. Xu, "Feature Extraction from Energy Distribution of Stamping Processes using Wavelet Transform," *J. of Vibration and Control*, **8**, pp. 1023–1032, 2002.

[8.9]   G. Zhang, M. Ge, H. Tong, Y. S. Xu, and R. Du, "Monitoring Stamping Operations Using Bispectral Analysis," *Engineering Applications of Artificial Intelligence*, **15**, pp. 97–104, 2002.

[8.10]  H. Y. M. Ng and R. Du, "Diagnosis of Sheet Metal Stamping Processes based on Thermal Energy Distribution," *Proc. of the 2005 IEEE Conference on Automation Science and Engineering*, August 1–2, 2005, Edmonton, Canada.

[8.11]  H. Y. M. Ng and R. Du, "Acquisition of Surface Thermal Distribution in Sheet Metal Forming," *Proc. of the 11th IEEE Int. Conf. on Mechatronics and Machine Vision in Practice*, Nov. 30 – Dec. 2, 2004, Macau SAR, China.

[8.12]  M. Yu and R. Du, "Sensitivity Analysis for Sheet Metal Stamping Based on Inverse Finite Element Modeling," *6th International Conference and Workshop on Numerical Simulation of 3D Sheet Forming Processes (NUMISHEET 05)*, Ann Arbor, MI, USA, Aug. 15–19, 2005.

[8.13]  Y. Huang and R. Du, "On the Development of Multi-Step Inverse FEM with Shell Model," *6th International Conference and Workshop on Numerical Simulation of 3D Sheet Forming Processes (NUMISHEET 05)*, Ann Arbor, MI, USA, Aug. 15–19, 2005.

[8.14]  Y. Huang, Y. P. Chen and R. Du, "A Study on the Effect of Punch Trajectory in Sheet Metal Stamping Using Implicit Dynamic Finite Element Simulation," *The Sixth World Congress on Computational Mechanics*, Beijing, China, Sept. 2004.

[8.15]  M. L. Yu, Y. P. Chen and R. Du, "Quality Prediction and Die Design Optimization for Sheet Metal Stamping Using Monte Carlo Simulation and Inverse Finite Element Modeling," *The Sixth World Congress on Computational Mechanics*, Beijing, China, Sept. 2004.

[8.16]  R. Du and W. Z. Guo, "The Design of a New Metal Forming Press with Controllable Mechanism," *Trans. of ASME, J. of Mechanical Design*, **125**(3), pp. 582–592, 2003.

[8.17]  W. Z. Guo, K. He, K. Yeung and R. Du, "A New Type of Metal Forming Press – Motion Control and Experiment Validation," accepted for publication in *Trans. of ASME, J. of Manufacturing Science and Engineering*, **127**(4), pp. 731–742, 2005.

[8.18]  S. Kalpakjian and S. R. Schmid, *Manufacturing Engineering and Technology*, 4th edition, Prentice Hall, 2001.

[8.19]  Y. Huang and R. Du, "A Study on the Tonnage Signal Using Finite Element Method," *2004 Japan-USA Joint Conference and Flexible Automation*, Colorado, USA, July 21 – 25, 2004.

[8.20]  R. Du, D. Xu and E. Liasi, "Snake Skeleton Graph – A New Method for Analyzing Signals that Contain Spatial Information," *Trans. of ASME, J. of Dynamic Systems Measurement and Control*, **125**(3), pp. 294–302, 2003.

[8.21]  V. N. Vapnik, "An Overview of Statistical Learning Theory," *IEEE Trans. on Neural Networks*, **10**(5), 1999.

[8.22]  C. Cortes and V. N. Vapnik, "Support Vector Networks", *Machine Learning*, (20), pp. 273–297, 1995.

[8.23]  B. Scholkopf, C. J. C. Burges and A. J. Smola, *Advances in Kernel Methods: Support Vector Learning*, MIT Press, Cambridge, MA, 1999.

[8.24]  J. L. Batoz, Y. Q. Guo and F. Mercier, "The Inverse Approach with Simple Triangular Shell Elements for Large Strain Prediction of Sheet Metal Forming Parts," *J. of Engineering Computations*, **15**(7), pp. 864–892, 1998.

[8.25]  C. H. Lee and H. Huh, "Blank Design and Strain Estimates for Sheet Metal Forming Processes by a Finite Element Inverse Approach with Initial Guess of Linear Deformation," *J. Materials Processing Technology*, **28**(1–3), pp. 145–155, 1998.

[8.26]  GOM, Gesellschaft für Optische Meßtechnik, Braunschweig, Germany, "Optical Metal Sheet Forming Analysis," www.gom.com/pub/publications/smf.pdf.

[8.27]  H. J. Antunez, "Thermo-Mechanical Modeling and Sensitivity Analysis for Metal-Forming Operations", *Computer Methods Application in Mechanical Engineering*, **161**, pp. 113–125, 1998.

[8.28]  W. E. Lorensen and H. E. Cline, "Marching Cubes: a High Resolution 3D Surface Construction Algorithm," *ACM Computer Graphics*, **21**(4), pp. 163–169, July 1987.

[8.29]  K. Y. K. Wong and R. Cipolla, "Structure and Motion from Silhouettes," *Proc. of 8th IEEE Int. Conf. on Computer Vision*, **2**, pp. 217–222, 2001.

# 9

# Robust State Indicators of Gearboxes Using Adaptive Parametric Modeling

Yimin Zhan and Viliam Makis

Department of Mechanical and Industrial Engineering
University of Toronto
Toronto, ON M5S 3G8, Canada
Emails: yimin.zhan@utoronto.ca, makis@mie.utoronto.ca

## Abstract

This chapter presents an in-depth study on the condition monitoring of rotating machinery using adaptive parametric modelling, focusing on the development of robust state indicators of gearboxes running from a brand new to breakdown state in a natural course, under varying load conditions. Three independent robust state indicators based on state-space representation of a time-varying autoregressive model and noise-adaptive Kalman filtering are proposed and compared with other state indicators considered in previous studies. The experimental validations make use of full lifetime vibration monitoring data of gearboxes under varying load conditions and analyze some critical properties of gear state indicators in real applications over the full lifetime horizon of gearboxes. The results show that the proposed three gear state indicators possess a highly effective and robust property in the state detection of a gearbox, which is independent of variable load conditions, as well as remarkable stability, early alarm for incipient fault and significant presence of fault effects. The proposed three gear state indicators can be directly employed by an online maintenance program as reliable quantitative condition covariates to make optimal maintenance decisions for rotating machinery.

## 9.1 Introduction

Proper machine condition monitoring procedures can result in lower maintenance costs and prolonged machine life. The most common family of machine condition monitoring methods is based on the analysis of vibration and acoustic signals, measured using a range of sensing techniques [9.1]. Rotating machines are used extensively in the manufacturing of industrial products. Gearboxes, as key rotating motion transmission components, play a critical role in the stable operation of rotating machinery and thus attract considerable research interest in the condition monitoring and maintenance of manufacturing equipment.

Gearbox condition signals often possess highly non-stationary properties due to the fact that defects and incipient failures often manifest themselves in the form of changes in the spectrum of a measured signal. This phenomenon has increasingly impelled the application of non-parametric joint time–frequency (T-F) methods to the analysis of non-stationary machine vibration signals since they are able to produce an overall view of the behavior of non-stationary signals by means of the so-called time-varying spectrum, which is defined in T-F space and represents the evolution of signal power as a function of both time and frequency [9.2]. However, the capacity to reveal power variations in T-F space as precisely as possible becomes a hard constraint for non-parametric T-F techniques when the aim is to monitor the occurrence of mechanical faults. At their early stages, faults start as almost impulsive events and determine a change in the 'signature' of the signal in T-F space [9.2]. Therefore, for early diagnosis, it is necessary to utilize methods with high temporal resolution, aiming at detecting spectral variations occurring in a very short time.

In comparison, the modern spectral analysis method is more effective. In modern spectral analysis, the techniques of time series modelling (AR, MA and ARMA, *etc.*), known as parametric spectrum analysis methods, have been applied to vibration signal analysis of rotating machinery by using time-invariant coefficients [9.3]–[9.5]. As a consequence, both accuracy and resolution can be significantly improved. Usually, the autoregressive model (AR) or a vector autoregressive model (VAR) are most preferred since they offer the best compromise between temporal representation and speed, efficiency and simplicity of algorithms enabling the estimation of model parameters. In fact, the spectrum of the ARMA process could even be represented purely in terms of the AR coefficients without resort to computing the MA coefficients [9.6]. Up to now, little attention has been focused on time-varying AR models where the evolution law of time-varying coefficients is assumed to be stochastic, and the parameter estimation of non-stationary time series models using advanced adaptive filtering theory for optimum condition-based maintenance (CBM) purposes in the sense of providing highly precise T-F domain information has rarely been investigated. Therefore, a state-space representation of a time-varying autoregressive model using a noise-adaptive Kalman filter (NAKF) will be presented in this chapter.

On the other hand, to associate a condition monitoring technique with operational maintenance decision analysis requires online calculation of independent quantitative indicators of the gearbox. Obviously, conventional state analysis techniques, like the two-dimensional spectral plot of fast Fourier transform (FFT) and three-dimensional time–frequency representation of wavelets, cannot meet this requirement.

Furthermore, development of a robust condition monitoring technique for detecting gearbox deterioration becomes far more complicated when it is subject to changeable operating condition. In most situations, varying operating conditions refer to variable load conditions or torque levels since it is the major source of contribution to the energy of the measured vibration signal. In manufacturing processes, many unexpected or uncertain sources can contribute to the fluctuation of load condition, *e. g.* non-uniformity of raw material in the machining process causing variation of load applied to a tooling system and then transmitting the varying load condition back to the output shaft of a gearing system. The vibration signal measured by vibration transducers mounted on the gearbox casing will inevitably represent, in their signal strength, the varying-load-induced energy variation. Under such circumstances, most conventional vibration monitoring techniques are unable to identify this non-deterioration energy variation. As a result, false alarms will be produced and then additional maintenance cost accrued. Therefore, an effective state detection technique should be insensitive to variable load condition.

Efforts in extraction of independent quantitative state indicators have been made in the literature considering a wide range of non-parametric methods. However, very few robust state indicators with respect to variable load condition have been obtained. Most state analyses are based on visual observation of a two-dimensional curve or three-dimensional contour plot, which is useful for machine condition monitoring but contributes little to condition-based maintenance.

Moreover, most research in the development of robust condition monitoring techniques was validated using limited vibration measurements picked up at a few specific inspections, e.g. [9.5][9.7][9.8]. The full lifetime performance of these methods has not been investigated. In other words, the stability of these methods over the full lifetime horizon of equipment is unknown. However, such stability is a vital necessity for online condition monitoring since strong variability of a state indicator will result in frequent false alarms and obscure the real indication of incipient fault. Therefore, this chapter is particularly concerned with the development and performance analysis of independent robust state indicators over the full lifetime horizon of gearboxes under varying load condition. In this chapter, three independent and robust state indicators of gearbox condition, based on the NAKF-based time-varying AR model, will be introduced and validated using complete lifetime vibration monitoring data of gearboxes running from a brand-new condition to breakdown.

The remainder of this chapter is organised as follows. Section 9.2 presents the NAKF-based state-space representation of a time-varying AR model and the theoretical aspects of three independent robust state indicators, termed the bispectral feature energy (BFE), the AR model residual-based state parameter (MRP), and the improved MRP (IMRP), respectively. The experimental set-up is described in Section 9.3. Sections 9.4, 9.5 and 9.6 present the performance analysis of BFE, MRP, and IMRP state indicators, respectively. Concluding remarks are given in Section 9.7.

## 9.2 Modeling

### 9.2.1 Noise-adaptive Kalman Filter-based Model

In this section, we present a state-space representation of a VAR model with time-dependent coefficients and provide recursive algorithms for the implementation of Kalman filtering. A VAR process is a discrete-time multivariate linear stochastic process given by

$$y_i = \sum_{k=1}^{p} A_k y_{i-k} + \varepsilon_i \qquad (9.1)$$

for $i = 1, 2, \ldots, N$, that is, the time series can be considered as the output of a linear all-poles filter driven by a white-noise signal with a flat spectrum, where $N$ is the sample size, $p$ the order of VAR model, $y_i$ the $i$th measurement vector of dimension $d\times1$, $A_k$ the $k$th $d\times d$ coefficient matrix and $\varepsilon_i$ a $d\times1$ sequence of zero-mean white Gaussian measurement noise. Considering the non-stationary property of vibration signatures we assume the coefficient matrices of the above VAR model to be time-varying

$$y_i = \sum_{k=1}^{p} A_k (i) y_{i-k} + \varepsilon_i \qquad (9.2)$$

To make use of the Kalman filtering algorithm, it is necessary to develop a state-space representation of the model (9.2). This can be achieved by rearranging the elements of the matrices of coefficients in a vector form using the vec-operator, which stacks the columns of a matrix on top of each other from left to right. Then, with the following notation:

$$a_i = vec([A_1 (i), A_2 (i) \ldots, A_p (i)]^T) \qquad (9.3)$$

$$Y_i = (y_i^T, y_{i-1}^T, \ldots, y_{i-p+1}^T) \qquad (9.4)$$

$$C_i = I_d \otimes Y_i^T \qquad (9.5)$$

where $I_d$ is a $d\times d$ identity matrix, $\otimes$ is the Kronecker product, an appropriate state-space representation of the VAR model with stochastic coefficients is given by

$$a_{i+1} = f_i(a_i) + v_i \qquad (9.6)$$

$$y_i = C_{i-1}^T a_i + \varepsilon_i \qquad (9.7)$$

where $a_i$ is the $pd^2\times1$ state vector, $v_i$ is a $pd^2\times1$ sequence of zero-mean white Gaussian state noise, uncorrelated with $a_1$ and $\varepsilon_i$, $\varepsilon_i$ is the same as in (9.1) and uncorrelated with $a_1$ and $v_i$, and the Equation (9.7) has an adaptive time-varying coefficient $C_{i-1}^T$ of dimension $d\times pd^2$. We also have $\hat{a}_0 (= E(a_0))$, the Gaussian $pd^2\times1$ initial state vector with covariance matrix $P_{0|0} (= Cov(a_0))$, and the noise covariance matrices:

$$E\{v_k v_i^T\} = Q_k \delta_{k,i} \qquad (9.8)$$

$$E\{\varepsilon_k \varepsilon_i^T\} = R_k \delta_{k,i} \qquad (9.9)$$

where $T$ denotes transposition, $\delta_{k,i}$ denotes the Kronecker delta, $\delta_{k,i} = 1$ if $k = i$ and 0 otherwise, $Q_k$ denotes the covariance matrix of the state noise term and $R_k$ denotes the covariance matrix of the measurement noise. For convenient implementation, $\hat{a}_0$ and $P_{0|0}$ are arbitrarily chosen in this model. Suppose that the evolution law of the state vector $a_i$ is a random walk process which results in the state-space representation

$$a_{i+1} = a_i + v_i \tag{9.10}$$

$$y_i = C_{i-1}^T a_i + \varepsilon_i \tag{9.11}$$

Thus, with the aid of a standard Kalman filter the following recursive prediction equations

$$P_{i|i-1} = P_{i-1|i-1} + Q_{i-1} \tag{9.12}$$

$$\hat{a}_{i|i-1} = \hat{a}_{i-1} \tag{9.13}$$

where $P_{i|i-1}$ is the one-step ahead prediction of the state covariance matrix, $P_{i-1|i-1}$ is the estimate (error) covariance matrix of the state vector, $\hat{a}_{i|i-1}$ is the one-step ahead predication of the state vector, $\hat{a}_{i-1}$ is the optimal filtering estimate of the state vector $a_{i-1}$ and updating equations

$$G_i = P_{i|i-1} C_{i-1} [C_{i-1}^T P_{i|i-1} C_{i-1} + \hat{R}_i]^{-1} \tag{9.14}$$

$$P_{i|i} = [I - G_i C_{i-1}^T] P_{i|i-1} \tag{9.15}$$

$$\hat{a}_{i|i} := \hat{a}_i \tag{9.16}$$

$$= \hat{a}_{i|i-1} + G_i(y_i - C_{i-1}^T \hat{a}_{i|i-1})$$

can be obtained for $i = 1, 2, \ldots, N$, where $G_i$ is the Kalman gain and the covariance matrix $Q_i$ of state noise is assumed to be known *a priori*. From the incoming measurement information $y_i$ and the optimal state prediction $\hat{a}_{i|i-1}$ obtained in the previous step, the innovations sequence is defined as

$$z_i := y_i - C_{i-1}^T \hat{a}_{i|i-1} \tag{9.17}$$

which leads to the estimate of $R_i$, using the $i$ most recent residuals, given by

$$\hat{R}_i = \frac{1}{i-1} \sum_{k=1}^{i} (z_k - \bar{z})(z_k - \bar{z})^T \tag{9.18}$$

where

$$\bar{z} = \frac{1}{i}\sum_{k=1}^{i} z_k \tag{9.19}$$

is the mean of the innovations up to the $i$th time instant. Note that the estimation approach (9.18) requires the white Gaussian property of $z_i$ [9.9]. Obviously, this property can be obtained if the Kalman filter operates in its optimum state. Therefore, an adaptive Kalman filter for estimating the state as well as the noise covariance matrices can be called a noise-adaptive Kalman filter (NAKF).

### 9.2.2 Bispectral Feature Energy

It can be shown that the parametric spectrum of the signal depends on the estimated parameters of the time series model. In fact, the relationship is given by

$$P(f) = \frac{P_N(f)}{|H(f)|^2} \tag{9.20}$$

that is, the signal power spectrum density (PSD) depends on what can be expressed as the product of $P_N(f)$, PDS of the white noise ($P_N(f) = P_{NO}$), and $H(f)$, frequency response of the linear filter. After estimate the coefficient matrices of the VAR model, an instantaneous estimate of the spectral density function, which in the multivariate case is a matrix valued function of frequency, can be given in terms of the VAR coefficient matrices

$$P_i(f) = [H_i^{-1}(f)]\hat{R}_i[H_i^{-1}(f)]^* \tag{9.21}$$

where the asterisk denotes the conjugate complex and

$$H_i(f) = I_d - \sum_{k=1}^{p} \hat{A}_k(i)e^{-j2\pi f T_s k} \tag{9.22}$$

The PSD function (9.21) is adaptive because each coefficient matrix $\hat{A}_k(i)$ is considered to be a stochastic process in order to fit the time-varying spectral characteristics of the vibration signals. In this manner, the model can, in principle, follow rapidly varying spectra because it has an inherently non-stationary structure. Accordingly, the PSD function, expressing the spectrum using the time-varying VAR parameters, becomes a function of two variables, time $i$ and frequency $f$ in the same way as for all T-F distributions [9.2].

Suppose the object of interest is a gear involved in meshing motion. To generalize the parametric time–frequency representation over each revolution of the target gear and then associate it with a bispectral detection statistic, the time-averaged spectrum should be calculated over each entire revolution and expressed as

$$X(f) = \frac{1}{M} \sum_{i=1}^{M} P_i(f)$$
(9.23)

where $M$ is the number of sampling points within each revolution, $P_i(f)$ is defined in Equation (9.21), and $X(f)$ can then be used by the following equations.

Among the most commonly used higher-order spectra is the complex-valued bispectrum associated with the third-order cumulant and defined as

$$B(f_1, f_2) = E[X(f_1)X(f_2)X^*(f_1 + f_2)]$$
(9.24)

where the asterisk denotes complex conjugation and $X(f)$ is the Fourier transform of $X(n)$. The bispectrum can be viewed as a decomposition of the third moment (skewness) of a signal over frequency. The expectation operator $E$ indicates an average over sufficient ensembles.

A critical concern about the bispectrum-based fault detection methodology is its sensitivity to the alternating load condition, such as the changes of output torque level. To overcome such a problem, the magnitude-squared bicoherence function obtained by normalization of the bispectrum is used and expressed as

$$\beta(f_1, f_2) = \frac{|B(f_1, f_2)|^2}{E[|X(f_1)X(f_2)|^2]E[|X(f_1 + f_2)|^2]}$$
(9.25)

Thus, the locally optimum state detection statistic $T_{lo}$ making use of bicoherence function is

$$T_{lo}(k) = \max_{1 \le l \le k} \sum_{m=l}^{k} (\beta_m - \beta_0)$$
(9.26)

where $\beta_0$ denotes the pre-change value of $\beta(f_1, f_2)$ reflecting the behavior of the machine under normal operating conditions and can be interpreted as a known quantity that can be readily estimated and $\beta_m$ denotes the value of $\beta(f_1, f_2)$ at the $m$th inspection [9.10]. The principle behind the condition monitoring by means of the state detection statistic $T_{lo}$ is that the large deviation from baseline indicates a fault. $T_{lo}(k)$ is actually a function of three variables given by $T_{lo}(k, f_1, f_2)$ where $k$ is time and $(f_1, f_2)$ is a bispectral frequency pair. To reduce computational requirements, $f_2$ is usually fixed at a characteristic frequency, say the gear meshing fundamental, which leads to a bivariate representation of $T_{lo}(k, f_1)$.

A bispectral feature energy index (BFE) was defined in [9.11] as follows

$$BFE(k) = \int_{f_s}^{f_e} T_{lo}(k, f_1) df_1 = \sum_{i=1}^{n} T_{lo}(k, f_{1,i})(f_{1,i+1} - f_{1,i})$$
(9.27)

where $f_{1,1}=f_s, f_{1,n}=f_e$ and $[f_s : f_r : f_e]$ is an equally-spaced frequency range by resolution $f_r$. The BFE($k$) provides a generalized indicator to track the state evolution of the target gear for a maintenance decision model. The predominant advantage of this hybrid diagnostic model is that it takes advantage of the improved parametric time–frequency analysis and then provides a robust state parameter, which is sensitive to variations of the state of the object of interest but insensitive to the changing operating regimes.

### 9.2.3 AR Model Residual-based State Parameter

Development of the AR model residual-based state parameter (MRP) [9.12] is motivated by recognition that the gear motion residual signal represents the departures of the time-synchronous average (TSA) from the average tooth-meshing vibration and usually shows evidence of faults earlier and more clearly than the TSA signal. Thus, fault-induced non-stationary features will be more significant in the gear motion residual signal, whereas the gear motion residual signal is usually stationary for healthy states of the gear of interest. Most importantly, since much of the energy in the healthy state of the gear of interest is concentrated at the meshing fundamental and its harmonics, the gear motion residual signal, obtained by removal of the regular gear meshing fundamental and its harmonics from the TSA signal, will be much less sensitive to varying load conditions than the TSA signal. Therefore, earlier alarms for incipient faults and less load dependence can be obtained simultaneously.

Further, an NAKF-based AR model is fitted to the stationary gear motion residual signal, which corresponds to the initial healthy state of the gear of interest. The resulting AR model coefficients will enable the model to give a consistent prediction for future stationary gear motion residual signals of the same family but yield significant prediction errors for any part that is outside, such as fault-induced non-stationarity or any frequency component not contained in the healthy state gear motion residual signal. Therefore, if the gear of interest remains in a healthy state, its model residuals will only represent the prediction error. However, if a localized fault is present, the prediction error will be more significant when the AR model established in the healthy state of the gear of interest is applied.

Most importantly, by selection of a proper model order for the healthy gear under various load conditions, significant enhancement in load-independence can be obtained in the AR model residuals. Consequently, a robust statistical measure, which takes the percentage of outliers exceeding the baseline three standard deviation limits can be applied to the AR model residuals to evaluate the state of the gear of interest.

Figure 9.1 illustrates in detail the proposed model. The model order is determined by selecting a number of gear motion residual signals from the initial period of each load condition under the healthy state of the gear of interest. During examination of each AR model order candidate, a matrix of state (or time-varying AR model coefficients) vectors with each column corresponding to each angular position (or phase) over the entire revolution of the gear of interest will be obtained by applying the NAKF to the gear motion residual signal at the first inspection. If the phase information is unavailable in the tests, the mean state vector of this state matrix should be calculated over the angular horizon and used as a constant state vector to calculate AR model residuals throughout the entire revolution for every

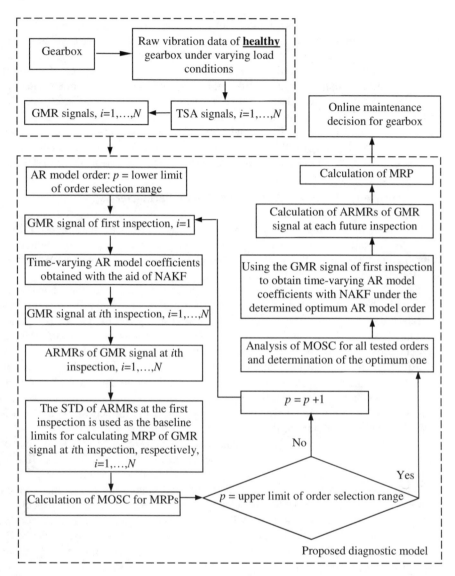

**Figure 9.1.** Diagram of the proposed diagnostic model. Notation: TSA – time synchronous averaging; AR – autoregressive; GMR – gear motion residual; NAKF – noise-adaptive Kalman filter; ARMRs – AR model residuals; STD – standard deviation; MRP – AR model residuals-based gear state parameter; MOSC – model order selection criteria

inspection performed during the full lifetime of the gear of interest. By doing so, the phase information embedded in the state matrix is lost. This usually would not invalidate the subsequent analysis since the difference among the columns of the state matrix estimated using the healthy gear's motion residual signals is predominantly induced by the geometric irregularities of the gear of interest, which is usually much less significant than fault-induced effects.

Extensive tests revealed that model order is the predominant parameter of the NAKF-based time-varying AR model resulting in notable variation of model performance under varying load-conditions. In order to select the optimum one, some statistical measures should be applied to evaluate the performance of the MRP under a certain model order. Unlike the conventional model order selection procedure based on, say, Akaike information criterion (AIC), the order selection criteria proposed in this study are aimed at a compromised model fitting with load-independence under varying load conditions. To be consistent with the improved MRP introduced in Section 9.2.4, eight AR model order selection criteria will be proposed therein, seven of which will be used to validate the performance of MRP.

### 9.2.4 Improved AR Model Residual-based State Parameter

Development of the improved AR model residual-based state parameter (IMRP) [9.13] is in general based upon the same theoretical background, but with three critical improvements. The reader is referred to Figure 9.1 for a general view of IMRP.

First, a number of gear motion residual signals extracted from the TSA signals obtained in the healthy state of the target gearbox but under different load conditions are selected. With the aid of an NAKF-based time-varying AR model, the AR model residuals can be calculated for each selected gear motion residual signal. Consequently, the AR model residuals with the largest standard deviation can be identified. Such a standard deviation is used in determining the baseline upper and lower limits, $\pm 3\sigma$, to calculate the percentage of outliers outside $[-3\sigma, +3\sigma]$ for the AR model residuals generated by applying the time-varying AR model to each selected gear motion residual signal. Such a percentage is termed the IMRP and is applied to evaluate the state of the target gear.

Second, the improved AR model order selection algorithm of IMRP consists of eight order selection criteria instead of seven as in [9.12] where the MRP is proposed, which incorporates four global criteria and four local criteria.

Let us assume that there are two different load conditions to be encountered during the lifetime of a gearbox, which cover at least $N$ and $M$ inspections respectively, and thus result in two IMRP sequences under a certain model order $p$, $IMRP(p, i)$ and $IMRP(p, j)$ corresponding with the two load conditions, respectively, where $i = 1, \ldots, N$ and $j = K+1, \ldots, K+M$ and $K \geq N$. Here, we assume that the $N+M$ inspections are carried out under the healthy state of the target gear.

Then, the ratio between the mean of $IMRP(p, i)$ and the mean of $IMRP(p, j)$, expressed as

$$R(p) = \frac{\max\left\{\dfrac{1}{N}\sum_{i=1}^{N} IMRP(p,i), \dfrac{1}{M}\sum_{j=K+1}^{K+M} IMRP(p,j)\right\}}{\min\left\{\dfrac{1}{N}\sum_{i=1}^{N} IMRP(p,i), \dfrac{1}{M}\sum_{j=K+1}^{K+M} IMRP(p,j)\right\}} \tag{9.28}$$

can be calculated, where it is assumed that we always assign the numerator the larger mean of IMRP and the denominator the smaller one. Thus $R(p)$ is always larger than 1 given that the numerator and the denominator are not equal, we can then obtain the first statistical measure

$$M_1(p) = R(p) - 1 \tag{9.29}$$

which is positive and should be close to zero when the optimum model order is obtained.

If a zero denominator is encountered for $R(p)$, $M_2(p)$ as an alternative is proposed and utilized to compare the means of $IMRP(p, i)$ and of $IMRP(p, j)$, which is defined as

$$M_2(p) = Abs\left[\frac{1}{N}\sum_{i=1}^{N} IMRP(p,i) - \frac{1}{M}\sum_{j=K+1}^{K+M} IMRP(p,j)\right] \tag{9.30}$$

which should be close to zero when the optimum model order is obtained.

The performance of IMRP is also expected to be stable and does not present strong variability. Therefore, the RMS, as an index of variability, is employed to characterize the dispersion among the IMRPs and is expressed as

$$s(p) = \sqrt{\frac{1}{N + M - 1}\left[\sum_{l=1}^{N+M}\left(IMRP(p,l) - \overline{IMRP}(p)\right)^2\right]} \tag{9.31}$$

where $IMRP(p, l) = IMRP(p, i) \cup IMRP(p, j)$, $l = 1, \ldots, N+M$, and $\overline{IMRP}(p)$ is the mean of $IMRP(p, l)$ under the order $p$, given by

$$\overline{IMRP}(p) = \frac{1}{N + M}\sum_{l=1}^{N+M} IMRP(p,l) \tag{9.32}$$

For consistency with $M_1(p)$, $M_2(p)$ and the following notation, $s(p)$ is replaced by $M_3(p)$, namely

$$M_3(p) = s(p) \tag{9.33}$$

In principle, $M_3(p)$ should achieve its minimum under the optimum model order.

In addition, the optimum model order should not give rise to significant outliers or abnormally high values of IMRP under the healthy state of the target gear. Therefore, the kurtosis can be employed to measure the severity of abnormal outliers and gives a statistical measure expressed as

$$M_4(p) = kurtosis\,(IMRP(p,l)) \tag{9.34}$$

The kurtosis function $kurtosis(\bullet)$ is provided by MATLAB®. The value of $M_4(p)$ should be close to 3 when the optimum model order is obtained.

Note that the above statistical measures, $M_1(p)$, $M_2(p)$, $M_3(p)$ and $M_4(p)$, are global ones since they take into account both load conditions. For local performance, we still need the same statistical measures to evaluate the performance of IMRP under each load condition so as to obtain acceptable local behavior of IMRP. In such a connection, the following four performance measures are proposed:

$$M_5(p) = \sqrt{\frac{1}{N-1}\left[\sum_{i=1}^{N}\left(IMRP\ (p,i) - \overline{IMRP}_1(p)\right)^2\right]} \tag{9.35}$$

and

$$M_6(p) = \sqrt{\frac{1}{M-1}\left[\sum_{j=K+1}^{K+M}\left(IMRP\ (p,j) - \overline{IMRP}_2(p)\right)^2\right]} \tag{9.36}$$

are the RMS strengths of $IMRP(p, i)$ and $IMRP(p, j)$, respectively, where $\overline{IMRP}_1(p)$ and $\overline{IMRP}_2(p)$ are the mean values of $IMRP(p, i)$ and $IMRP(p, j)$ under the first and second load conditions and given by $\overline{IMRP}_1(p) = \frac{1}{N}\sum_{i=1}^{N} IMRP(p,i)$ and $\overline{IMRP}_2(p) = \frac{1}{M}\sum_{j=K+1}^{K+M} IMRP(p, j)$, respectively. Similar to $M_3(p)$, $M_5(p)$ and $M_6(p)$ should be as small as possible under the optimum model order.

Furthermore,

$$M_7(p) = kurtosis\ (IMRP\ (p,i)) \tag{9.37}$$

and

$$M_8(p) = kurtosis\ (IMRP\ (p, j)) \tag{9.38}$$

are the kurtoses of $IMRP(p, i)$ and $IMRP(p, j)$, respectively. Similar to $M_4(p)$, the values of $M_7(p)$ and $M_8(p)$ should be close to 3 when the optimum model order is obtained.

It is natural to think and also confirmed by experiments that the four global statistical measures of IMRP, $M_1(p)$, $M_2(p)$, $M_3(p)$ and $M_4(p)$, are the major criteria for selecting a model order leading to load-independent model fitting, while the four local statistical measures, $M_5(p)$, $M_6(p)$, $M_7(p)$ and $M_8(p)$, are auxiliary ones. If the exact model order selection cannot be made via the global measures, the auxiliary ones should be examined. In the case of fewer inspections available under a certain load condition, the corresponding local statistical measures are less reliable and thus could be ignored. By means of these eight statistical measures together, one can conveniently evaluate the performance of IMRP under a certain model order and thus select the optimum one. It must be noted that, most of the time, the eight statistical measures cannot simultaneously achieve their target values. Therefore, a compromised decision must be made.

Third, five indexes are proposed in this study to assess the performance of IMRP and its counterparts. The first one is to evaluate the variability of a state indicator in the normal state of a gearbox. This evaluation can be simply conducted by visual inspection and thus will be denoted by using the grading system of *Least, Less, Moderate* and *Strong*. The second evaluates the load-independent property of a state indicator. Such an evaluation can also be accomplished by visual inspection and thus will be marked *Yes* or *No*. The third performance index is the data file number of the first alarm which is critical for the emanation of an early alarm of an incipient fault. It is obvious that the smaller the data file number, the better performance a state indicator demonstrates. Two more auxiliary performance indexes are proposed for quantitative comparison. The first index, which takes the ratio of value of each state indicator corresponding to the data file of the first alarm to its maximum value in the healthy state of the gearbox, is utilized to evaluate the significance level of an early alarm. Apparently, a higher alarm significance index generates a more convincing alarm for incipient gearbox faults. The second index, which is defined as the ratio of mean of each state indicator after an incipient fault to its mean before the incipient fault within the entire data file range of each test run, is used to evaluate the enhancement of fault-induced effects. Similarly, a higher fault enhancement index indicates a stronger enhancement of fault-induced effects.

## 9.3 Experimental Set-up

The mechanical diagnostics test bed (MDTB) as shown in Figure 9.2 was utilized in this study to provide data on a commercial transmission as its health progresses from new to faulted, and finally to failure. The vibration data used here were obtained from the Applied Research Laboratory at the Pennsylvania State University [9.14]. The MDTB is functionally a motor-drive train-generator test stand [9.15]. The gearbox is driven at a set input speed using a 30 Hp, 1750 rpm AC drive motor, and

**Figure 9.2.** Mechanical diagnostic test bed

the torque is applied by a 75 Hp, 1750 rpm AC absorption motor. The maximum speed and torque are 3500 rpm and 225 ft-lbs respectively. Variation of the torque is accomplished by a vector unit capable of controlling the current output of the absorption motor. The MDTB is highly efficient because the electrical power that is generated by the absorber is fed back to the driver motor. The mechanical and electrical losses are sustained by a small fraction of wall power. The MDTB has the capability of testing single and double reduction industrial gearboxes with ratios from about 1.2:1 to 6:1. The gearboxes are nominally in the 5–20 Hp range. The system is sized to provide the maximum versatility to speed and torque settings. The motors provide about 2 to 5 times the rated torque of the selected gearboxes, and thus the system can provide good overload capability. Torque limiting clutches are used on both sides of the gearbox to prevent the transmission of excessive torque as could occur with gear jam or bearing seizure. In addition, torque cells are used on both sides of the gearbox to directly monitor the efficiency and the loads transmitted [9.15].

For each test-run analyzed in this study, there were a number of unequally-spaced inspections performed during the lifetime of each gearbox. Each inspection provides a data file collected in a 10-second window at set times, which cover 200,000 sampling points in total, triggered by accelerometer RMS thresholds. The sampling frequency is 20 kHz. The signals of the MDTB accelerometers are all converted to digital data format with the highest resolution to which the accelerometers are accurate [9.15]. To ensure that the accuracy of the accelerometers is preserved, 16-bit data acquisition boards have been used for most of the measurements. When instrumentation was determined to have an accuracy below 16 bits, a 12-bit data acquisition card is used. To increase the percentage of useful

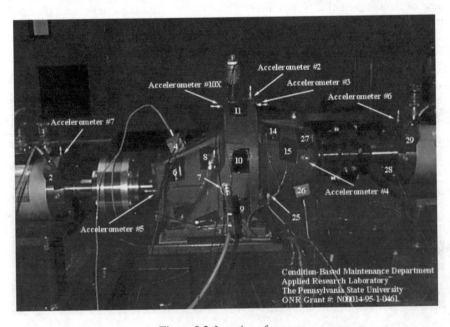

**Figure 9.3.** Location of sensors

information in the collected data and decrease the amount of storage required, a window of data is buffered in memory first. The buffered data is not stored immediately on the hard disk; it is first pre-processed to determine if it contains any new or noteworthy information about the system. The data is stored in an unconditioned raw format with a header and index. The header identifies the general information about the data such as sampling rate, type and number of accelerometers, test condition, *etc*. An index points to the beginning of data blocks that could represent different accelerometer readings or time intervals. Converting programs are also written so that the data can be imported into other desired databases. Following the test-run, the data are moved to CD-ROM and/or digital tape [9.15].

Figure 9.3 shows the location of accelerometers. It is mentioned in [9.16] that the single axis shear piezoelectric accelerometer data A03 for axial direction presents the best quality data for state diagnosis of a gearbox. Therefore, data recorded by this accelerometer are selected in this study to investigate the performance of the proposed model. As annotated in the original MDTB data CDs, the target gear is always the driven gear since its failure is the major factor causing the test bed to shutdown. The pinion gear is not investigated in this study since it is in a healthy state throughout the lifetime of each test-run.

Test-runs 6, 5 and 14 are analyzed for BFE, MRP and IMRP, respectively, in this study. The gearbox specifications of these three test-runs are shown in Table 9.1. The gear pair geometry and dimension information are shown in Table 9.2.

**Table 9.1.** Gearbox specifications for test-runs 6, 5 and 14

| Gearbox ID | DS3S0150 + 06, 05 | DS3S034014 |
|---|---|---|
| Make | Dodge APG | Dodge APG |
| Model | R86001 | R86005 |
| Rated Input Speed | 1750 rpm | 1750 rpm |
| Maximum Rated Output Torque | 528 in-lbs | 555 in-lbs |
| Maximum Rated Input Hp | 10.0 | 4.66 |
| Gear Ratio | 1.533 | 3.333 |
| Contact Ratio | 2.388 | 2.565 |
| Gear Mesh Fundamental (Hz) | 875 | 612.5 |
| Gear Mesh Fundamental (Hz) | 875 | 612.5 |

## 9.4 Performance Analysis of BFE

The full lifetime vibration data of test-run 6 is used in the validation of BFE. The gearbox in this test was driven until there were ten broken teeth and two root cracks on the driven gear. There were a total of 31 inspections carried out during the lifetime of this test-run. The output torque level was 540 in-lbs before data file 13 and switched to 1620 in-lbs from data file 13. Signals collected by accelerometers A02 and A03 are used to construct a bivariate time series. It was found that the

**Table 9.2.** Gear pair geometry and dimension information of test-runs 6, 5 and 14

| Gear | Test-runs #6 and #5 | | Test-run #14 | |
|---|---|---|---|---|
| | **Driven Gear** | **Pinion Gear** | **Driven Gear** | **Pinion Gear** |
| Number of Teeth | 46 | 30 | 70 | 21 |
| Normal Diametrical Pitch | 21.000 | 21.000 | 25.400 | 25.400 |
| Standard Pitch Diameter | 2.262 | 1.475 | 2.846 | 0.854 |
| Operating Pitch Diameter | 2.264 | 1.476 | 2.876 | 0.863 |
| Base Circle Diameter | 2.038 | 1.393 | 2.564 | 0.770 |
| Outside Diameter | 2.358 | 1.572 | 2.938 | 0.961 |
| Root Diameter | 2.134 | 1.348 | 2.753 | 0.775 |
| Standard Helix Angle | 14.478 | 14.478 | 14.478 | 14.478 |
| Normal Circular Tooth Thickness | 0.0726/0.0710 | 0.0730/0.0716 | 0.0654/0.0637 | 0.0724/0.0710 |
| Whole Depth of Tooth | 0.11190 | 0.11190 | 0.09252 | 0.09252 |

white Gaussian assumption of innovations is generally valid for the bivariate TSA signals of test-run 6 and optimum filter behavior is generally obtained under order $p=30$. Thus, the AR model order of 30 is determined for this bivariate time series.

In order to facilitate the following analysis we quote a graph of the fault growth parameter (FGP) as shown in Figure 9.4(d) proposed by Miller, who investigated the same test-run 6 in his work [9.16]. The normalized plots of BFE are illustrated in Figures 9.4(a) and (b) for two variates corresponding to accelerometers A02 and A03, respectively, which show the BFE progressions evaluated over 31 data files. As can be seen in either Figure 9.4(a) or (b), the BFE shows a very slight variation between the two dashed lines where the alternating of two operating regimes takes place from data file 12 to 13. In comparison with the conventional RMS method of accelerometer A02 as shown in Figure 9.4(c), which was used in test-run 6 to track the state evolution and trigger the shutdown, the BFE method, which is not sensitive to the alternating operating regimes, demonstrates a highly robust property, whereas the RMS method presents an abrupt jump between data file 12 and 13 due to the alternating of operating regimes, which may produce a false alarm. However, it is difficult to define the time of occurrence of a faulty state from the BFE plot of accelerometer A02, although the first notable variation of BFE takes place from data file 15 to 16. This predicament is greatly ameliorated as shown in Figure 9.4(b) where the BFE plot of accelerometer A03 shows a very remarkable jump from data file 16 to 17 but presents stable performance before data file 16. This is consistent with the FGP as shown in Figure 9.4(d) where the FGP goes up gradually after data file 16 but presents stable behavior with low-level variations before data file 16. However, it must be noted that the jump given by the BFE of accelerometer A02 from data file 16 to 17, as shown in Figure 9.4(b), indicates that it gives a much more significant alarm for an incipient fault than FGP, and thus conveys a more certain warning of an incipient fault to maintenance personnel. Thus, the fault-induced effect is enhanced by BFE. Further, the FGP shows evidence of a load

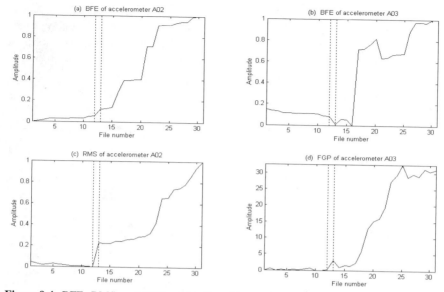

**Figure 9.4.** BFE, RMS and FGP versus data file number for test-run 6. The two vertical dashed lines denote the switch of load condition from 540 in-lbs in data files 1–12 to 1620 in-lbs for data files 13–31

switch-induced jump between data files 12 and 13 in Figure 9.4(d). In addition, one can see that the RMS method signifies the occurrence of a faulty state much later than both BFE and FGP.

Therefore, we conclude that the proposed BFE index, using accurate T-F domain information generated by the NAKF-based model, is superior to the conventional RMS method and corroborates well with the findings of Miller [9.16]. First, the RMS method is very sensitive to alternating operating conditions but the BFE index is not and thus eliminates the possibility of false alarms caused by alternating operating conditions. Second, the BFE index is able to give early warning of incipient faults in rotating machinery. This is very significant for industrial applications since it provides sufficient warning to schedule timely maintenance activities and therefore avoid catastrophic failures.

## 9.5 Performance Analysis of MRP

The full lifetime vibration data of test-run 5 was used in the validation of MRP. The gearbox in this test was driven until there was a break in two adjacent teeth of the driven gear. There was a total of 83 data files collected during the lifetime of this test. The output torque level was 540 in-lbs before data file 13 and switched to 1620 in-lbs from data file 13 on.

A comparative investigation was carried out between the proposed MRP and FGP, AR model residuals kurtosis and gear motion residuals kurtosis. For convenience, the abbreviations RMS1, RMS2, RMS3, MRK1 and MRK2 will be used to stand for the RMS strengths of AR model residuals, gear motion residuals, TSA signal, AR model residuals kurtosis and gear motion residuals kurtosis, respectively. Also, the standard deviation of the baseline AR model residuals,

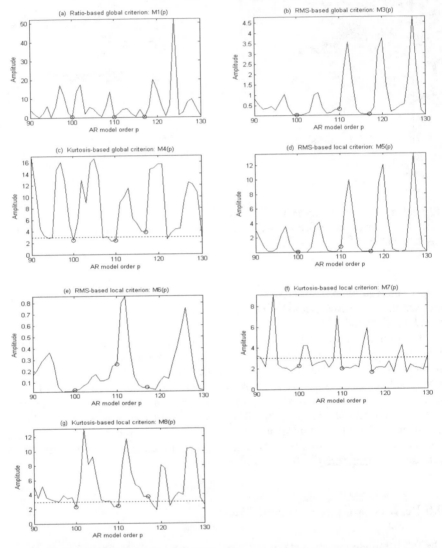

**Figure 9.5.** Statistical measures of MRP for AR model order selection for test-run 5. Orders marked by circles are 100, 110 and 117, respectively. (a) $M_1(100)$=0.2600, $M_1(110)$=0.1484, $M_1(117)$=0.1654;    (b)    $M_3(100)$=0.0304,    $M_3(110)$=0.3087,    $M_3(117)$=0.0516;    (c) $M_4(100)$=2.4505, $M_4(110)$=2.3905, $M_4(117)$=3.8183; (d) $M_5(100)$=0.0137, $M_5(110)$=0.7059, $M_5(117)$=0.0277;    (e)    $M_6(100)$=0.0333,    $M_6(110)$=0.2578,    $M_6(117)$=0.0562;    (f) $M_7(100)$=2.2560, $M_7(110)$=1.9611, $M_7(117)$=1.5913; (g) $M_8(100)$=2.2948, $M_8(110)$=2.3852, $M_8(117)$=3.6814

determined at the beginning of this test by using the gear motion residual signal of the first inspection, is a constant throughout the whole test.

For this test, 12 gear motion residual signals under the first load condition of 540 in-lbs and the first 40 gear motion residual signals under the second load condition of 1620 in-lbs are used to determine the optimum AR model order. The AR model order selection criteria $M_1(p)$ and $M_3(p)$ to $M_8(p)$ (not $M_2(p)$ proposed for IMRP in Section 9.2.7) will be used in this test.

An initial inspection of $M_1(p)$, $M_3(p)$ and $M_4(p)$ as shown in Figures 9.5(a), (b) and (c), respectively, indicates that the optimum AR model order could be selected from 100, 110 and 117. In detail, Figure 9.5(a) shows that the order of 110 gives the smallest value of $M_1(p)$. Figure 9.5(b) shows that the order of 100 is the best when using $M_3(p)$, and in Figure 9.5(c), both $M_4(100)$ and $M_4(110)$ are less than 3, but $M_4(117)$ gives a kurtosis value of 3.818 which is the largest among the kurtosis values of these three orders. Thus, the order of 117 is excluded. Further examination of $M_7(p)$ and $M_8(p)$ in Figures 9.5(f) and (g) shows no notable difference between order 100 and 110, whereas $M_5(100)$ and $M_6(100)$ have smaller RMS values in comparison with $M_5(110)$ and $M_6(110)$ as shown in Figures 9.5(d) and (e), respectively. Therefore, an AR(100) model is selected for this test.

In Figures 9.6 and 9.7, the two vertical dashed lines in each subplot denote the switch of load condition from data file 12 to 13. The load condition versus data file number is shown in Figure 9.6(a). As shown in Figures 9.6(b) and (c), both RMS1 and RMS2 indicate that the time instant corresponding to the presence of an incipient fault is around file 74. Unfortunately, RMS3, as shown in Figure 9.6(d), not only fails to reveal the time of occurrence of the incipient fault, but also shows a significant load-dependent variation around file 12. Moreover, load dependence is

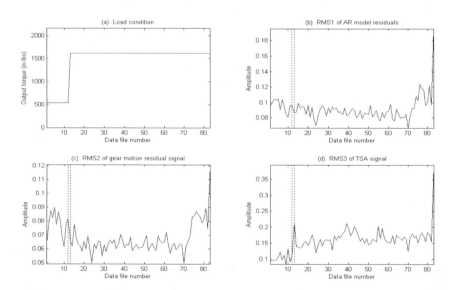

**Figure 9.6.** Load condition, RMS1, RMS2 and RMS3 versus data file number for test-run 5. The two vertical dashed lines in subplots (b), (c), and (d) denote the switch of load condition between data file 12 and 13

also observed in RMS2, as clearly shown around file 12 in Figure 9.6(c). Therefore, RMS1 shows the best performance when compared with RMS2 and RMS3.

Figure 9.7 shows the comparison between the proposed MRP and other state parameters. As shown in Figure 9.7(a), the variation of MRP is at very low level between data files 1 and 71. It is independent of varying load conditions as indicated by the arrow line shown in Figure 9.7(a), and there is no varying-load-induced jump when the switch of load condition takes place between the two vertical dashed lines. After the incipient fault initiates around data file 73, the MRP jumps to a higher level. We remark here that a justifiable evaluation of the performance of FGP should not take into account condition 1 (540 in-lbs output torque level) since the FGP could be highly load-dependent as shown in Figure 9.7(b). Therefore, correct diagnosis by the FGP should assess the state of the driven gear by checking the deviation from the baseline pattern of FGP when the driven gear is in its healthy state under the second load condition.

Further, in view of the load independence of the MRP, we evaluate its performance under the healthy state of the driven gear by considering all data files before the occurrence of the incipient fault. Thus, further investigation denotes that mean values of the MRP within data file ranges [72, 82] and [1, 71] are 1.799% and 0.1608% respectively, which result in a ratio of 11.19. On the other hand, the FGP, as shown in Figure 9.7(b), demonstrates very serious load-dependent behavior and a dramatic jump is clearly present as indicated by the arrow at the load switch between data file 12 and 13. Even though the FGP goes up quickly to a level higher than the MRP during the intermediate and advanced faulty states within data file

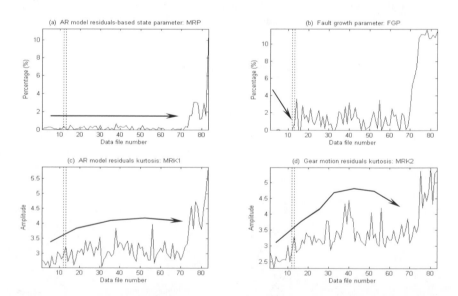

**Figure 9.7.** MRP, FGP, MRK1 and MRK2 versus data file for test-run 5. The two vertical dashed lines in each subplot denote the switch of load condition between data file 12 and 13. (a) The arrow line denotes the general trend of MRP; (b) the arrow mark denotes the load switch-induced jump of FGP; (c) the arrow curve denotes the general trend of MRK1; (d) the arrow curve denotes the general trend of MRK2

range [72, 82] by comparing Figures 9.7(a) and 9.7(b), the mean values of FGP within data file ranges [72, 82] and [13, 71], $i. e.$ 10.26% and 1.079% respectively, result in a ratio of 9.510, which is less than the 11.19 of the MRP. Thus, this phenomenon implies that the fault-induced feature embedded in the AR model residuals is further amplified in comparison with the decomposed gear motion signal proposed by Wang and McFadden [9.17]. Additionally, considerable variability is present in the traces of MRK1 and MRK2 as shown in Figures 9.7(c) and 9.7(d), respectively. Moreover, at some data files, the kurtosis-based MRK1 and MRK2 corresponding to the healthy state of the driven gear are equal to or higher than the values corresponding to the intermediate faulty state of the driven gear due to their strong sensitivity to outliers. This is inconsistent with the actual state evolution of the driven gear. Further comparison reveals that MRK1 shows less variability than MRK2 within data file range [1, 71]. The reason is that AR model residuals are much less load dependent than gear motion residual signals. Therefore, the proposed MRP possesses remarkable advantages over the other three state parameters in this case.

## 9.6 Performance Analysis of IMRP

The full lifetime vibration data of test-run 14 was used in the validation of IMRP. The gearbox used in test-run 14 contains a 70-tooth driven gear and a 21-tooth pinion gear as shown in Table 9.2. This test-run was shut down due to seven broken teeth of the driven gear at the end of this test. There was a total of 323 data files collected in this test-run. The output torque level was 550 in-lbs before data file 173 and switched to a sinusoid condition from file 173 on. The highest load reached 300% nominal torque level, while the lowest went down to 50% nominal torque level.

Data files within file ranges [1, 60] under the constant and [173, 242] under the sinusoid load conditions are used to determine the optimum AR model order. Figure 9.8 shows the plots of $M_1(p)$ to $M_8(p)$. It is obvious that order 85 is the only order which gives zero $M_2(p)$, $M_3(p)$ and $M_6(p)$, as shown in Figures 9.8(b), 9.8(c) and 9.8(f), respectively. Unlike the previous case, test-run 5, which contains a lot of orders with zero $M_2(p)$, $M_3(p)$, $M_5(p)$ and $M_6(p)$, it is believed that order 85 in this test is the optimum AR model order since it is the unique order that gives the lowest $M_2(p)$, $M_3(p)$ and $M_6(p)$ based on Figures 9.8(b), 9.8(c) and 9.8(f), respectively. Therefore, order 85 is determined for this test.

Similar to Section 9.5, a comparative investigation will be presented among the RMS1, RMS2 and RMS3, which is followed by a comparative investigation among the proposed IMRP and three other gear state indicators, $e. g.$ an improved FGP (FGP1) proposed by Lin $et al.$ [9.18], FGP and MRK1.

Figure 9.9 shows the constant to sinusoid load condition and plots of RMS1, RMS2 and RMS3. It is obvious that both RMS1 and RMS2 show load-dependent behavior within the data file range of sinusoid load condition as well as the same data file number 280 for incipient fault as shown in Figures 9.9(b) and 9.9(c), respectively. RMS3 in Figure 9.9(d) provides no helpful information and thus is not addressed here.

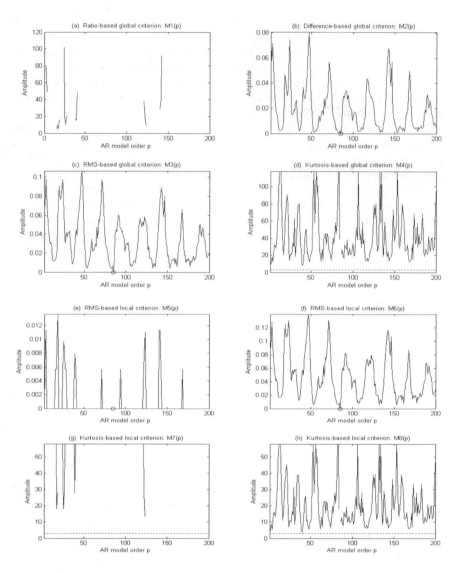

**Figure 9.8.** Statistical measures of IMRP for AR model order selection for test-run 14. Order marked by circle in each subplot is 85. (a) $M_1(85)$=NaN; (b) $M_2(85)$=0; (c) $M_3(85)$=0; (d) $M_4(85)$=NaN; (e) $M_5(85)$=0; (f) $M_6(85)$=0; (g) $M_7(85)$=NaN; (h) $M_8(85)$=NaN

Figure 9.10 shows plots of IMRP, FGP1, FGP and MRK1. It is obvious that IMRP shows the least variability in the healthy period of the target gear within data file range [1, 267] as shown in Figure 9.10(a). FGP1 in Figure 9.10(b) shows less variability compared to FGP in Figure 9.10(c). In addition, MRK1 in Figure 9.10(d) shows less variability than FGP but more than FGP1. Apparently, MRP is independent of load condition as shown in Figure 9.10(a), but FGP1 and FGP are

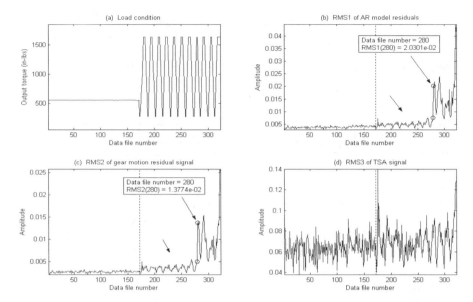

**Figure 9.9.** Load condition, RMS1, RMS2 and RMS3 versus data file number for test-run 14. The vertical dashed line in each subplot of (b), (c) and (d) denotes the position of the first data file, 173, collected under the sinusoidal load condition

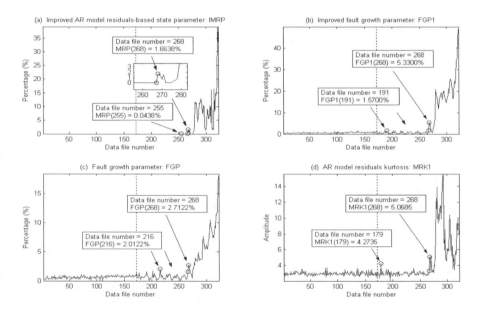

**Figure 9.10.** IMRP, FGP1, FGP and MRK1 versus data file number for test-run 14. The vertical dashed line in each subplot of (b), (c) and (d) denotes the position of the first data file, 173, collected the under sinusoidal load condition

not. All state indicators give the first alarm for incipient fault at data file 268. Furthermore, both the alarm significance index and fault enhancement index of IMRP significantly exceed those of FGP1, FGP and MRK1 as reported in Table 9.3. Therefore, IMRP significantly outperforms FGP1, FGP and MRK1 in this test.

**Table 9.3.** Performance statistics of state indicators in test-run 14

| State Indicator | IMRP | FGP1 | FGP | MRK |
| --- | --- | --- | --- | --- |
| Variability in normal state | Least | Less | Strong | Moderate |
| Load independence | Yes | No | No | Yes |
| Data file number of first alarm | 268 | 268 | 268 | 268 |
| Alarm significance index | 38 | 3.3949 | 1.3479 | 1.1860 |
| Fault enhancement index | 4.0626e+03 | 3.1140e+01 | 9.1205 | 2.5561 |

## 9.7 Conclusions

In this chapter, three robust state indicators of gearboxes operating under varying load conditions have been proposed based on an NAKF-based state-space representation of a time-varying AR model. The three state indicators have been validated using full lifetime vibration monitoring data of gearboxes.

The first gear state indicator BFE was developed by associating the NAKF-based model with a robust bispectral deterioration detection statistic. The experimental validation shows that the BFE index is independent of load condition and yields a significant alarm for an incipient gear fault in comparison with other state indicators proposed in relevant studies. However, the stability of BFE during the healthy period of the gearbox is not as satisfactory as those of MRP and IMRP.

The second gear state indicator MRP was developed by considering a highly effective combination of gear motion residual signals, compromised AR model fitting with a specific AR model order selection method for selecting a robust order value with respect to variable load condition, and a statistical measure taking the percentage of outliers exceeding the baseline three standard deviation limits. Experimental validation shows that the MRP is superior to kurtosis-based state indicators and other recently proposed gear state indicators in the literature in both stability during the healthy period of the gearbox and the significant presence of gear fault-induced effects.

The third gear state indicator IMRP improves MRP by considering three more crucial enhancements. First, instead of using the first standard deviation as in MRP, the largest standard deviation of AR model residuals, which correspond to a certain gear motion residual signal among all the gear motion residual signals selected for compromised AR model fitting, is used in the three standard deviation limits in IMRP. Second, a more practical AR model order selection method is proposed in IMRP, considering eight AR model order selection criteria. Third, a systematic performance assessment method has been proposed to evaluate five critical properties of gear state indicators. Experimental validation shows that IMRP significantly outperforms other newly proposed gear state indicators.

Most importantly, the proposed three gear state indicators can be directly employed as condition covariates by an online maintenance program for optimal maintenance decision-making.

## Acknowledgment

We are most grateful to the Applied Research Laboratory at Penn State University and the Department of the Navy, Office of the Chief of Naval Research (ONR) for providing the data used to develop this work. We thank Bob Luby at PricewaterhouseCoopers and Professor Andrew Jardine, Murray Wiseman, Dr Daming Lin and Dr Dragan Banjevic in the CBM Lab at the University of Toronto for their support. The authors also wish to thank the Natural Sciences and Engineering Research Council of Canada (NSERC), Material and Manufacturing Ontario of Canada (MMO) and the CBM Consortium companies for their financial supports.

## References

[9.1]   Samimy, B., Rizzoni, G., 1996, "Mechanical Signature Analysis Using Time-Frequency Signal Processing: Application to Internal Combustion Engine Knock Detection," In *Proceedings of the IEEE 84*, pp. 1330–1343.

[9.2]   Conforto, S., D'Alessio, T., 1999, "Spectral Analysis for Non-stationary Signals From Mechanical Measurements: A Parametric Approach," *Mechanical Systems and Signal Processing*, **13**(3), pp. 395–411.

[9.3]   Baillie, D.C., Mathew, J., 1996, "A Comparison of Autoregressive Modeling Techniques for Fault Diagnosis of Rolling Element Bearings," *Mechanical Systems and Signal Processing*, **10**(1), pp. 1–17.

[9.4]   Dron, J.P., Rasolofondraibe, L., Bolaers, F. and Pavan, A., 2001, "High-resolution Methods in Vibratory Analysis: Application to Ball Bearing Monitoring and Production Machine," *International Journal of Solids and Structures*, **38**(24–25), pp. 4293–4313.

[9.5]   Wang, W.Y., Wong, A.K., 2002, "Autoregressive Model-Based Gear Fault Diagnosis," *Journal of Vibration and Acoustics*, **124**(2), pp. 172–179.

[9.6]   Naidu, P.S., 1996, *Modern Spectrum Analysis of Time Series*, CRC Press.

[9.7]   Lin, J., Qu, L., 2000, "Feature Extraction Based on Morlet Wavelet and Its Application for Mechanical Fault Diagnosis," *Journal of Sound and Vibration*, **234**(1), pp. 135–148.

[9.8]   Roan, M.M., Erling, J.G. and Sibul, L.H., 2002, "A New, Non-linear, Adaptive, Blind Source Separation Approach to Gear Tooth Failure Detection and Analysis," *Mechanical Systems and Signal Processing*, **16**(5), pp. 719–740.

[9.9]   Moghaddamjoo, A., Kirlin, R.L., 1989, "Robust Adaptive Kalman Filtering with Unknown Inputs," *IEEE Transactions on Acoustics, Speech, and Signal Processing*, **37**(8), pp. 1166–1175.

[9.10]  Parker JR, B.E., Ware, H.A., Wipe, D.P., Tompkins, W.R., Clark, B.R., Larson, E.C. and Poor, H.V., 2000, "Fault Diagnostics Using Statistical Change Detection in the Bispectral Domain," *Mechanical Systems and Signal Processing*, **14**(4), pp. 561–750.

[9.11]  Zhan, Y. M., Jardine, A. K. S., 2004, "An On-line Diagnostic System for a Gearbox Subject to Vibration Monitoring Based on Adaptive Modeling," In *Proceedings of the 4th International Conference on Intelligent Maintenance Systems*, paper no. 029.

[9.12]  Zhan, Y. M., Makis, V., 2006, "A Robust Diagnostic Model for Gearboxes Subject to Vibration Monitoring," *Journal of Sound and Vibration*, **290**(3–5), pp. 928–955.

[9.13]  Zhan, Y. M., Makis, V. and Jardine, A. K. S., 2006, "Adaptive State Detection of Gearboxes under Varying Load Conditions Based on Parametric Modeling," *Mechanical Systems and Signal Processing*, **20**(1), pp. 188–211.

[9.14]  Condition-Based Maintenance Department, Applied Research Laboratory, 1998, *MDTB data (Data CDs: test-runs #6, #5 and #14)*, The Pennsylvania State University, 1998.

[9.15]  Byington, C. S., Kozlowski, J. D., 2000, "Transitional Data for Estimation of Gearbox Remaining Useful Life," *Mechanical Diagnostic Test Bed Data*, Condition-Based Maintenance Department, Applied Research Laboratory, The Pennsylvania State University.

[9.16]  Miller, A. J., 1999, "A New Wavelet Basis for the Decomposition of Gear Motion Error Signals and Its Application to Gearbox Diagnostics," *Master of Science Thesis*, The Graduate School, The Pennsylvania State University.

[9.17]  Wang, W. J., McFadden, P. D., 1995, "Decomposition of Gear Motion Signals and Its Application to Gearbox Diagnostics," *Journal of Vibration and Acoustics*, **117**(3A), pp. 363–369.

[9.18]  Lin, D., Wiseman, M., Banjevic, D. and Jardine, A. K. S., 2004, "An Optimal Condition-Based Maintenance Program for Gearboxes Subject to Tooth Failure," *Mechanical Systems and Signal Processing*, **18**(5), pp. 993–1007.

# 10

# Signal Processing in Manufacturing Monitoring

C. James Li

Rensselaer Polytechnic Institute
Troy, NY 12180, USA
Email: lic3@rpi.edu

**Abstract**
This chapter outlines three generic types of signatures encountered in monitoring of manufacturing process and equipment, *i. e.* periodic type including modulation, transient type, and dynamics changing type, and then summarizes signal processing algorithms and their suitability for various types of signatures under four categories, *i. e.* time domain, frequency domain, time–frequency distribution and model based methods. Additionally, decision-making strategies are discussed.

## 10.1 Introduction

The constant drive for lower production cost, higher quality, more flexibility, better safety and more environmentally friendly, and the ever-increasing per-dollar computing power of microprocessors are some of the forces behind process/equipment monitoring and control. Sensing, which transduces a physical variable to an electrical signal, is the first component in monitoring. Signal processing is the next, which accentuates and/or extracts relevant features from the signal. Decision-making is the last step where the features are interpreted and proper actions are formulated. This chapter will only describe signal processing and, to a lesser extent, decision-making. Sensing is a whole different topic by itself.

Roughly speaking, applications in manufacturing monitoring can be divided into machine health monitoring, tool condition monitoring (*e. g.* wear, breakage), workpiece inspection (*e. g.* geometry) and process monitoring. The four are so different that each deserves a chapter by itself. Additionally, manufacturing processes and equipment are numerous and vastly different, as one can see in the contrast between, say, metal cutting and semiconductor fabrication. New equipment and processes are created constantly. Obviously, there is no universal physical variable(s) and signal processing that would be appropriate for all applications. From space considerations, it is not practical to summarize signal processing for all applications. To keep it general, this chapter summarizes three generic types of signatures encountered in manufacturing equipment/process monitoring, and then discusses signal processing in terms of time domain, frequency domain, time–frequency distributions and model-based distributions and their suitability with respect to the various types of signatures.

Complexity in signal processing also depends on sensor subsystems. Budgetary and technology constraints limit the type and number of sensors one can allocate for a machine and a process. Mounting strategy also affects the quality of signals. In some cases, one measures a physical variable of interest, such as tool wear, directly. In other cases, one measures an indirect variable, such as vibration or acoustic emission (AE), and infers the state of the tool from it. Generally speaking, sensing cost decreases at the expense of increased signal processing as one moves from direct measurement to indirect measurement.

## 10.2 Types of Signatures

Simply speaking, signal processing is used to accentuate features associated with the state of a machine/process to facilitate identification of the state. Therefore, it would be wise to consider signature-generating mechanisms. In broad terms, there are three types of signature-generating mechanism, which can occur individually and simultaneously.

The first type produces periodic signals through periodic phenomena such as periodic forces resulting from inbalance and misalignments or periodic tool–workpiece contacts. Frequency domain techniques such as spectrum analysis are the most appropriate because the signals are usually characterized by a set of pure tones whose frequencies can be predicted from a reasonable knowledge of the process/equipment. A somewhat more complicated phenomenon, i. e. modulations, occurs when the magnitude and/or phase of a periodic signal, i. e. the carrier, is modulated by another phenomenon (which is frequently periodic too). For example, periodic tool–workpiece contact could be modulated by spindle run out. Modulations are characterized by sidebands in the spectrum.

The second type produces a transient signal resulting from an abrupt event such as an impact. The abrupt event can be one-time only due to tool breakage or periodic due to a localized bearing defect. Spectrum analysis would not be effective because the energy is spread over a wide range of frequencies and therefore difficult to detect. On the other hand, advanced signal processing such as wavelet transform and time–frequency distributions, would be more effective.

The third type is a response to changes in the input–output relationship of a system or a process, such as machine deterioration. Such changes are usually difficult to detect or quantify by observing the signals independent from one another. One usually has to identify the underlying dynamics driving all of the machine/process variables from a set of signals to determine if a change in machine/process dynamics has occurred.

Additionally, the signal picked up by a sensor is distorted by the transmitting path between the source and the sensor. The path may attenuate important features significantly and its nonlinearities usually make the signal more complicated than the original (e. g. clipping of a single tone signal produces a family of harmonics). Multiple paths is another issue.

## 10.3 Signal Processing

Signal processing methods can be roughly divided into time domain, frequency domain, time–frequency domain, and model-based.

### 10.3.1 Time Domain Methods

When a signal is processed in the time domain without being transformed into another domain, such as the frequency domain, the processing technique is classified as a time domain method. The simplest among them includes magnitude and energy. Peak to valley, average, area under the curve and slope are standard forms.

#### *Statistical Parameters*

Statistical moments are frequently used. A number of examples of statistical parameters that have been used as diagnostic parameters are given below. For a discrete signal, $x(n)$, let $N$ and $\bar{x}$ denote the length and the mean, respectively. Then

$$Peak\text{-}to\text{-}valley = \max(x(n)) - \min(x(n)) \tag{10.1}$$

$$RMS = \sqrt{\frac{1}{N}\sum_{i=1}^{N}(x(i)-\bar{x})^2} \tag{10.2}$$

$$Crest\ Factor = \frac{Peak\text{-}to\text{-}valley}{RMS} \tag{10.3}$$

$$Kurtosis\ (normalized\ 4th\ moment) = \frac{\frac{1}{N}\sum_{i=1}^{N}(x(i)-\bar{x})^4}{RMS^4} \tag{10.4}$$

Probability density function

$$p(x \le x(n) \le x+\Delta x) = \frac{No.\ of\ x(n)\ between\ x\ and\ (x+\Delta x)}{N} \tag{10.5}$$

With the exception of autocorrelation, a common drawback of time domain techniques is that they don't provide information about periodicities. RMS is signal energy, whose changes might signify faults or can be related to some process characteristics although, by itself, it provides no further diagnostic information. For example, most mechanical faults increase the RMS of machine vibration eventually, and the RMS of acoustic emission has been used for monitoring process stages in chemical mechanical processes (CMP) and the machining of brittle materials [10.1]. However, RMS could be affected by many things and therefore is not specific. It is also not sensitive to a few short-lived, not very large impulses, which could be better signs of trouble in some cases. Peak-to-valley is sensitive to impulses such as tool breakage, and crest factor is its improvement in terms of reduced sensitivity to changes in operating conditions such as loading and speed. However, a single noise spike can throw both off. The quadruple power of kurtosis emphasizes larger amplitudes and is therefore more sensitive to impulses than RMS. The shape of the probability density function can be used to characterize changes in waveform.

In general, the success of these parameters depends on finding a frequency band that is dominated by the signature. Otherwise, they could be signifying something else. Generally speaking, there is no theoretical guideline in choosing frequency band and the band may not even stay constant over time. Some trials are usually necessary.

### Event Counting

This kind of methods look at the number of signal peaks above a certain threshold during a fixed length of time. For example, it has been used for AE tool wear monitoring as reviewed in [10.2] and detection of localized defects in rolling element bearings, i. e. the so called "shock pulse counting method".

### Energy Operator

In communication theory, the energy operator is used to decode the information on an amplitude modulated or frequency modulated signal [10.3]. In a similar way, the energy operator can be used to extract the amplitude or frequency modulation from a signal. In the discrete form, the energy operator is defined as

$$\psi(x(n)) = x^2(n) - x(n+1)x(n-1) \tag{10.6}$$

If the signal is a single harmonic,

$$x(n) = a(n)\cos(\phi(n)) \tag{10.7}$$

then the energy operator can be shown to be related to the amplitude and frequency as follows:

$$\Psi(x(n)) = a^2(n)\dot{\phi}^2(n) \tag{10.8}$$

Obviously, modulations in amplitude and/or frequency will be revealed in the energy operator. Note that this algorithm requires narrowband filtering of the signal so that the signal can be approximated by a single harmonic.

Figure 10.1(a) shows a gear signal average which was obtained from 200 rotations of a 20-tooth input pinion in a three-stage reduction gearbox. There are two artificially seeded defects of different sizes at teeth 11 and 18 of the pinion, respectively, simulating fractured teeth at different stage. The average is plotted as a function of angular position (tooth number). It is obvious that the advanced chipping on tooth 11 can be immediately detected from the signal average even by the naked eye. The larger amplitude of vibration around that tooth indicates reduced meshing stiffness due to the defect. However, the smaller defect on tooth 18 has little noticeable effect on the signal average, presenting a tougher challenge for detection algorithms. Figure 10.1(b) shows the energy operator which highlights both defects.

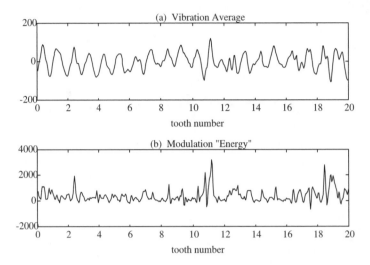

**Figure 10.1.** Gear vibration average

## *Short-time Signal Processing*

At times, one deals with non-statinoary signals whose properties such as energy and mean, change slowly with time. This leads to a variety of "short-time" processing methods in which short segments of the signal are isolated and processed as if they were short segments from a sustained signal with fixed properties. This is repeated periodically as often as desired. Often these short segments overlap one another. The result of the processing on each frame may be either a single number, or a set of numbers. Therefore, such processing produces a new time-dependent sequence which can serve as a representation of the signal.

$$O(n) = \sum_{m=-\infty}^{\infty} f(x(n)w(n-m)) \qquad (10.9)$$

where $f(\ )$ represents an operation applied to the signal segment and $w$ is a window function. For example, a rectangular window is $w(n)=1/N$ if $0 \leq n \leq N\text{-}1$, and 0 otherwise ($N$ is the window width). If $f(\bullet)=\bullet$, Equation (10.9) represents a moving average calculated from a sliding window as it moves along the time. It actually perform lowpass filtering with a cutoff frequency depending on the width of the window. If $f(\bullet)=|\bullet|$, Equation (10.9) calculates a short-time magnitude. If $f(\bullet)=|\bullet|^2$, Equation (10.9) calculates a short-time energy. Figure 10.2 shows a bearing vibration (sampled at 50 kHz) characterized by periodical ringing excited by roller–defect impacts. Using a window width of one quarter of the interval between neighboring ringings, short-time energy is calculated as shown. The figure shows that short time energy can be considered as a general envelope estimator. If $f(\bullet)=(1/2N)(sgn(x(n)-sgn(x(n-1))))$, Equation (10.9) calculates a short-time zero crossing which can be used to reveal fluctuations in frequecy composition of a signal. For practical application examples, please refer to [10.4].

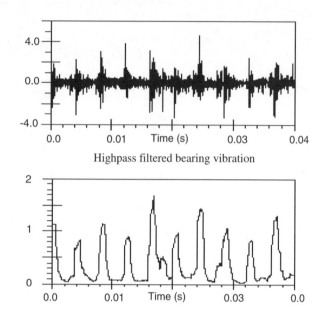

Highpass filtered bearing vibration

**Figure 10.2.** Short-time energy

### Synchronized Averaging

This algorithm is frequently used as a pre-processor to enhance the signal to noise ratio of a periodic signal. The technique consists of ensemble averaging consecutive segments of the signal, each one fundamental cycle long, *e. g.* one rotation or one part cycle (Figure 10.3). It is assumed that a sensor, *e. g.* an encoder, is used to clock the sampling so that samples are taken at the same position cycle after cycle. Each cycle will generate a fixed number of samples and we denote them as $x_i(n)$, where $i$ is the cycle number and $n$ is the sample number. The point by point average over $M$ cycles can then be calculated as

$$x_{ave}(n) = \frac{1}{M}\sum_{i=1}^{M} x_i(n)$$

(10.10)

$M$ should be large enough that a synchronized average would remain almost the same even if a larger $M$ is used. Synchronized averaging attenuates components whose periods are not compatible with the cycle period. If the clocking sensor is not available or practical, a once-per-revolution tachometer signal is usually acquired in order to carry out some kind of order-tracking algorithm.

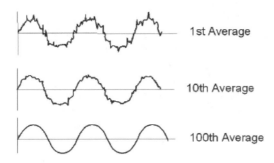

1st Average

10th Average

100th Average

**Figure 10.3.** Synchronous averaging

## 10.3.2 Frequency Domain Methods

The most basic and well-known method in this group is spectral analysis based on the discrete Fourier transform (DFT). It has become very popular since the birth of the fast Fourier transform (FFT) algorithm, which enabled hardwired analyzers and efficient computer coding. The power spectrum reveals how energy is distributed over frequencies and therefore is very useful in identifying periodic phenomena and gauging their strength. Because a large number of forcing functions in rotating machines are proportional to a fundamental frequency such as the rotating frequency, spectra sometimes are plotted against multiples (integer or fractional) of rotational speed, which are called orders. Such a spectrum is often called an order spectrum and analysis performed on the amplitude and phase of the rpm harmonics are called order analysis.

Filtering is accomplished by applying different weights to different frequencies. Common ones include highpass, lowpass, bandpass/stop filters. Issues involved in spectrum analysis (resolution, variances, windowing, aliasing, *etc.*) and filtering (linear phase, FIR, IIR, rolloff, distortion, *etc.*) are discussed in numerous books such as [10.5] and will not be elaborated further here.

While DFT uses a set of harmonics as the basis function, other basis functions have also been employed. For example, Haar functions (square waves) are the basis functions for the Haar transform, which has been used for processing a stamping tonnage signal to identify and monitor fault-sensitive parameters [10.6].

In the following, a number of frequency domain methods, mostly based on the Fourier transform, will be summarized.

### *Cepstrum Analysis*

Let $F$ be the Fourier transform operator, and $F^{-1}$ the inverse transform. Cepstrum is defined in a couple of different ways:

$$x_c(\tau) = F\left[\ell n |X(\omega)|^2\right]; \quad \text{original definition}$$

$$or = F^{-1}\left[\ell n[X(\omega)]\right] \ where X(\omega) = F(x(t)) \tag{10.11}$$

In its original definition, cepstrum is sometimes called "spectrum of the logarithmic power spectrum". The difference between the two definitions is that phase information is lost in the original definition. The cepstrum is useful in detecting spectrum periodicity such as families of harmonics or sidebands found in the spectra. It reduces a whole family of harmonics into a single cepstral line. This makes it much easier to distinguish among different families. Filtering with the cepstrum (like filtering with the spectrum) can separate different families of harmonics. The logrithmic operator also makes it sometimes possible to remove the effect of the signal transfmission path and recover a signal in the presence of echoes.

An application example of the cepstrum is gear signal processing. Gear vibrations contain rotating frequency harmonics, meshing harmonics, which are integer multiples of the rotating frequency, and sidebands due to modulation at gear rotating frequency, e. g. a crack in one of the teeth – a once per revolution modulation. Hence, the sidebands have the same spacing of the rotating frequency. Since all of the above have the spacing of rotating frequency, i. e. a spectrum periodicity equal to the rotating frequency, the cepstrum (spectrum of a spectrum) will have a peak at the modulation period, i. e. rotating period (which is usually different from one gear to the next). Because a damaged gear frequently increases the level of sidebands, cepstrum peaks can be indicative of the condition of the gear.

### *Hilbert Transform*

The Hilbert transform of a signal $x(t)$ and its inverse transform are defined as below.

$$\tilde{x}(t) = H\{x(t)\} = x(t) * \frac{1}{\pi t} = \frac{1}{\pi} \int_{-\infty}^{\infty} x(\tau) \frac{1}{t-\tau} d\tau \tag{10.12}$$

$$x(t) = H^{-1}\{\tilde{x}(t)\} = -\frac{1}{\pi} \tilde{x}(t) * \frac{1}{t} = -\frac{1}{\pi} \int_{-\infty}^{\infty} \tilde{x}(\tau) \frac{1}{t-\tau} d\tau$$

The transform can be considered as filtering with a filter whose impulse response is $1/\pi t$. The filter's frequency response has unity magnitude, and phase $-90$ degrees for positive frequency and 90 degrees for negative frequency. The amplitude and instantaneous phase are calculated as

$$|x(t)| = \sqrt{x^2(t) + \tilde{x}^2(t)} \tag{10.13}$$
$$\theta_i(t) \quad = \quad \tan^{-1}\left[\tilde{x}(t)/x(t)\right]$$

One can also form an analytical signal

$$x_a(t) = x(t) + j\tilde{x}(t) \tag{10.14}$$

whose magnitude and phase are given in Equation (10.13). When signal $x(t)$ is a harmonic function with time-varying amplitude and phase (modulations), it can be shown that Equation (10.13) calculates the time-varying amplitude and phase. This is why the Hilbert transform has been widely used for amplitude and phase demodulations for a narrowband signal. (However, this is not true when $x(t)$ is not a harmonic function, which is why the signal usually has to be bandpass filtered.)

The Hilbert transform is not the only signal processor that reveals modulations. A number of signal processors that accentuate modulations will be discussed in the next three sub-sections [10.7].

## *SB Ratio*

The SB ratio is the ratio of sideband power to carrier power. It has been applied to track gear damage as a damaged gear frequently increases sideband level.

## *Residual*

The signal is first bandpass filtered around a dominant carrier. The residual is then obtained by subtracting the carrier from the filtered signal, and therefore only contains sidebands. It was shown that the envelope of the resulting signal is directly related to both amplitude and phase modulations. In the case of gear monitoring, the largest meshing harmonic is usually chosen as the carrier about which the filter band is centered.

## *FM0, FM4, NA4, NB4*

The figures of merit are some of the best known gear condition indices (although they are definitely suitable for other applications where similar fault signature-generating mechanisms are found). These figures of merit [10.8] were designed to detect certain kinds of gear faults from gear vibrations. Except FM0, they all accentuate modulations through sideband extractions. FM0 is an indicator of major faults in a gear mesh and is defined as

$$FM0 = \frac{peak - to - valley}{\text{sum of RMS of meshing harmonics}} \tag{10.15}$$

When a tooth breaks, the peak-to-valley level tends to increase and thus FM0 increases. However, a single noise spike can throw it off.

FM4 is calculated by removing the tooth-meshing harmonics and their first-order side-bands from the vibration and taking the normalized kurtosis of the signal that remains (the so called "difference signal").

$$FM4(M) = \frac{\frac{1}{N}\sum_{i=1}^{N}(d_i - \bar{d})^4}{\sigma^4} \tag{10.16}$$

where $d$ is the difference signal, $\bar{d}$ the mean value of the difference signal, $N$ the total number of data points in the signal, $\sigma$ the standard deviation of the signal, $M$ the current time record number in the run ensemble, and $i$ the data point number in the time record. FM4 indicates the amount of localized damage, such as pitting or small cracks.

NA4 is similar to FM4, however, a residual signal is constructed by removing meshing frequency components from the vibration signal. The fourth statistical moment is divided by the current run time averaged variance of the residual signal, resulting in the quasi-normalized kurtosis given below:

$$NA4(M) = \frac{\dfrac{1}{N}\sum_{i=1}^{N}(r_i - \overline{r})^4}{\left\{\dfrac{1}{M}\sum_{j=1}^{M}\left[\dfrac{1}{N}\sum_{i=1}^{N}(r_{ij} - \overline{r})^2\right]\right\}^2} \tag{10.17}$$

where $r$ is the residual signal, $\overline{r}$ the mean value of the residual signal, and $j$ the time record number in the run ensemble.

With this quasi-normalized kurtosis method, the change in the residual signal is compared to a weighted baseline for the specific system in the "good" condition until the average of the variance itself changes.

NB4 is similar to NA4, except it uses the envelope of the signal bandpassed about the dominant meshing harmonic. The envelope is an estimation of the amplitude modulation present in the signal and is most often due to transient variations in the loading. A few damaged teeth will cause transient load fluctuations and thus the amplitude will modulate.

Some studies have confirmed that NA4 and NB4 react to the onset and growth of pitting well. However, they are not as sensitive when it comes to tooth fracture.

### Bicoherence

Bicoherence, as defined below, quantifies the phase dependency among two arbitrary frequency components and the third frequency component at the sum frequency of the two. Since the third frequency is the sum of the first two, it is usually plotted on a bi-frequecncy plane of the first two. When the first two are harmonic components of a periodic signal (and therefore the third one too because it is the sum frequency), bicoherence value depends on the strength of the periodic signal since harmonic components of a periodic signal have a fixed phase relationship with one another.

$$b^2(f_1, f_2) = \frac{\left|E\left[X(f_1)X(f_2)X^*(f_1 + f_2)\right]\right|^2}{E\left[\left|X(f_1)X(f_2)\right|^2\right]E\left[\left|X(f_1 + f_2)\right|^2\right]} \tag{10.18}$$

where $E[\bullet]$ denotes the expected value, and $X(f_i)$ denotes the complex frequency component from the Fourier transform.

Because bicoherence is based on phase, it works even when the periodic signal is less than dominant. For example, Figure 10.4 shows bearing vibrations of a good bearing and a bearing with a roller defect. Obviously, one cannot easily tell them apart because the periodic ringing excited by defect–raceway impacts is masked by noise. However, the periodic ringing creates a family of harmonics with a fundamental frequency equal to the defect–raceway impact frequency, *i. e.* the characteristic defect frequency. These harmonics are the Fourier components of a periodic signal and therefore are correlated in phase. Consequently, bicoherence should show increased magnitude at frequency pairs $(f, f)$, $(2f, f)$, $(3f, f)$ ..., $(2f, 3f)$, $(2f, 4f)$, etc., where $f$ is the characteristic defect frequency. Figure 10.5 shows bicoherence magnitudes of the two bearing vibrations at some of the aforementioned frequency pairs where $f_{rs}$ is the defect–raceway impact frequency. Obviously, one can tell the two bearings apart with bicoherence, although one cannot see the repeated ringing patterns in the damaged bearing signal [10.9].

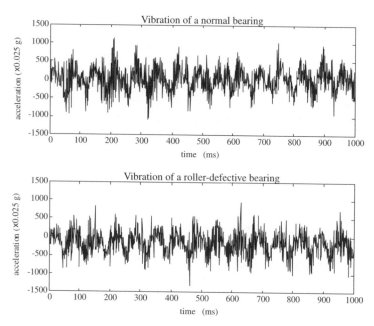

**Figure 10.4.** Bearing vibrations of a good and a roller-damaged bearing (sampling frequenncy 1,000 Hz)

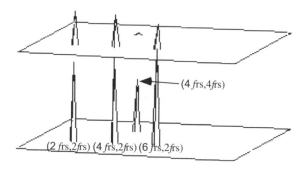

**Figure 10.5.** Bicoherence values of a good bearing and roller-damaged bearing

Generally speaking, phase information has not received much attention from the monitoring and signal processing community. Its utility should be exploited further with higher priority for possible payoffs.

## Cyclostationary

Cyclostationary refers to the phenomenon that the statistical moments of a time series are periodic. A signal $x(t)$ is said to be cyclostationary of order $n$ if its $n$th moment is a periodic function of time $t$. The fundamental frequency $\alpha$ of the periodicity is the cyclic frequency. Given a signal $x(t)$ cylostationarity of order 2 can be measured by the spectral coherence function (SCF).

$$C^\alpha(f) = \frac{E\left[X(f-\alpha/2)X^*(f+\alpha/2)\right]}{\sqrt{\left|S\ (f-\alpha/2)\right|} \times \sqrt{\left|S\ (f+\alpha/2)\right|}} \tag{10.19}$$

where $S$ is the power spectrum. According to the expression, SCF measures the normalized correlation between frequency lines centered about $f$ and separated by $\alpha$. SCF is typically plotted on the bifrequency plane of $f$ and $\alpha$.

SCF has been applied to gear signal processing. In the presence of a faulty tooth, a rough mesh occurs once per revolution and this cyclic event leads to side-bands around the gear meshing harmonics with a spacing of the gear rotating speed $f_s$ [10.10]. When SCF is calculated for a faulty gear with $\alpha = f_s$ and $f$ being one of the meshing harmonics, elevated levels are expected due to increased cyclostationarity with a cyclic frequency of $f_s$ (and its multiples). A similar situation can also happen when one of the cutting inserts is chipped in a face milling tool.

### 10.3.3 Time–frequency Methods

Manufacturing equipment and processes frequently produce non-stationary signals whose distribution of energy over frequencies changes over time. Such changes may be periodic or aperiodic. Obviously, traditional spectral analyses are not adequate for such signals for they assume stationary signals. Time–frequency methods, in general, are designed to provide information about how the distribution changes over time.

## Spectrogram

By applying FFT to short segments of a signal and arranging the spectra in chronogical order, one gets a spectrogram which reveals how the energy is distributed along both the frequency and time axes. It is defined as

$$x_n(k) = DFT[x(m)w(n-m)]$$
$$= \sum_{m=n-N+1}^{n} x(m)w(n-m)e^{-j\frac{2\pi}{N}mk} \tag{10.20}$$

where $w(\ )$ is the window function like the one described for short-time signal processing. (In fact, a spectrogram is the combination of short-time signal processing and DFT and it is therefore, sometimes, called short-time DFT.)

Figure 10.6 shows the spectrogram of a damaged bearing. The vibration consists of periodic spikes signaling the periodic impacts between rollers and the outer-race defect. The spectrogram not only shows the energy fluctuation but also how the energy is distributed over frequencies.

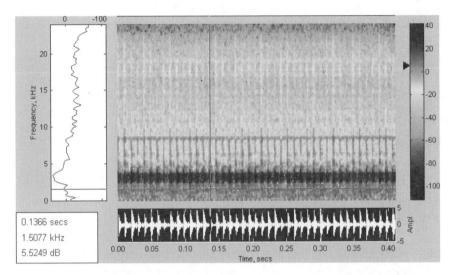

**Figure 10.6.** Spectrogram of an outer-race-damaged bearing signal

Window width or segment length is the key parameter in calculating spectrogram because it simultaneously determines the spectral resolution and time localization. The wider the window, the better the frequency resolution. On the other hand, a wider window means the spectrogram obtained is less localized and less sensitive to time-localized transient events. A compromise between frequency resolution and time localization is inevitable in choosing a window width. Usually, the compromise yields a window width that offers neither enough frequency resolution at the low frequency end of the spectrum nor adequate time localization at the high frequency end.

*Wavelet Transform (WT)*

By adjusting window width according to frequencies, the wavelet tranform mini-mizes the aforementioned shortcoming of the spectrogram. For a continuous signal $x(t)$ the transform and inverse transform are given as

$$W_x(a, b) = \int g^{*(a,b)}(t)x(t)dt \qquad (10.21)$$

$$x(t) = \frac{1}{c_g} \iint g^{(a,b)}(t)W_x(a,b)\frac{1}{a^2}da\,db \qquad (10.22)$$

where $g^{(a,b)}(t)$ is the shifted and dilated wavelet, defined by

$$g^{(a,b)}(t) = \frac{1}{\sqrt{a}} g\left(\frac{t-b}{a}\right) \qquad (a>0) \tag{10.23}$$

and the asterisk denotes the complex conjugate. The mother wavelet $g(t)$ must be oscillating and short-lasting.

Let

$$h^{(a)}(t) = -\frac{1}{\sqrt{a}} g^*\left(-\frac{t}{a}\right) \tag{10.24}$$

then, Equation (10.21) becomes a convolution

$$W_x(a,b) = \int h^{(a)}(t)\, x(b-t)\, dt \tag{10.25}$$

Clearly, for a given dilation $a$, $W_x(a,b)$, as a function of $b$, can be viewed as the output of a filter whose impulse response is $h^{(a)}(t)$ and input is $x(t)$. The dilation scales the frequency response of the filter along the frequency axis. For example, if the frequency response of the mother wavelet has a passband $[\omega_{min}, \omega_{max}]$, the baby wavelet $h^{(a)}(t)$ has a passband $[\omega_{min}/a, \omega_{max}/a]$.

On the other hand, when $b$ is fixed, if $g(t)$ is taken to be a finite-length function which is zero outside an interval around the origin, say $[-\tau, \tau]$, Equation (10.21) says that the value of $W_x(a,b)$ depends on the values of $x(t)$ in the interval $[b-a\tau, b+a\tau]$. Smaller $a$ makes the interval shorter and gives better localization in the time domain. This adaptive window width makes the wavelet transform a powerful tool for detecting and locating discontinuities or abrupt changes [10.11] and rapid transient of small amplitude.

An example mother wavelet is the exponential decaying sinusoidal, which is similar to the impulse response of a mass–spring–damper system.

$$\begin{aligned} g(t) &= \exp(-\sigma t)\ \sin \omega_0 t \quad && t \geq 0 \\ g(t) &= -g(-t) && t < 0 \end{aligned} \tag{10.26}$$

### Wigner–Ville Distribution (WVD)

As one moves along time axis, a time-indexed autocorrelation function can be computed about each instant. The Fourier transform of such an autocorrelation function is an instantaneous power spectrum for the instant. One gets a time–frequency distribution by lining up these instantaneous spectra along the time axis. Given that $t$ is time and $\omega$ is frequency, the mathematical definition of WVD is as follows:

$$W(t,\omega) = \frac{1}{2\pi} \int x\left(t+\frac{1}{2}\tau\right) x^*\left(t-\frac{1}{2}\tau\right) e^{-j\tau\omega} d\tau$$

$$= F\left[x\left(t+\frac{1}{2}\tau\right) x^*\left(t-\frac{1}{2}\tau\right)\right]$$

(10.27)

The limitation of WVD is that it produces spurious peaks due to the cross-terms between two frequency components.

### *Choi–Williams Distribution (CWD)*

Using an exponential kernel function, the Choi–Williams time–frequency distribution [10.12] has better cross-term properties than the WVD and finer frequency resolution at high frequencies than the wavelet transform, although at higher computational cost. It is defined as

$$E_f(t,\omega) =$$

$$F\left[\int_\mu \frac{1}{\sqrt{4\pi\tau^2/\sigma}} e^{-\frac{(\mu-t)^2}{4\tau^2/\sigma}} x\left(\mu+\frac{\tau}{2}\right) x^*\left(\mu-\frac{\tau}{2}\right) d\mu\right]$$

(10.28)

In order to get sharper auto-term resolution, $\sigma$ should be large. On the other hand in order to reduce the effect of cross-terms, $\sigma$ should be small. Therefore, a large $\sigma$ (> 1.0) is recommended for signals whose amplitude and frequency change quickly, and a small $\sigma$ ($\leq$ 1.0) is recommended for signals whose amplitude and frequency change relatively slowly. It is different from WVD in that each time-indexed autocorrelation function is calculated from a set of neighboring samples with weights determined by an exponential kernal function.

Li and Tzeng [10.13] used CWD to process the torsional vibration of a milling machine spindle and identified the time–frequency "pixels" that are most correlated with tool wear via statistical analysis. A wear estimator then uses pixel magnitude to estimate the amount of tool wear.

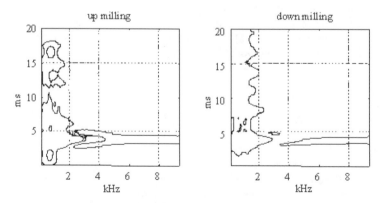

**Figure 10.7.** Choi–Williams distributions of milling measurements

### 10.3.4 Model-based Methods

#### Time Series Analysis

Simply speaking, linear difference equations such as AR, ARMA are used as parametric representations of a signal or the dynamics of a process/machine. The model parameters are found by minimizing modeling error. The model parameters or variables derived from them, such as damping ratio, have been used as features [10.14]. Model residuals and prediction errors have also been used to reveal abnormalies [10.15]. The analysis has been extended to include nonlinear models as well.

#### Wideband Demodulation

When modulations occur on a periodic signal which is more than just a harmonic function (*i. e.* A periodic signal contains a family of harmoincs), demodulations based on the Hilbert transform should only be done after bandpass filtering about one of the harmonics to guarantee that the signal is a single harmonic function. However, bandpass filtering eliminates a large amount of useful information. This issue is unique to mechanical signal processing because the carrier used in communication is mostly a harmonic function.

The wideband demodulation method employs a nonlinear programming method and a signature model to identify amplitude and phase modulations from the wideband signal. No information has to be eliminated and the interactions among sidebands of neighboring harmonics are accounted for. An example of the signature model is given below:

$$y(n) = \sum_{k=0}^{K} [1 + a_k(n)] X_k \cos[kNn + \phi_k + b_k(n)] \qquad (10.29)$$

where $X_k \cos(\ )$ are harmonics of a periodical signal, and $a_k, b_k$ are amplitude and phase modulations to be identified.

The algorithm has been shown to be more sensitive to gear tooth defects than narrowband demodulation [10.16]. However, it is computationally more expensive than the narrowband demodulation.

#### Virtual Sensor

Frequently, it is impractical or impossible to directly meaure a desired physical variable or place a sensor right at the source of the signal. For example, one may have to settle for a vibration transducer somewhere on the spindle while the cutting force at the tool–workpiece interface is really what one is interested in. If the dyanmic relationship between a sensor measurement and what one is interested in is relatively constant, it is possible to pre-calibrate a dynamic model between the two using empirical data and analytical/numerical models. A model that estimates a physical variable from measurent of other physical variables is called a virtual sensor.

To estimate milling tool wear, Li and Tzeng [10.13] developed a virtual sensor using nonlinear dynamic modeling to identify the relationship between translation and torsional vibrations from simultaneous measurements of both. Once this relationship is available, an approximation to the torsional acceleration can be constructed by knowing the translation vibration.

The model used was a second-degree nonlinear output-error model. An orthogonal method was employed to identify the significant terms and their appropriate coefficients. Additionally, a Kalman filter is another viable alternative for virtual sensing [10.17].

### Embedded Models

Some mechanical faults or process stages introduce nonlinearities. For example, a transverse crack in a structure modulates local stiffness as the crack breathes during vibration. Input–output models, as a black-box representation, are not capable of isolating/identifying such nonlinearities. Embedded modeling embeds a parametric representation of the unknown nonlinearity into a first principles model, and employs appropriate solution methods to identify the unknown by matching the model response to the measured response.

For example, a model for a one-degree-of-freedom mass–spring system is

$$m\ddot{x} + kx = f(t) \tag{10.30}$$

where $m$ is mass, $k$ stiffness and $f(t)$ forcing function.

If the spring has a crack whose breathing during vibration modulates its stiffness, one could embed a periodic stiffness function, e.g. $k = a^*\cos(x)$ to identify the stiffness and then estimate the size of the crack. The method has found many applications, such as detection and size estimation of a transverse crack on a vibration beam [10.18], gear tooth [10.19] and flexible rotor [10.20].

## 10.4 Decision-making Strategy

### 10.4.1 Simple Thresholds

Some features have been proposed as single-shot diagnostic variables that have an absolute threshold and therefore do not need historical data for trending. For example, kurtosis is said to indicate a damaged bearing for any value significantly higher than 3, say 4.5. Realistically, such variables are just too simple to offer single-shot diagnosis under most circumstances.

Typically, a threshold has to be set in a case-by-case manner. It can be set absolutely, or relatively to, for example the maximum of an exemplar signal (or the average of a number of them). For example, an absolute maximum can be set for machine life threatening events such as collision or spindle bearing seizure (e. g. Th1 in Figure 10.8), or a threshold can be set at, say, 200% of the maximum of an exemplar signal. Usually, there is more than one threshold for different purposes, such as tool fracture and a missing tool. Response time is scheduled according to the seriousness and urgency of an event. In some cases, immediate action is necessary while, in others, a slower response is more appropriate. When the exemplar is the previous cycle, the threshold is no longer fixed and is therefore sometimes called a floating threshold.

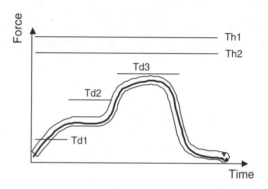

**Figure 10.8.** Simple and time-dependent thresholds and a part signature

### 10.4.2 Statistical Process Control (SPC)

When part dimensions are measured by sensors, the control chart can be used directly. When indirect sensing is involved, an indirect method has to be used. Gong *et al.* [10.21] proposed a two-step method integrating an online sensor and a control chart for such an application. A manufacturing process is monitored by a sensor and a sample of parts is inspected whenever a warning is issued by the sensor. If the sample mean is outside the predetermined control limits, the process is suspended to conduct a diagnosis to determine the cause for possible corrective actions.

### 10.4.3 Time/Position-dependent Thresholds

Due to the highly nonstationary nature of most manufacturing process cycles, a constant threshold is frequently too limiting. Finite duration thresholds can be placed in or across different time or position segments of a cycle to enable more refined monitoring (*e. g.* Td1–3 in Figure 10.8) [10.22].

### 10.4.4 Part Signature

A part signature can be produced from a single observation but more frequently from repeated observations of a cycle by averaging. The signature can be represented by step functions or more flexible parametric or nonparametric curves through curve fitting or nonparametric profiling such as principal curves [10.23]. Envelopes can then be placed over and under the signature to monitor process stage or faults (Figure 10.8). Timing variations between the part signature and the current cycle must be considered to minimize missed detection or false alarms. More sophisticated classification on the local or global deviations from the part signature has also been developed.

### 10.4.5 Waveform Recognition

In some tool monitoring systems [10.24], a number of force waveform patterns of tool breakage and chipping are stored in the system. For example, when a carbide tool breaks the cutting force suddenly rises and then drops to zero while the force mostly just drops to zero when breakage occurs in ceramic tools. The system continuously matches the incoming signal to the stored waveform patterns. Waveform recognition could be devised not to be sensitive to variables such as cutting conditions and material properties.

### 10.4.6 Pattern Recognition

With the features extracted by signal processing, one can take advantage of a number of pattern classification methods such as linear discriminant function, fuzzy logic, neural net, fuzzy neural net, decision tree, support vector machine, *etc.*, to monitor the process stage or machine/process faults. Typically, exemplars are made available through empirical means, and features providing the most discriminant power are identified through statistical analyses. A classifier is then "trained" to optimally classify the exempliers in terms of some objective functions. In mulit-sensor configurations, sensor fusion is frequently done in a similar way.

Due to the proliferation of feature extraction and classification tools, this is an area where much research has been done, sometimes at the cost of more fundamental research from which more insights could have been gained. No further elaboration will be given here because the subject is secondary to signal processing and the literature is just too vast. However, the community will benefit greatly if further research is directed at the development of theoretical guidelines and systematic utilization instead of just trying different combinations of the tools made available by the AI community.

### 10.4.7 Severity Estimator

Once a fault is detected, it is usually desirable to determine its severity. For example, it is desirable to track tool wear or a gear tooth crack to know when one has to change the tool or perform a maintenance action. A severity estimator is a mathematical representation of the relationship between some features and the actual severity. It would be best if an analytical model could be established, for that will minimize the need for empirical data over an actual failure trajectory. Reference [10.19] established a model between tooth meshing stiffness and tooth crack size to estimate the size of a tooth crack. When an analytical model is unavailable or infeasible, a black box model calibrated by empirical run-to-fail data can be used instead. For example, reference [10.13] used a neural network calibrated by empirical data to estimate tool wear from the time–frequency distribution of a virtual sensor output. Furthermore, discrete-time time-invariant Markov models have been used to describe a machine's deterioration trajectory [10.25].

## 10.5 Conclusions

It is difficult to summarize in a short chapter the numerous and diverse signal processing algorithms that have been developed for manufacturing equipment/ process monitoring. Begining with three generic types of signature encountered in manufacturing processes and equipment, this chapter summarizes signal processing algorithms under four categories, *i. e.* time domain, frequency domain, time–frequency distribution and model-based methods. Then, decision-making strategies are briefly discussed. For more information, see [10.1] for AE in machining monitoring, [10.24] for commericial tool condition monitoring systems, [10.7] for gear condition monitoring, and [10.26] for bearing condition monitoring.

Signal processing is a field that has evolved rapidly in recent years and the evolution will accelerate as the driving forces behind it are getting stronger every day. Despite much research into advanced signal processing such as time–frequency distribution and model-based methods, current commercial monitoring systems still use only the simplest forms, such as RMS and filtering. More development work is needed to "ruggedize" monitoring algorithms so that they can be used reliably on the shop floor. On the other hand, the requirement for "prognosis" will promote more fundamental research into the real physics behind signals. Blindly applying signal processing algorithms and empirically setting thresholds, which may work well with detection, is not suitable for severity assessment, or for extrapolating past and current states into the future.

## References

[10.1]    Dornfeld, D. A., Lee, Y. and Chang, A., 2003, "Monitoring of Ultraprecision Machining Processes," *International Journal of Advance Manufacturing Technology*, **21**, pp. 571–578.

[10.2]    Iwata, K. and Moriwaki, T., 1977, "An Application of Acoustic Emission Measurements to In-process Sensing of Tool Wear," *Annals CIRP*, **26**(1), pp. 21–26.

[10.3]    Maragos, P., Quatieri, T. F. and Kaiser, J. F., 1993, "On Amplitude and Frequency Demodulation Using Energy Operators," *IEEE Transactions on Signal Processing*, **41**(4), pp. 1532–1550.

[10.4]    Li, C. J. and Wu, S. M., 1989, "On-line Detection of Localized Defects in Bearings by Pattern Recognition Analysis," *ASME J. of Engineering for Industry*, **111**(4), pp. 331–336.

[10.5]    Oppenheim, A. V. and Schafer, R. W. 1975, *Digital Signal Processing*, Prentice Hall, NJ.

[10.6]    Koh, C. K. H., Shi, J. and Williams, W., 1995, "Tonnage Signature Analysis Using the Orthogonal (Harr) Transforms," *NAMRI/SME Transactions*, **23**, pp. 229–234.

[10.7]    Li, C. J., 2001, "Diagnostics: Gear vibrations and diagnostics," *Encyclopedia of Vibration*, Editors: Braun, S. G., Ewins, D. J. and Rao, S. S., Academic Press.

[10.8]    Choy, F. K., Braun, M. J., Polyshchuk, V., Zakrajsek, J. J., Townsend, D. P. and Handschuh, R. F., 1994, "Analytical and Experimental Vibration Analysis of a Faulty Gear System," *NASA Technical Memorandum*, 1.15:106689.

[10.9]   Li, C. J., Ma, J., Hwang, B. and Nickerson, G. W., 1991, "Pattern Recognition Based Bicoherence Analysis of Vibrations for Bearing Condition Monitoring," *Proceedings of Symposium on Sensors, Controls and Quality Issues in Manufacturing*, ASME Winter Annual Meeting, Atlanta, GA, PED-**55**, Edited by T. I. Liu, C. H. Meng and N. H. Chao, ASME, New York, pp. 1–11.

[10.10]  Capdessus, C., Sidahmed, M. and Lacoume, J. L., 2000, "Cyclostationary Processes: Application in Gear Faults Early Diagnosis," *Mechanical Systems and Signal Processing*, **14**(3), pp. 371–385.

[10.11]  Grossmann, A., 1988, "Wavelet Transforms and Edge Detection", in *Stochastic Processes in Physics and Engineering*, S. Albeverio *et al.*, eds., D. Reidel Publishing Co., Dordrecht, Holland, pp. 149–157.

[10.12]  Choi, H. and Williams, W. J., 1989, "Improved Time-frequency Representation of Multicomponent Signals Using Exponential Kernels", *IEEE Transactions on Acoustics, Speech, and Signal Processing*, **37**(6), pp. 862–871.

[10.13]  Li, C. J. and Tzeng, T. C., 2000, "Muliti-Milling-Insert Wear Assessment Using Nonlinear Virtual Sensor, Time-Frequency Distribution and Neural Networks," *Mechanical Systems and Signal Processing*, **14**(6), pp. 945–957.

[10.14]  Szabo, B., Wu, S. M., 1991, "On-line techniques for tool condition monitoring in milling operations", *IFAC Symposia Series - Proceedings of a Triennial World Congress*, **5**, pp. 59–64.

[10.15]  Li, C. J. and Limmer, J. D., 2000, "Model-Based Condition Index For Tracking Gear Wear and Fatigue Damage," *Wear*, **241**, pp. 26–32.

[10.16]  Ma, J. and Li, C. J., 1996, "Gear Defect Detection through Model-Based Wideband Demodulation of Vibrations," *Mechanical Systems and Signal Processing*, **10**(5), pp. 653–665.

[10.17]  Park, S. S. and Altintas, Y., "Dynamic Compensation of Spindle Integrated Force Sensors with Kalman Filter", *ASME Journal of Dynamic Systems, Measurement and Control*, in press.

[10.18]  Batayneh, W. M., 2005, "Beam Crack Diagnosis using Embedded Modeling", *Ph.D. thesis*, Rensselaer Polytechnic Institute, Troy, New York.

[10.19]  Li, C. J. and Lee, H., 2005, "Gear Fatigue Crack Prognosis Using Embedded Model, Gear Dynamic Model and Fracture Mechanics," *Mechanical Systems and Signal Processing*, **19**(4), pp. 836–846.

[10.20]  McKee, K. and Li, C. J., 2003 "Rotor Transverse Crack Detection and Diagnosis Using Embedded Modeling", *Proceedings of ASME IMECE*, 3, Paper No, IMECE 43504.

[10.21]  Gong, L., Jwo, W. and Tang, K., 1997, "Using On-line Sensors in Statistical Process Control," *Management Science*, **43**(7), pp. 1017–1029.

[10.22]  1997, "Tool Monitoring for the Protection and Unsupervised Operation of Machine Tools", brochure of Nordmann Company, Köln, Germany.

[10.23]  Hastie, T. and Stuetzle, W., 1989, "Principal Curves," *Journal of the American Statistical Association*, **84**, pp. 502–516.

[10.24]  K. Jemielniak, 1999, "Commercial Tool Condition Monitoring Systems," *International Journal of Advanced Manufacturing Technology*, **15**, pp. 711–721.

[10.25]  White, C., 1977, "A Markov Quality Control Process Subject to Partial Observation," *Management Science*, **23**, pp. 843–852.

[10.26]  Li, C. J. and McKee, K., 2001, "Diagnostics: Bearing (Roller & Journal) Vibrations and Diagnostics," *Encyclopedia of Vibration*, Editors: Braun, S. G., Ewins, D. J. and Rao, S. S., Academic Press.

# 11

# Autonomous Active-sensor Networks for High-accuracy Monitoring in Manufacturing

Ardevan Bakhtari and Beno Benhabib

Computer Integrated Manufacturing Laboratory
Department of Mechanical and Industrial Engineering, University of Toronto
5 King's College Road, Toronto, Ontario M5S 3G8, Canada
Email: bakhtar@mie.utoronto.ca

**Abstract**
In manufacturing, information acquired through integrated sensors can be used to increase flexibility, reliability, and accuracy of autonomous robotic systems. Furthermore, use of such sensors, as a means by which to implement flexible automation, can potentially diminish costs by reducing the need for customized and complex tooling often needed in non-programmable automation.

Robotic sensors can be categorized into three groups: medium-range and short-range proximity sensors (typically, for object recognition and/or position/orientation estimation) and contact sensors (typically, for force/torque measurements). This chapter focuses on the use of medium-range sensors; more specifically, the objective is the review of the state-of-the-art in autonomous active sensor networks.

## 11.1 Sensor Networks

A sensor network, also commonly referred to as a distributed sensor system, consists of multiple similar and/or dissimilar sensors that work in tandem for a common surveillance task. Sensors in such networks may differ in type, intrinsic parameters (sensor's internal configuration such as focus and aperture), and extrinsic parameters (sensor's external configuration, *i.e.* position and orientation). There are three general types of sensors that comprise a network [11.1]:

1. *Complementary Sensors* perceive different features using separate sensors. The data is combined to produce more detailed information than would be available from any single sensor.
2. *Cooperative Sensors* operate in a synergistic manner, providing information that cannot be derived from any one sensor. A common application of cooperative fusion is stereo vision.
3. *Competitive Sensors*, commonly referred to as *redundant sensors*, perceive the same feature(s) in the environment in order to (*i*) reduce uncertainty of

the observed data and, thus, increase the accuracy of the surveillance system, and/or (*ii*) achieve a degree of system fault tolerance through detection and omission of data from malfunctioning sensors.

In this section, below, some of the key aspects of sensor networks will be discussed in more detail.

### 11.1.1 Sensor Fusion

Sensor (or data) fusion refers to the combination of outputs from multiple sources to yield a single sensory data [11.2]. The objective is to collect data that would provide information with lower uncertainty and greater detail than would be available from any individual sensor. There are three general types of sensor fusion: low-level, mid-level, and high-level [11.3].

Low-level fusion, also referred to as pixel fusion, is the combination of raw data (images) from multiple sensors into a single image (*e. g.* [11.4]). The single image is then processed for feature extraction and decision making. Although pixel-level fusion is highly robust, it is a complex procedure, especially, when dealing with dissimilar sensors or sensors with different intrinsic and extrinsic configurations.

Mid-level fusion, or feature-level fusion, is the combination of features extracted from each item of sensory data. Decisions are made based on the fused features (*e. g.* [11.5]). This type of fusion is simpler to implement than pixel fusion and, thus, is commonly used in parameter estimation applications, such as localization and tracking.

*High-level fusion*, or *decision fusion*, occurs when independent decisions from individual sensors as well as *a priori* information regarding the environment are merged by a central fusion module to determine a global decision[1] (*e. g.* [11.6]).

### 11.1.2 Sensor Selection

A multi-sensor surveillance system does not necessarily have to utilize all its sensors for every given task; rather, it is often necessary for a system to dynamically select a subset of its sensors and assign them to a particular *Object of Interest* (OoI) such that the surveillance task is accomplished and the effectiveness of the sensing system is maximized [11.7]. In general, there are two types of systems that give rise to two distinct sensor-selection problems; those that utilize cooperative sensors (non-overlapping) and those that use competitive sensors (overlapping).

Sensor selection for a surveillance system with multiple *non-overlapping* (complementary or cooperative) sensors requires the system to simply determine which sensor has a view of the OoI. This is commonly accomplished by offline calculation of the area in the workspace that is visible to each sensor. For example, the system proposed in [11.8] discretizes the workspace into a number of sectors and, once the OoI enters a sector, the sensors assigned to the sector provide

---

[1] In a decentralized multi-sensor system, on the other hand, each node (consisting of local sensors) performs data fusion based on local observations and the information communicated between other nodes.

synchronous information about the OoI. It should be noted that the system assumes there is only one OoI within the workspace. The system proposed in [11.9], on the other hand, uses multiple non-overlapping cameras to reconstruct the path taken by an OoI, in a multi-object environment, as it passes the view of several cameras. The system assumes that an OoI will follow one of the several predefined paths with given probabilities. The first step is to run a motion-detection and recognition algorithm on each video stream. This returns probabilities of detecting the OoI for each camera. Then, a probabilistic model is created through a Bayesian network to find the path taken by the OoI. This is compared to the initially expected paths to ensure that all transitions and transition times are plausible.

The more difficult problem of sensor selection for systems with multiple *overlapping* (competitive) sensors is a relatively new area of research and, thus, has not yet been given the same degree of attention as the non-overlapping case. In this case, the system must determine which sensors are best suited for the surveillance of the OoI. For example, the system presented in [11.10] tracks a single OoI in a multi-object environment through images taken by multiple overlapping cameras. The use of competitive sensors allows the system to overcome occlusions and reduce uncertainty through sensor fusion. This system also requires each camera's visible region (*i. e.* a region within the cameras field of view) to be predefined; however, the system can detect occlusions and consequently redefine the visible region. When an OoI is detected, the sensor planner selects all cameras that have a clear line of sight to the target to perform surveillance.

### 11.1.3 Sensor Modeling

Sensor modeling aims to define and model what is being observed, how it is observed, and the accuracy of those measurements [11.11]. In this section, the problem of modeling sensor-observation uncertainty is addressed.

*11.1.3.1. Modeling Sensor-observation Uncertainty*

Observation errors are the result of unexpected changes or unmeasured variances in the sensor itself. The two main types of observation errors are statistical and calibration errors, respectively.

Statistical errors, or *random errors*, are caused by white noise in the sensory observations. Random errors are properties of the sensor and, thus, cannot be altered; however, they can be modeled as a stochastic process. The model most commonly used is to assume that the sensor's parameter estimation $x_e$ is the correct value corrupted by a random disturbance $H$:

$$x_e = x + H(\Omega) \tag{11.1}$$

where $H$ is a function of some known factors $\Omega$ (such physical sensor parameters and environmental conditions).

Calibration errors, or *systematic errors*, are errors caused by variations between the model and the sensor. When a mathematical model of a sensor is created the form of the model must be selected and the model must be calibrated through sensor measurements. The form of the model usually is a simplification of the actual sensor and, thus, often ignores certain environmental factors. Furthermore, during calibration, there always exists systematic and random errors in the sensor-calibration devices; however, by acquiring many observations the negative effects of systematic and statistical errors can be minimized. Lastly, uncontrollable physical factors may change the properties of the sensor during run-time and, thus, introduce additional calibration errors.

Calibration errors are much more difficult to deal with than statistical errors. Although, they are typically not included in the sensor model, they can be considered as another corruption of the sensor reading by a fixed error, $h$:

$$x_e = x + H(\Omega) + h \tag{11.2}$$

*11.1.3.2. An Example: Modeling Uncertainty in Object Localization*

In this section, an example of sensor modeling is considered. The sensors are CCD cameras that are used for independent estimation of the position and orientation (pose) of a marked target. There are six variances in sensor measurements, three for the target position ($x$, $y$, $z$) and three for orientation ($n_x$, $n_y$, $n_z$), which define the uncertainty in object localization.

The first step is to determine which environmental factors have an impact on the uncertainty in sensor measurements. Through variational analysis, it is determined that only two controlled parameters affect the measurements significantly: the Euclidean camera-to-target distance $d$, and the camera's bearing $\theta$ (Figure 11.1). Uncontrolled parameters, such as illumination, are not considered in this study.

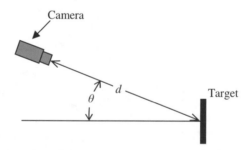

**Figure 11.1.** Camera distance $d$ and bearing $\theta$

Next, two-factorial experiments are performed to determine the relationship between each measurement variance and the two controlled parameters, $d$ and $\theta$. As an example, the results for two variances (estimation along the $y$-axis and one surface normal) are shown in Figure 11.2.

**11.1.4 An Example of a Multi-sensor Network**

The system proposed in [11.12] combines information from both complementary and cooperative sensors for improved target detection and tracking. A fixed

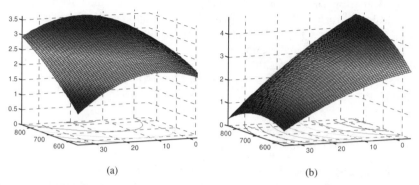

(a)                                    (b)

**Figure 11.2.** Response surfaces of variances in (a) orientation estimation about $z$-axis, $\sigma_{n_z}^2$ and (b) $y$-axis position, $\sigma_y^2$

wideangle camera monitors the entire workspace, detecting and tracking target gross motion. Information from this camera is used to select and conFigure a set of active binocular cameras based on *a priori* assigned camera workspaces. Through a two-level fusion strategy, the system combines information from the binocular cameras and the wide-angle static camera for more accurate and robust target tracking (Figure 11.3).

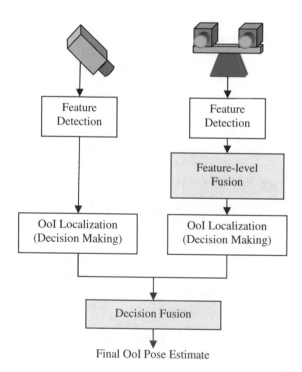

**Figure 11.3.** Multi-stage fusion used to integrate information from multiple vision systems

First, data from each of the binocular cameras are combined with a feature-level (mid-level) fusion algorithm based on triangulation method to yield a 3D estimate of the target's pose. This information is, then, combined with the pose estimate from the wide-angle camera using a decision-level (high-level) fusion. Finally, information regarding the target's trajectory from previous observations are merged with this estimate using a Kalman filter (KF) estimator (*i. e.* high-level fusion) [11.13].

## 11.2 Active Sensors

An *active-sensor* is one that may be dynamically reconfigurable, in some or all of its intrinsic and/or extrinsic parameters, in order to improve the sensing performance by perceiving the target from preferred viewing poses in terms of maximizing feature visibility and minimizing occlusions. At any given instant during run-time, an online planner is used to determine the optimal *sensing-system configuration* based on our latest knowledge of the environment, historical information regarding motion of the objects, the models of the sensors, as well as the specific required sensing task at hand.

### 11.2.1 Active-sensor Networks for Surveillance of Moving Objects in Static Environments

Traditionally, sensor planning has been utilized for determining the configuration of a set of sensors in *static* surveillance environments. Sensor planning in a static environment has been categorized as either a generate-and-test synthesis, or expert system [11.14].

*11.2.1.1 Generate and Test*

In generate-and-test systems, sensor-placement plans are developed by searching through discretized sensor configurations based on the task constraints. Advantages of generate-and-test systems include [11.14]:

- their use of established techniques for space discretization and searching;
- use of multiple constraints to efficiently yield regions in the search space that can be intersected to find a feasible region; and
- easy detection of occlusions.

However it should be noted that traditional generate-and-test approaches have the disadvantage of computational cost due to fine discretization coupled with an exhaustive search. There have been a number of methods proposed for reducing the computational cost of discretization such as variable space discretization and iterative discretization (*e. g.* [11.15], [11.16]). Furthermore, by using intelligent search algorithms, the computational cost can be significantly reduced; however, guarantee of a global optimal solution may be lost.

*Examples of Generate-and-Test Systems*

The system presented in [11.17] uses a virtual sphere, created around the OoI, representing all the possible poses for the sensor (Figure 11.4). The sphere is, then, discretized and poses that are unoccluded and fit within the workspace of the robot are selected. The sphere is centered around the OoI with the radius serving as a simple resolution and focus constraint. The sphere is, then, tessellated with icosahedrons that are iteratively subdivided by splitting each triangular space into 4 new faces, thus, achieving the level of detail required by the sensing task at the cost of increasing search space. Next, each discretized space is evaluated for occlusions by projecting a ray from the center of the facet being processed to the edges of the features being imaged. A distance-based criterion is used to rank all occlusion-free regions. The idea is that discretized regions that are farthest from occluded facets are preferred. The final step is to search for the highest ranked regions that would be achievable by the sensor (*i. e.* a camera mounted on an industrial robot).

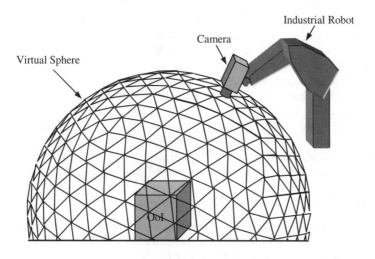

**Figure 11.4.** A virtual sphere generated around the OoI

Similarly, in [11.18], the sensing planner tries to find the minimum number of viewpoints that would allow observation of all the features on an OoI. However, in contrast to the system in [11.17], not only the virtual sphere around the object is discretized, but the surface of the OoI itself is also discretized (Figure 11.5). At each sensor location, the system finds the sensor orientation that would result in the maximum number of visible surface points. The planner, then, determines a set of sensor locations that would observe every discretized point on the object with minimum number of sensor poses.

*11.2.1.2 Synthesis*

Synthesis methods determine sensor configurations by using the analytical relationship between task requirements and the sensor parameters. This can be

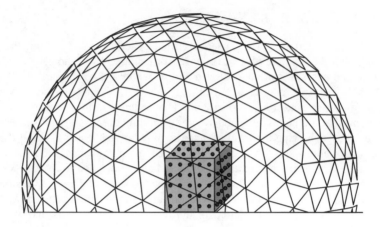

**Figure 11.5.** Discretizing the OoI and the sensor's workspace

a difficult task due to the dimensionality of a function space [11.14]. The dimension of a function space would depend on the degree of freedom (dof) of the sensor (for example, a sensor with 3D mobility as well as 3 intrinsic parameters would have a 9-dimensional function space). However, the dof of the sensor may be restricted by the physical and task constraints, thus, reducing the dimensionality of the functions.

The synthesis approach requires an understanding of and formulation between the parameters to be planned and the surveillance task. This requirement makes the system highly application specific. However, synthesis methods are in general less computationally costly than generate-and-test methods that systematically search for a solution. Furthermore, by exactly calculating sensor parameters, the accuracy of synthesis approaches is higher than generate-and-test methods.

*Examples of Systems Using the Synthesis Systems*

In [11.19], the sensor planner synthesizes a region of viewpoints by first imposing a 3D bound on the position of the camera by each of the task constraints (*e. g.* field of view, focus, and resolution) (Figure 11.6). The intersections of these bounds are considered to be regions of acceptable viewpoints. As above-mentioned, the general sensor-planning problem involves solving for all the dof of the sensor; however, in this work a subset of the sensor parameters (*i. e.* 3D position) is considered and other parameters such as orientation, focal length, and aperture are assumed to be a function of the sensor position with respect to the OoI. For example, the orientation of the sensor is chosen such that the optical axis is along a unit vector from the sensor's position to the center of the OoI. By solving for these parameters the dimensionality of the problem is greatly reduced.

The system presented in [11.20] determines sensor parameters that satisfy the sensor constraints (occlusion free, in focus, and within field-of-view) based on information provided through a CAD model of the OoI as well as detailed camera models. The sensor parameters are: 3D position (*x, y, z*), orientation (pan and tilt), back principle point to image distance, focal length, and aperture. All task constraints are formulated analytically resulting in a region of admissible values in the 8D hyperspace of sensor configurations. The intersection of these regions is, then, determined to find sensor configurations that satisfy all task constraints simultaneously. The center of the intersecting hyperspace is selected as the desirable configuration for the sensor. This is presented as an optimization problem with the task constraints as constraints of optimization and the objective function as the distance between a generalized configuration (*i. e.* a point in the 8D hyperspace) and the bounding hypersurfaces.

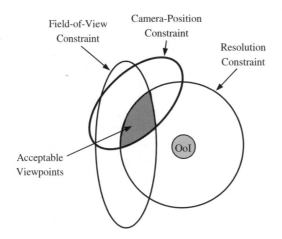

**Figure 11.6.** The intersection of geometric bounds

### 11.2.1.3 Expert Systems

The sensor-planning systems in the final category considered herein use *expert knowledge* for viewing and illumination, in a rule-based environment. When information regarding the particular OoI being observed is available, the expert system provides advice on the appropriate sensor configuration [11.14]. These systems address high-level aspects of the problem in a qualitative manner, such as if a certain object should be viewed from the back or front [11.21]. The qualitative advice needs to be extended with a synthesis or generate-and-test approach in order to determine the exact sensor configuration. For example, the system presented in [11.22] uses an expert system to suggest a possible viewing configuration given detailed information regarding the OoI (specular, diffuse, opaque, transparent, translucent, *etc.*), as well as the surveillance task at hand (type and size of features to be detected).

### 11.2.2 Online Sensor Planning for Surveillance of Dynamic Environments

Recently, there has been interest in sensor planning in dynamic environments, (*e. g.* [11.23]–[11.25]). Most current systems address this problem by utilizing methods developed for sensor planning in static environments, such as those mentioned above. In order to improve the surveillance performance, online planning can be used to reconfigure an active-sensor network (*i. e.* active-sensing system) in real-time, based on the current and estimated future states of the environment. For example, the system proposed in [11.26] optimizes sensor configurations offline by discretizing time and treating each time instant as a static case to be able to utilize the sensor-planning method presented in [11.18]. However, by relying solely on an offline plan, the surveillance system cannot react to unexpected variations in the environment. In order to address this problem, the system presented in [11.24] uses an offline heuristic (expert) method to determine sensor motions in 2D space based on an *a priori* known object trajectory and an online controller to readjust sensor motions to account for some deviations in actual object trajectory from the expected.

*Real* online sensor planning systems do not rely on offline plans and, therefore, do not require any *a priori* knowledge about the object's trajectory making them more robust to unexpected variations in the environment. For example, the system in [11.25] combines sensor-placement constraints and the shape and current pose of the OoI via a Bayesian network for task-specific sensor planning. The Bayesian network is reconstructed continuously to reflect changes in the pose of the target as determined by the active sensors. The derived sensing action maximizes the amount of target visible while minimizing the sensing cost (sensor movement).

## 11.3 Agent-based Approach to Online Sensor Planning

As the number of sensors increases, the use of a centralized planner for sensor selection and configuration becomes more difficult. Therefore, recently, a number of agent-based approaches have been proposed to address the problem of online sensor planning in order to decrease complexity and increase robustness and scalability (*e. g.* [11.27][11.28]).

Agent-based systems are described as a collection of interacting, autonomous, flexible, and goal-driven components that when properly combined, through cooperation, competition, and negotiation, exhibit an emergent behavior able to deal with complex systems that would otherwise be impractical to manage using a centralized controller [11.29].

### 11.3.1 Agents

An agent is a physical or virtual entity with the following principal properties [11.29]:

- possessing its own resources and capable of acting autonomously in an environment and communicate directly with other agents;
- driven by a set of tendencies (individual objective or satisfaction function to optimize) and whose behavior tends towards satisfying the objectives, while obeying the laws of the environment.

Agents are categorized as either reactive or symbolic. Reactive agents rely on simple behaviors to make decisions based on the information available about the current status of the surroundings and the status of other agents. Symbolic agents make decisions based on explicit global representation and logical reasoning. Although reactive agents are robust and understandable at the single-agent level, they are generally inefficient and extremely complex at the multi-agent level, especially when global behavior is required [11.30]. On the other hand, symbolic agents are generally more predictable and their global behavior is easier to understand. However, they are less robust and computationally more expensive as the environment must be symbolically represented.

### 11.3.2 Advantages and Drawbacks of Multi-agent Systems

The main advantage of agent-based systems is reduced computational complexity in sensor planning, since each agent searches for a solution (*e. g.* sensor pose with respect to the target) in order to optimize its own surveillance performance. The second advantage of agent-based systems is the increased robustness realized through fault tolerance. In a centralized surveillance system, the failure of one unit (the sensor planner) means failure of the whole system. On the other hand, in multi-agent-based systems, each sensor is considered as a separate module and, thus, the failure of one module does not lead to the failure of other modules. Agent-based planning also offers the advantage of flexibility and scalability as different sensors of varying types and abilities can be added or removed from the system with minimum impact on the system's planning strategy.

The main disadvantage of a multi-agent-based system is loss of global optimality. It may be useful here to introduce the notion of local versus global performance. Local performance is the sensor's recorded, or estimated, performance. Global performance, on the other hand, refers to the performance of the system as a whole. Agents utilizing global performance must receive recorded or estimated system performance from a centralized unit. In general, most multi-agent systems make decisions based on a combination of local and global performance metrics. The more this balance leans towards local performance the less computationally expensive the decision-making process becomes, however, at the expense of system performance optimality.

### 11.3.3 Examples of Agent-based Sensor-planning Systems

In this section, two systems are considered; the first system uses an agent-based sensor planning method to continuously track an OoI, while in the second system surveillance of the OoI is performed at discrete time instants (demand instants). Discrete surveillance allows the system to predict the status of the environment at future demand instants in order to optimize the performance of the surveillance system for the OoI's trajectory.

#### *11.3.3.1 A Sensor-planning System for Continuous Surveillance*

The system proposed in [11.23] and [11.24] uses a set of cooperative active-vision sensors for continuous tracking and surveillance of multiple targets (Figure 11.7). Each camera is controlled individually through an agent-based architecture. An agent is a network-connected computer with an attached PTZ (pan, tilt, zoom) camera. Furthermore, it is assumed that all cameras have overlapping visual fields:

1.  The system starts with each agent independently scanning the workplace for a target[1]. An agent searching for a new target is referred to as a *freelancer*.
2.  If an agent detects a target, it navigates the gaze of all other agents towards the target in order to estimate its 3D position. This requires the agent to share information regarding the detected target with all other agents.
3.  Each agent must, then, decide on its own if it should search for a new target or join the group of agents (the *agency*) tracking the specific target. Each agent is also responsible for dynamic reconfiguration of the extrinsic (pan and tilt) and intrinsic (focus and zoom) parameters of the associated camera.
4.  An agency is dynamically maintained; thus, agents that lose visibility of the target leave the agency and new agents join in their place.

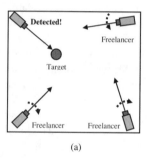 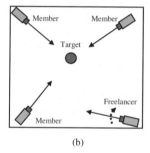 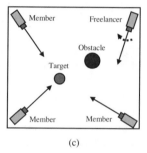

       (a)                            (b)                           (c)

**Figure 11.7.** An example surveillance scenario of the system in [11.23]: (a) a freelancer detects the target, (b) three agents join in to track the target, and (c) occlusion forces an agent to leave the agency; a new agent joins in its place

*11.3.3.2 A Sensor-planning System for Discrete-time Surveillance*

In [11.31], an agent-based sensor-planning method is used to maximize the effectiveness of the surveillance system, which is used to provide estimates of object parameters (pose) at predetermined times along its trajectory. These predetermined times are referred to as demand instants, $t_i$. If the sensing system comprises multiple redundant sensors, a subset of these may be sufficient to satisfy the sensing requirements of a demand. Namely, a sensor fusion process does not need to combine the information from all sensors in the system. Instead, a subset of sensors may be selected to survey the object at a particular demand instant, allowing other sensors to be configured in anticipation of future use.

*Quality of the Sensing Data*

As discussed in Section 11.1.3.2, integral to sensor planning is an estimate of the quality of data that each sensor can provide for the demand instant at hand and for

---

[1] An active background-subtraction method is used for target detection.

the span of a rolling horizon of several demand instants, given the sensor's current pose in the workspace and its motion capabilities. This estimate is used to select sensors for inclusion in the surveillance of the OoI at a demand instant and assess the desired pose of each sensor during surveillance. A visibility measure that is inversely proportional to the measurement uncertainty may be used to assess the fitness of a single sensor, at one or several demand instants. The model-based visibility measure provides a more robust basis for sensor selection than would simple distance measures or determining if the OoI is visible to a sensor (in a *line-of-sight* sense) or not. The visibility measure for the $j$th sensor servicing the $i$th demand instant is defined herein as

$$
v_{ji} = \begin{cases} \dfrac{1}{\|\mathbf{R}\|} & \text{If target is unoccluded} \\ 0 & \text{else} \end{cases},
$$

(11.3)

where $\mathbf{R}$ is the covariance matrix associated with the sensor measurement. For example, $\mathbf{R}$ is a function of six variance parameters: three for the Cartesian position of the target $(x, y, z)$ and three for its orientation $(n_x, n_y, n_z)$. Our variance analysis experiments, Appendix A, led to the conclusion that only two controlled sensor parameters significantly affect the measurement variances: the Euclidean camera-to-target distance $d$, and the bearing of the camera with respect to the target, $\theta$:

$$
d = \|\boldsymbol{o}_p - \mathbf{s}_p\|
$$

(11.4)

where $\boldsymbol{o}_p$ is the OoI's position and $\mathbf{s}_p$ is the sensor position, and

$$
\theta = \cos^{-1}(\boldsymbol{o}_n \circ \boldsymbol{\tau})
$$

(11.5)

where $\boldsymbol{o}_n$ is the unit vector corresponding to the OoI orientation and $\boldsymbol{\tau}$ is the viewing direction of the sensor,

$$
\boldsymbol{\tau} = \frac{\boldsymbol{o}_p - \mathbf{s}_p}{\|\boldsymbol{o}_p - \mathbf{s}_p\|}
$$

(11.6)

The visibility measure for a sensor over the span of a rolling horizon is defined as

$$
v_j = \sum_{i=1}^{m} a_i v_{ij}
$$

(11.7)

where $m$ is the number of demand instants in the rolling horizon and $a_i$ is the weight of the $i$th demand instant. The weight factor is constant for all sensors and represents the uncertainty in predictions of the future OoI poses. Although the weight factor $a_i$ is user specified and depends on the surveillance system at hand, in general the value of $a_i$ decreases as predictions of target poses are made further into the future.

*Coordination Strategy*

Sensor planning is accomplished using two complementary strategies: A co-ordination strategy to determine how many and specifically which sensors should be used at each demand instant in order to optimize the performance of the surveillance system over the span of the rolling horizon; and, a positioning strategy to determine the optimal pose of each sensor for any demand point being serviced.

The proposed agent-based system consists of multiple *sensor* agents, a *referee* agent, and a *judge* agent. Each sensor agent tries to maximize its own performance over the span of the rolling horizon. Although not directly controlled by a centralized controller, the sensor agents must abide by the external rules of the environment monitored and enforced by two virtual agents. The rules are set to ensure the collective behavior of the sensor agents exhibits the desired system behavior.

*Sensor Agent*:  The sensor agent is responsible for choosing the demand instants that the associated sensor will service and determining its optimum pose in order to maximize the sensor's performance metric (*i. e.* visibility) over the span of the rolling horizon. If a demand instant is not serviced, the sensor would have zero visibility for that demand instant, however, it would allow more time for the sensor to manoeuvre for the next demand instant.

Each sensor agent searches through all possible combinations (for example, [1, 1, 1] is a combination referring to servicing all demand instants in a 3 demand-instant horizon) using a depth-first approach. The total search space for a sensor agent is $2^m$, where $m$ is the number of demand-instants in the rolling horizon. At each combination searched, the sensor determines the best achievable poses to service the selected demand instants through the positioning strategy outlined below. Using the optimum poses and the object locations the sensor agent determines the expected achievable visibility for each combination. The sensor agent then evaluates the combinations searched to determine acceptable solutions. The acceptable solutions are constrained by the following internal rules:

- A demand instant cannot be serviced if it is occluded.
- Combination [0, 0, 0, ... , 0], representing a  sensor not being assigned to any demand instant, is only considered if all other combinations are occluded[1].

Next, the sensor ranks all acceptable combinations in descending order of combined visibilities. The *r*th ranked acceptable solutions for the *j*th sensor is denoted herein as $S_{jr}$. The sensor agent sends the first ranked acceptable solution, $S_{j1}$, to the referee agent.

*Referee agent*: The referee agent is a virtual agent that monitors the intentions of the sensor agents and ensures the external rules are not violated. The external rules would depend on the surveillance task at hand. In this work, the following external rule is defined, in order to ensure the sensors are well distributed among the demand instants of the rolling horizon:

- At least one sensor must be assigned to each demand instant.

---

[1] The [0, 0, ... , 0] combination allows the sensor to simply follow the target so that it may be used in the future.

Once the referee agent detects a violation of the external rules it initiates the judge agent in order to resolve the conflict.

*Judge Agent*: Upon initiation, the judge agent sends a command to each sensor agent requesting the agent's second ranked acceptable solutions, $S_{j2}$. Along with their alternate solutions, each sensor agent also sends its expected loss in visibility if it is forced to select its second ranked solution.

The judge agent uses a *depth-first* approach to search through the possible permutations of first and second ranked solutions for a combination that would resolve the conflict (an example combination is $\{S_{11}\ S_{21}\ S_{32}\ S_{41}\}$). The judge agent, then, selects the combination that would yield the lowest visibility drop, and informs the sensor agents of its decision.

If no acceptable combination is found the judge agent increases the search space by requesting the sensors' third ranked solutions. This process is repeated until an acceptable combination is found or allowable search time has elapsed. In the event that no acceptable combination is found within the allowable search time, the first ranked solutions are used.

*Positioning Strategy*

In a single-target multi-object environment, the positioning strategy is performed not only based on the trajectory of the OoI (*i. e.* the target) but also based on other objects that are not of interest but which may act as occlusions. The first step in determining the best achievable pose is to determine the occluded regions in the workspace. In order to accomplish this, the pose of each object (target or obstacle) is predicted for the demand instant. Next, each object is modeled as a single geometric primitive (*e. g.* a sphere or cylinder) rather than as a collection of 3D polyhedra in order to decrease computational complexity.

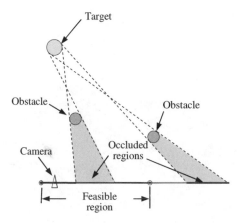

**Figure 11.8.** 2D example of occluded regions

Occluded regions of a sensor's workspace are determined by modeling the target as a light source and calculating geometric shadow volumes [11.32], cast by the obstacles, using the pose and size of each object in the workspace (Figure 11.8). The algorithm subsequently determines the region of the workspace that the sensor can travel to before the target reaches the demand instant, referred to herein as feasible region. This region is defined by the sensors' dynamic motion capabilities such as maximum velocity and acceleration. Lastly, the algorithm determines an optimal sensor pose that would yield maximum visibility, which is both feasible and unoccluded (*i. e.* acceptable regions). This is done by discretizing the acceptable region into a pre-specified number of positions. An optimum pose is selected by evaluating the visibility metric at each discrete position.

## 11.4 An Active-sensor Network Example for Object Localization in a Multi-object Environment

### 11.4.1 Experimental Set-up

The experimental set-up presented in this section uses the agent-based sensor planning method described in Section 11.3.3.2. This system uses four mobile cameras to determine the pose of a single target, represented by a circular marker, manoeuvring through the workspace on a planar trajectory (Figure 11.9). Due to system limitations only stationary obstacles, positioned within the workspace at predefined locations, could be considered during the experiments. A stationary overhead camera is utilized to obtain gross estimates of the target motion and location of obstacles. These estimates are used to plan the motion of the cameras that have one or two degrees of freedom (dof) – all have one-dof rotational capability (pan), while two of the cameras can also translate linearly (see Table 11.1 for component list).

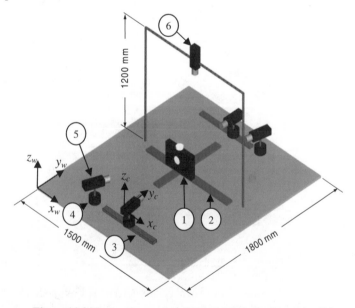

**Figure 11.9.** System layout (see Table 11.1 for hardware details)

**Table 11.1.** Hardware specifications

| Part No. | Hardware | Characteristic |
|---|---|---|
| 1 | Target | Matte black aluminium plate marked with white circular marker (Diameter = 25 mm). |
| 2 | x-y Table | Range: 500 mm (x)/200 mm (y) <br> Positional Accuracy: 48 μm (x)/24 μm (y) <br> Maximum Velocity: 0.3 m/s |
| 3 | Two Linear Stages | Range: 300 mm <br> Positional Accuracy: 30 μm <br> Maximum Velocity: 0.3 m/s |
| 4 | Four Rotary Stages | Positional Accuracy: 10 arc s <br> Maximum Velocity: $\pi/6$ rad/s |
| 5 | Four Dynamic CMOS Cameras | Resolution: 640×480 pixels <br> Lens Focal Length: 25 mm |
| 6 | One Static CCD Camera | Resolution: 640×480 pixels <br> Lens Focal Length: 12 mm |

## 11.4.2 Experiments

In order to illustrate the implementation of a sensor-planning algorithm in an active-vision system, an experiment is briefly discussed here. In general, this and other experiments performed using the described set-up verified that the performance of a surveillance system can be tangibly improved with the use of an effective sensor planning strategy, primarily due to (i) increased robustness of the system (*i. e.* its ability to cope with *a priori* unknown target trajectories and the presence of obstacles), (ii) decreased uncertainty associated with estimating the target's pose through sensor fusion, and (iii) increased reliability through sensory fault tolerance.

*Procedure*

In the specific example discussed herein, the performances of two systems (*i. e. fast* and *slow*) are compared. The speed of the target in both systems is set to 5 mm/s, following the trajectory shown in Figure 11.10. The target is a 25 mm circular marker, while the two 60 mm obstacles are positioned at each side of the target trajectory (Figure 11.10).

System evaluation is carried out using the visibility metric discussed in Section 11.3.2.2: Target visibility by a sensor is calculated using the expected variance in the measurements that is a function of the sensor's Euclidean distance to the target and its bearing (*i. e.* the angle the camera's local axis makes with the normal of the target's surface).

The system performance was also evaluated by a *post-process system evaluation* module to determine the exact errors in the real-time estimation of the target's pose. The first of these errors is the *absolute error in position estimation*, $e_{position}$, defined as the Euclidean distance between the true target position, $o_p = (x_t, y_t, z_t)$, and the system's estimate of the target's position, $\chi_{fused} = (x_e, y_e, z_e)$:

$$e_{position} = \left\| \mathbf{o}_p - \chi_{fused} \right\|$$ (11.8)

Similarly, the absolute error in surface normal estimation, $e_{orientation}$, is the angle between the true target orientation, $\mathbf{o}_n$, and the estimated surface normal, $\mathbf{o}_{n_{fused}}$:

$$e_{orientation} = \cos^{-1} \left( \mathbf{o}_n \cdot \mathbf{o}_{n_{fused}} \right)$$ (11.9)

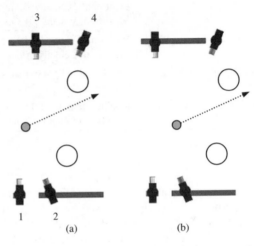

**Figure 11.10.** Initial sensor poses for (a) slow system and (b) fast system

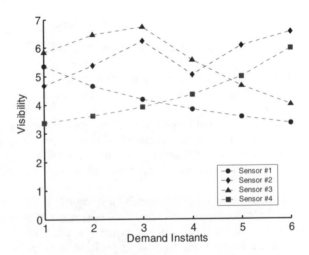

**Figure 11.11.** Observed visibilities of each sensor of the fast system

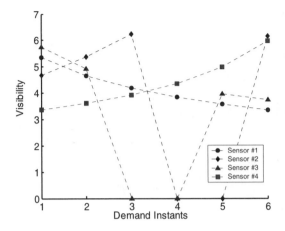

**Figure 11.12.** Observed visibilities of each sensor of the slow system

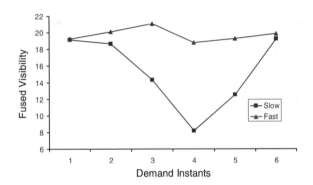

**Figure 11.13.** Fused visibilities of both systems

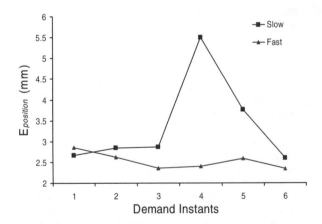

**Figure 11.14.** Absolute errors in target position estimates

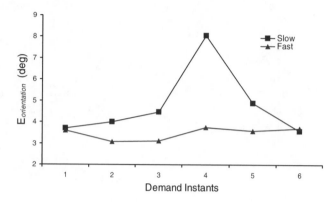

**Figure 11.15.** Absolute errors in target's surface normal estimates

*Results*

The pose of the moving target was estimated at six distinct locations (demand instants). The visibilities of each sensor at every demand instant for both systems are given in Figures 11.11 and 11.12, respectively, and the fused visibilities for both systems are shown in Figure 11.13. As noted, the visibilities of the target by the fast system are tangibly higher than those of the slow system. The corresponding absolute position errors associated are shown in Figure 11.14 and absolute errors in surface-normal estimations are given in Figure 11.15. Despite the presence of random noise in both systems, the data confirms the tangible improvement of system performance through the use of the fast system.

## Acknowledgment

The authors would like to acknowledge the support of Natural Sciences and Engineering Research Council of Canada (NSERC).

## References

[11.1]    H. Durrant-White, "Sensor Models and Multisensor Integration," *The International Journal of Robotics Research*, **7**, pp. 97–113, 1998.
[11.2]    R. Luo and M. Kay, "Multisensor Integration and Fusion for Intelligent Machines and Systems," in *Data Fusion in Robotics and Machine Intelligence* (M. Abidi and R. Gonzalez, Eds.), San Diego, CA: Acadamic Press, pp. 7–136.
[11.3]    Belur V. Dasarathy. "Information Fusion as a Tool in Condition Monitoring," *Trans. on Information Fusion*, **4**, pp. 71–73, 2003.
[11.4]    W. Ross, A. Waxman, W. Streilein, M. Aguiiar, J. Verly, F. Liu, M. Braun, P. Harmon and S. Rak, "Multi-Sensor 3D Image Fusion and Interactive Search," *IEEE Int. Conf. on Information Fusion*, pp. TUC2:10–17, Paris, France, 2000.
[11.5]    X. Zhang, "An Information Model and Method of Feature Fusion", *IEEE Int. Conf on Signal Processing*, pp. 1389–1392, Beijing, China, 1998.

[11.6]   Y. Yuan and M Kam, "Distributed Decision Fusion with a Random-Access Channel for Sensor Network Applications," *IEEE Trans. on Instrumentation and Measurement*, **53**(4), pp. 1339–1334, 2004.

[11.7]   S. Abrams, P. Allen and K. Tarabanis, "Computing Camera Viewpoints in an Active Robot Work Cell," *The International Journal of Robotics Research*, **19**, pp. 267–285, 1999.

[11.8]   B. Horling, R. Vincent, J. Shen, R. Becker and K. Rawlins, "V. Lesser: SPT Distributed Sensor Network for Real Time Tracking," Technical Report 00–49, University of Massachusetts, Amherst, MA, 2000.

[11.9]   V. Kettnaker and R. Zabih, "Bayesian Multi-Camera Surveillance," *IEEE Int. Conf. on Computer Vision and Pattern Recognition*, **2**, pp. 253–259, Los Alamitos, CA, 1999.

[11.10]  T. Ellis "Multi-Camera Video Surveillance," *IEEE Int. Conf. on Security Technology*, pp. 228–233, Atlantic City, NJ, 2002.

[11.11]  G. D. Hager, *Task-Directed Sensor Fusion and Planning*, Kluwer Academic Publishers, Boston, MA, 1990.

[11.12]  P. Piexoto, J. Batista and H. Araujo, "Integration of Information from Several Vision Systems for a Common Task of Surveillance," *J. of Robotic and Autonomous Systems*, **31**, pp. 99–108, 2000.

[11.13]  R. E. Kalman, "A New Approach to Linear Filtering and Prediction Problems," *Transactions of the ASME—Journal of Basic Engineering*, **82**(D), pp. 35–45, 1961.

[11.14]  K. A. Tarabanis, P. K. Allen, and R. Y. Tsai, "A Survey of Sensor Planning in Computer Vision," *IEEE Transactions on Robotics and Automation*, **11**(1), pp. 86–104, 1995.

[11.15]  R. Munos and A. Moore, "Variable Resolution Discretization in Optimal Control," *Transaction of Journal of Machine Learning*, **49**(2–3), pp. 291–323, 2002.

[11.16]  T. Yairi, K. Hori and S. Nakasuka, "Sensor Space Discretization in Autonomous Agent Based on Entropy Minimization of Behavior Outcomes," *IEEE Conference on Multisensor Fusion and Integration for Intelligent systems*, pp. 111–116, Taipei, Taiwan, 1999.

[11.17]  S. Sakane, T. Sato and M. Kakikura, "Model-Based Planning of Visual Sensors Using a Hand-Eye Action Simulator: HEAVEN," *IEEE Int. Conf. on Advanced Robotics*, pp. 163–174, Versailles, France, 1987.

[11.18]  G. H. Tarbox and S. N. Gottschlich, "Planning for Complete Sensor Coverage in Inspection," *Trans. on Computer Vision and Image Understanding*, **61**(1), pp. 84–111, 1995.

[11.19]  C. K. Cowan and P. D. Kovesik, "Automated Sensor Placement from Vision Task Requirements," *IEEE Transactions on Pattern Analysis and Machine Intelligence*, **10**(3), pp. 407–416, 1988.

[11.20]  A. Tarabanis, R. Y. Tsai and P. K. Allen, "The MVP Sensor Planning System for Robotic Vision Tasks," *IEEE Transactions on Robotics and Automation*, **2**(1), pp. 72–79, 1995.

[11.21]  B. G. Batchelor, D. A. Hill and D. C. Hodgson, *Automated Visual Inspection*. IFS Ltd, Bedford, UK, 1985.

[11.22]  B. G. Batchelor, "A Prolog Lighting Advisor," in *Proc. of SPIE Intelligent Robots*, **1193**, pp. 295–302, 1989.

[11.23]  T. Matsuyama, T. Wada and S. Tokai, "Active Image Capturing and Dynamic Scene Visualization by Cooperative Distributed Vision," *Trans on Advanced Multimedia Content Processing*, **11**(4), pp. 252–288, 1999.

[11.24]  T. Matsuyama, N. Ukita, "Real-time Multi-Target Tracking by a Cooperative Distributed Vision System," *Proc. of the IEEE Journal*, **90**(7), pp. 1136–1150, 2002.

[11.25]  H. Zhou and S. Sakane, "Sensor Planning for Mobile Robot Localization Using Bayesian Network Inference," *Journal of Advanced Robotics,* **16**(8), pp. 751–771, 2002.

[11.26]  R. Niepold, S. Sakane and Y. Shirai, "Vision Sensor Set-up planning for a Hand-Eye System Using Environmental Models," In *Proceedings of the Society of Instrument and Control Engineers of Japan,* **7**(1), pp. 1037–1040, Hiroshima, Japan, 1987.

[11.27]  J. R. Spletzer and C. J Taylor, "Dynamic Sensor Planning and Control for Optimally Tracking Targets," *Int. Journal of Robotic Research,* **22**(1), pp. 7–20, 2003.

[11.28]  M. Kamel and L. Hodge, "A Coordination Mechanism for Model-Based Multi-Sensor Planning," *IEEE Int. Symp. on Intelligent Control,* pp. 1143–1149, Vancouver, 2002.

[11.29]  J. Liu and J. Wu, *Multi-Agent Robotic Systems,* CRC Press, Washington, D. C., 2001.

[11.30]  E. Spier, "From Reactive Behaviours to Adaptive Behaviours: Motivational Models for Behaviour in Animals and Robots," *Ph.D. Thesis,* Oxford University, UK, 1997.

[11.31]  A. Bakhtari and B. Benhabib, "Agent-Based Active-Vision System Reconfiguration for Autonomous Surveillance of Dynamic, Multi-Object Environments," *IEEE Int. Conf. on Intelligent Robots and Systems,* Edmonton, Canada, 2005.

[11.32]  K. Thakur, F. Cheng and K. T. Miura, "Shadow Generation Using Discretized Shadow Volume in Angular Coordinates," *IEEE Conf. on Computer Graphics and Applications,* pp. 224–233, Canmore, Canada, 2003.

# 12

## Remote Monitoring and Control in a Distributed Manufacturing Environment

Lihui Wang, Weiming Shen, Peter Orban and Sherman Lang

Integrated Manufacturing Technologies Institute
National Research Council of Canada
London, ON N6G 4X8, Canada
Email: lihui.wang@nrc.gc.ca

### Abstract

Remote monitoring and control are crucial in decentralized manufacturing environments. This is evidenced by today's distributed shop floors where agility and responsiveness are required to maintain high productivity and flexibility. However, there exists a lack of an effective system architecture that integrates remote condition monitoring and control of automated equipment. Addressing this problem, this chapter introduces a web-based and sensor-driven technique that bridges this missing link. A framework of *WISE-SHOPFLOOR* (Web-based integrated sensor-driven e-shop floor) was designed to realize such a concept. The conceptulization, architectural design, and system implementation are discussed in detail and two case studies on robot control and remote machining are presented. Enabled by Java and web technologies, *WISE-SHOPFLOOR* demonstrates significant promise of intelligent distributed manufacturing.

## 12.1 Introduction

Distributed manufacturing shop floors characterized by a large variety of products in small batch sizes are typical for global companies and outsourced small-to-medium sized enterprises. To stay competitive, they demand dynamic control and real-time monitoring capabilities that are responsive and adaptive to rapid changes in daily production to achieve the best product quality and efficiency. These become achievable when distributed manufacturing is combined with the e-manufacturing concept. Since 1993 shortly after the debut of the Web, a number of methods and frameworks have been proposed for solving problems in distributed manufacturing. However, most of them are developed for collaborative design, web-based rapid prototyping, and project management, *e. g.* WebCADET [12.1] for distributed design support, CyberCut [12.2] for web-based rapid machining, and NegotiationLens [12.3] for conflict resolution. Unfortunately, most of today's manufacturing equipment does not have built-in capability to transmit and receive data. Few of the available web-based systems are designed for remote real-time monitoring and control. Some related systems listed below in the area of event monitoring are limited in their functionality and platform requirements.

The latest *Cimplicity* from GE Fanuc Automation (USA) allows users to view their factory's operational processes through an XML-based *WebView* screen, including all alerts on every *Cimplicity* system [12.4]. The *FactoryFlow* from Unigraphics Solutions (USA) is an offline factory-floor layout planning, material handling, and simulation package [12.5]. By most estimates, the number of CNC machines capable of linking to the Internet is less than 10% of the installed base [12.5]. Seeking the opportunity to link CNC machines with the Internet, MDSI (Ann Arbor, MI, USA) uses *OpenCNC* [12.6], a Windows-based software-only machine tool controller with real-time database, to automatically collect and publish machine and process data on a network. In 1999, Hitachi Seiki (Japan) introduced *FlexLink* [12.7] to its turning and machining centers. Working together with *PC-DNC Plus* from Refresh Your Memory (USA), *FlexLink* is able to do in-process gauging, machine monitoring, and cycle-time analysis. Since 1998, Mazak (Japan) has operated its high-tech *Cyber Factory* concept [12.8] at its headquarters in Oguchi, Japan. The fully networkable *Mazatrol Fusion controllers* allow Mazak machines to communicate over wireless factory networks for applications including real-time machine tool monitoring and diagnostics. In addition, Japan-based Mori Seiki introduced a *CAPS-NET* system that polls machine tools on Ethernet at settable increments, usually five-second or longer, for engineers to get updates on machine tools' run-time status in production [12.9]. To bring legacy machine tools with only serial ports online, e-Manufacturing Networks Inc. (now Memex Division in Hamilton, Ontario, Canada) introduced its *ION Universal Interface* and *CORTEX Gateway* [12.10] to help the old systems go online, and to monitor information flow and the status of the CNC machine tools on the network.

As summarized in Table 12.1, despite all the accomplishments, the available systems are either for offline simulation or for monitoring only. Most systems require a specific application to be installed instead of a standard web browser, which reduces a system's portability. Remote shop floor monitoring and control remain impractical as web-based applications due to security concerns and real-time constraints. Reducing network traffic, increasing system performance, and overcoming security barriers are the major concerns in web-based manufacturing system developments.

To bridge the gap, we propose a web-based and sensor-driven approach for real-time monitoring and control in distributed manufacturing. It serves as the glue in factory automation, and to bring machines, robots and sensors together through an integrated *WISE-SHOPFLOOR* framework. This chapter discusses in detail the concept, architecture design, and implementation issues through two case studies on robot control and remote machining.

## 12.2 WISE-SHOPFLOOR Concept

The *WISE-SHOPFLOOR* framework [12.11] is designed to provide users with a web-based and sensor-driven intuitive environment where real-time monitoring and control are undertaken. Instead of camera images (usually large in data size),

Table 12.1. Available systems/products and their functionalities

| Product or System | Online | | Offline | | |
|---|---|---|---|---|---|
| | Monitoring | Control | Planning | Analysis | Simulation |
| WebCADET | | | √ | | √ |
| CyberCut | | | √ | | |
| NegotiationLens | | | | | √ |
| Cimplicity | √ | | | | |
| FactoryFlow | | | √ | √ | √ |
| OpenCNC | √ | | | √ | |
| FlexLink | √ | | | √ | |
| Cyber Factory | √ | | √ | √ | √ |
| CAPS-NET | √ | | | | |
| ION Universal Interface | √ | | | | |
| CORTEX Gateway | √ | | | | |

a physical device of interest (*e. g.* a robot or a machine) can be represented by a scene graph-based Java 3D model with behavioral control nodes embedded. Once downloaded from its application server, the 3D model is rendered by a local CPU and can work on behalf of its remote counterpart showing real behavior for visualization at a client side. It remains alive by connecting with the physical device (via Java Servlets) through low-volume message passing (sensor data or user control commands). The 3D model provides users with increased flexibility for visualization from various perspectives such as walk-through or fly-around that are not possible by using stationary optical cameras. In addition to motion data, other sensory data including temperature, vibration and force can also be transmitted via the network and shown in colors or contour lines on the 3D model for condition-based monitoring. The largely reduced network traffic makes real-time monitoring and remote control practical for dispersed users through a shared *Cyber Workspace* [12.12].

Figure 12.1 shows how intelligent sensors are relevant to the *WISE-SHOPFLOOR*. By combining virtual 3D models with real devices through synchronized real-time data communications, this sensor-driven approach allows shop floor personnel to assure optimal operations as well as web-based trouble-shooting – particularly useful when they are off-site.

Although the *WISE-SHOPFLOOR* framework is initially designed as an alternative to camera-based monitoring systems, an off-the-shelf web-ready camera can easily be switched on remotely to capture unpredictable (or un-modeled) real scenes for diagnostic purposes, whenever it is needed. In addition to real-time monitoring and remote control, the framework can also be extended and applied to distributed process planning, dynamic scheduling, collaborative design, remote diagnosis, and virtual machining. It is tolerant to hostile, invisible or non-accessible environments, *e. g.* inside a nuclear reactor or outside a space station, where placing optical cameras is sometimes not feasible.

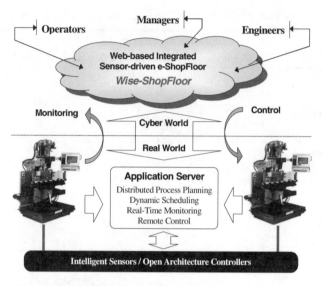

**Figure 12.1.** *WISE-SHOPFLOOR* concept

## 12.3 Architecture Design

Targeting remote monitoring and control in distributed manufacturing, the *WISE-SHOPFLOOR* is designed to use the popular client–server architecture and the VCM (View-Control-Model) design pattern with built-in secure session control. The proposed solution to meet both the user requirements of rich data sharing and the real-time constraints are listed below.

- using interactive sensor-driven scene graph-based Java 3D models instead of bandwidth-consuming camera images for visualization;
- transmitting only the sensor data and control commands between virtual models and real devices for real-time monitoring and remote control;
- providing remote users with thin-client graphical user interface (GUI) for model manipulation and shop floor navigation; and
- deploying the core control logics in a secure application server using Java Servlets.

As shown in Figure 12.2, the mid-tier application server handles major security concerns, such as session control, viewer registration, data collection/distribution, and real device manipulation. A central *SessionManager* is designed to look after the issues of user authentication, session control, and sensitive data logging. All initial transactions need to pass through the *SessionManager* for access authorization. In a multi-client environment like *WISE-SHOPFLOOR*, different clients may require different sets of sensor data for device monitoring. It is not efficient to have multiple clients sharing the same model talk with the same device directly at the same time. Instead, a publish-subscribe design pattern is adopted to collect sensor data and distribute them to the right clients at the right time, efficiently. For this purpose, a server-side module, *SignalCollector*, is responsible for sensor data collection from networked physical devices. The collected data are then passed to another server-side module called *SignalPublisher* that in turn multicasts the data to the registered subscribers (clients) through applet-servlet communication. A *Registrar* is designed to maintain a list of subscribers with the requested sensor data. A Java 3D model can thus communicate indirectly with sensors and controllers no matter where the clients reside. HTTP streaming is chosen as the data communication protocol for the best combination between applets and servlets and for by-passing firewall protection. For security reasons, a physical device is controllable only by the *Commander* that resides in the application server. Another server-side component called *DataAccessor* is designed to separate logical and physical views of data. It encapsulates JDBC (Java Database Connectivity) and SQL codes, and provides standard methods for accessing data.

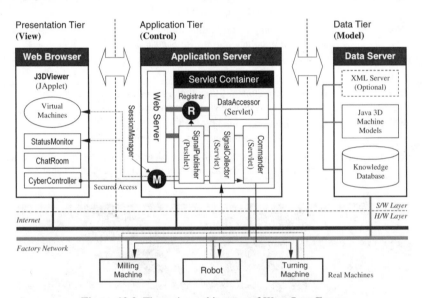

**Figure 12.2.** Three-tier architecture of *WISE-SHOPFLOOR*

Although the global behaviors of Java 3D models are controlled by the server based on real-time sensor signals, users still have the flexibility of monitoring the models from different perspectives, such as choosing machine, changing viewpoint, and selecting control mode, through *J3DViewer* at client side. Authorized users can submit control commands through *CyberController* to the application server. The *Commander* at server-side then takes over the control for real device manipulations. Another client-side module *StatusMonitor* can provide end users with a view of run-time status of the controlled device. For the purpose of collaborative trouble-shooting, a *ChatRoom* is included in the framework for synchronized messaging among connected users.

Table 12.2 summarizes the functionality of each module designed into the *WISE-SHOPFLOOR* framework. Figure 12.3 shows one typical shop floor represented by

Java 3D models and driven by sensor signals. More details on data communication
and prototype implementation are provided in the following sections.

**Table 12.2.** Modular functionality of *WISE-SHOPFLOOR* framework

| | |
|---|---|
| DataAccessor | Server-side component. Responsible for retrieving Java 3D models and recording critical device/sensor data for trouble-shooting. |
| SignalCollector | Server-side component. Responsible for sensor signal collection from physical devices based on clients' requests. |
| SignalPublisher | Server-side component. Responsible for sensor data distribution to the registered subscribers (clients) via multicasting. |
| Commander | Server-side component. Responsible for controlling a physical device on behalf of an end user to meet enhanced security requirements. |
| SessionManager | Server-side component. Authorized to check clients' authentication and assign appropriate access rights back to the clients. |
| Registrar | Server-side component. Maintaining a list of subscribers for sensor data distribution in collaboration with SignalPublisher. |
| J3DViewer | Client-side component. Providing means for end users to control the view of Java 3D models, individually. |
| CyberController | Client-side component. Providing a set of widgets for end users to manipulate a physical device through Cyber Workspace. |
| StatusMonitor | Client-side component (JTable). Used to monitor the run-time status of a physical device, with sensor data served by application server. |
| ChatRoom | Client-side component. A shared workspace designed for collaborative trouble-shooting. Messages are coordinated by SessionManager. |

**Figure 12.3.** A shop floor modeled in Java 3D

## 12.4 Data Collection and Distribution

### 12.4.1 Information Flow

Acquiring data from a machine for real-time monitoring, control, and inspection is limited by the available bandwidth for the data transmission. Broadcasting data about all machines to every clients would require sending more messages than necessary, slowing down the transmission of each message, and reducing the application's ability to display data and images in real time. Polling initiated by a client requires two-way communication, while only the information sent to the server from the client is of any use. The best solution to reducing network congestion and ensuring quick data transmission is to have data multicast to only the clients requesting the data, with an open connection established for data streaming, and sending data whenever the data is updated. To this end, the *WISE-SHOPFLOOR* implements a *Publish-Subscribe* design pattern. A client subscribes to information pertaining to a specific machine, leaving an open connection to receive data. When a new sensor signal for that machine is posted, it is published only to those clients who have subscribed to it. In the *WISE-SHOPFLOOR*, this data communication is handled by a modification of the *Pushlet Framework* [12.13]. Figure 12.4 shows the communication pathway for data to and from clients and real machines.

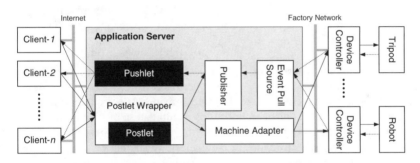

**Figure 12.4.** Information flow between clients and devices

### 12.4.2 Applet–Servlet Communication

As shown in Figure 12.5, a client-side applet of the *WISE-SHOPFLOOR* communicates with *Pushlet*, an HTTP servlet developed by Just van den Broecke, of JustObjects [12.13]. Invoking the *Pushlet* with an HTTP "Get" request with a "subject" parameter allows a client to subscribe to that subject. When receiving a subscription request, the *Pushlet* leaves the connection to the client open, allowing data to be streamed in without reopening a connection for each data communication.

On the client-side of the *Pushlet* is the *JavaPushletClient*. The *JavaPushletClient* sends the request for the *Pushlet* subscription, and opens an input stream from the socket. The *Pushlet* client then loops continuously and checks for data in the data stream. If the publisher has written new data to the stream, the *Pushlet* client overwrites the next most recent data with the new data, ensuring that only the most recent information is used to update a Java3D model. The actual update of the model, however, comes from a different loop. Java 3D provides an interface called InputDevice, which can be registered to the Java 3D Physical Environment [12.14]. Once registered, a schedule is created to call a Polling and Processing method from the InputDevice. In the *WISE-SHOPFLOOR*, this schedule is designed such that the method is called each time a frame is rendered, so that each frame renders a machine with only the most recent information about the machine. Figure 12.5 shows the pathway of applet–servlet communication.

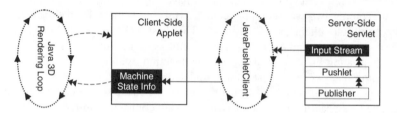

**Figure 12.5.** Streaming based applet–servlet communication

The Publisher (or *SignalPublisher* in Figure 12.2) sends information through the connection established by the *Pushlet*. This data is found by the *JavaPushletClient* loop, and is pushed into a client-side storage location. On a different thread, the Java 3D rendering loop retrieves the data and updates the on-screen Java 3D model for machine condition monitoring.

The *Pushlet Framework* also provides a *Postlet* servlet, used by clients to "Post" events to the application server (or *Commander* in Figure 12.2). When a client wishes to control a machine, he/she needs to seek permission from the *SessionManager* in the application server and enters into the control mode. At any given time, only one client can be granted the control authority for manipulation of a given machine. The client-side applet then connects to the *Postlet* servlet, sending an HTTP "Get" request with the desired instructions as parameters. When the *Postlet* passes the data to the *Commander*, the connection is closed, while the *Commander* sends the data to the appropriate machine controller, on behalf of the client, for remote control.

### 12.4.3 Sensor Signal Collection and Distribution

On the physical machine side, data collection is slightly different. There are many different types of machines that usually have different types of controllers. The *Pushlet* package provides an Adapter, called Event Pull Source [12.13] (see Figure 12.4), which can be extended to obtain data from a required source (real device). Events (sensor signals) are "pulled" from an Event Pull Source at a regular interval, which can be set to a desired increment to approximately replicate the sampling rate for real-time monitoring.

In the collection of sensor data from real machines, the server containing the *Pushlet* servlet actually acts as a client of the machine controllers, establishing a socket connection and working with the machine interface of each controller. The concrete implementation of the Event Pull Source is one adapter between the interface of a machine controller and the interface required by the *Pushlet* framework. However, the communication to the real machine must be in both directions to achieve control, although the Event Pull Source communicates in only a single direction – from the machine to the application server (*SignalCollector* in particular, as shown in Figure 12.2). A machine controller is not able to interpret the *Pushlet* event, and thus will not be a client of the *Pushlet*. Another *Pushlet* adapter, the Machine Adaptor (see Figure 12.4), is required to take information from the *Postlet* (*i.e.* from the client), and send it to a machine controller in the required format. As the *Pushlet Framework* does not provide this functionality, the WISE-SHOPFLOOR uses a wrapper for the *Postlet*, which determines whether data is destined for the publisher or the machine, and thus directs it appropriately.

It should be noted that Java security policy prevents an applet from talking directly with a real device through socket communications. Besides, the socket communications have the drawback of inability to traverse firewalls. This is why a more complex data communication model is introduced here, adopting HTTP streaming as upper-level communication protocol between clients and server while using TCP (Transmission Control Protocol) for lower-level communication with machines. Figure 12.6 shows the three-tier *WISE-SHOPFLOOR* concept and Figure 12.7 illustrates comprehensive data streaming, from a different perspective.

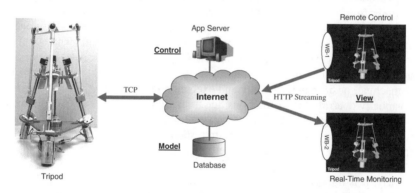

**Figure 12.6.** Combined communication protocols in *WISE-SHOPFLOOR*

## 12.4.4 Virtual Control versus Real Control

There are several differences between the communication required for real control of a physical machine and the communication for virtual control of a machine model. Virtual control mode is used for simulation such as robot trajectory planning or tool path optimization. With virtual control, there is no real machine available to send information about its current status. Instead, a control command is first interpreted on the client side, and the resulting status is sent to the *Postlet*, where it is immediately published without any processing. In other words, the server-side module *Commander* is bypassed. In reality, machines work differently. Instructions must be sent to the machine controllers, which execute the instructions, and then return information regarding its run-time status. Thus, the data packets that a client sends for virtual control and real control do not follow the same format.

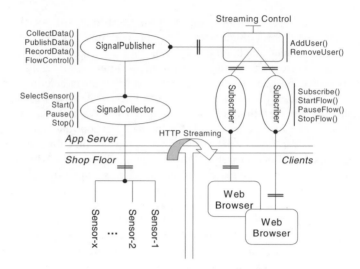

**Figure 12.7.** Upper-level data communication with streaming control

Virtual control considers the environment and control scenario to be "perfect". Instructions from a client are always executed. The "machine" only resides in the computer of the client with control privilege, and data does not need to be processed by another module before being sent to other clients for monitoring. The client with control privilege does not need to subscribe to the data because the data for monitoring is unaltered from the control instructions. However, when controlling a real machine, positioning accuracy and execution delay exist. A machine operator may manually override the instructions due to malfunctions or other exceptions. For true monitoring purposes, the data received for monitoring must come from the actual machine, representing its run-time status. Thus, redirection, as well as interpretation of the different data packet types, must be handled by the wrapper of the *Pushlet*.

Clearly, it is important to avoid any security disasters of hardware at shop floor level while developing a web-based system in real control mode. This concern is addressed in more detail in the next section.

## 12.5 Shop Floor Security

According to an NCMS report [12.15], there is a growing consensus that linking shop floor hardware to the Internet will become the backbone technology for collaborative manufacturing. However, a major concern of implementing Internet or web-based collaborative manufacturing systems is the assurance that proprietary information about the intellectual property owned by the organization or information about the company's operations is available only to authorized individuals. Any web-based collaborative systems must accommodate privacy of the individuals and organizations involved in collaborative activities. In the *WISE-SHOPFLOOR* environment, information about the operations or the information provided by individuals or organizations should only be shared by those involved. Web-based remote monitoring and control typically involve sharing information in the form of detailed run-time operations, as well as real-time and mission-critical hardware controls. For general acceptance of the *WISE-SHOPFLOOR* framework, the secrecy of the proprietary information must be properly maintained. To meet security requirements, our approach depends on a security infrastructure built into the Java platform. This security infrastructure consists of byte-code verification, security policies, permissions, and protection domains. In addition to the Java security infrastructure, other security and privacy issues are considered in the framework for implementation, including digital rights management for information access and sharing, data encryption, and process confidentiality protection [12.16].

Figure 12.8 shows how a remote user can get access indirectly to the real shop floor without violating shop floor security policy. All data communication between the end user and a shop floor device goes through the application server, and is processed by a server-side module before passing the data to its receiver. As mentioned in Section 12.3, only the server-side modules are allowed to collect sensor data or manipulate devices within their limits. On the other hand, all end users are physically separated from the real shop floor by using segmented networks (Intranet/Internet, and Factory Network) with the application server as a gateway.

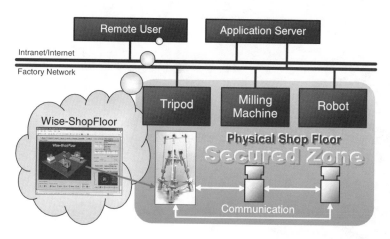

**Figure 12.8.** Indirect secure access to a physical shop floor

In what follows, we will demonstrate the effectiveness of the *WISE-SHOPFLOOR* prototype system through two case studies on remote robot control and web-based machining, with brief performance evaluations.

## 12.6 Case Study 1: Remote Robot Control

Realizing that a parallel structure is much stiffer, researchers and engineers started developing parallel kinematic machines (PKMs) in the past decade, and these are now considered another milestone in NC machine tool history. One of the advantages of PKM is that no motor has to carry another. This feature offers the potential in the machine tool industry to develop lightweight machines running at high speeds with high accuracy.

The testing device used in this section as the first case study is a 3-dof (degrees of freedom) parallel kinematic machine, a *tripod*, developed at the authors' lab. In order to conduct web-based remote monitoring and control in the *WISE-SHOPFLOOR* environment, a kinematic model and a geometric model of the tripod are needed. The kinematic model formulates the link–joint relationship of the tripod and it is then embedded into the geometric model for 3D graphics rendering.

### 12.6.1 Constrained Kinematic Model

The tripod shown in Figure 12.9(a) is based on the fixed-length-leg configuration with linear motion component actuators. A tool can be mounted on the moving platform (the top plate) of the tripod. The movement of the moving platform is controlled by sliding the three fixed-length legs along their respective guideways. Details on kinematic/dynamic analysis and structure optimization of this tripod can be found in [12.17].

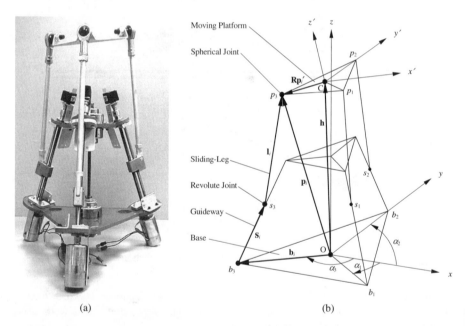

(a)                                                        (b)

**Figure 12.9.** A 3-dof parallel kinematic machine: (a) tripod prototype, (b) kinematic model

As shown in Figure 12.9, the tripod consists of three kinematic chains, including three fixed-length legs with identical topology driven by ball screws, which connect the moving platform to the fixed base. Although the tripod under study may rotate around $x$ and $y$ axes and move along the $z$ axis, only three variables are independent, while the other three are dependent [12.18]. The kinematic equation for the position of the $i$th spherical joint ($i = 1, 2, 3$) is given as

$$\mathbf{p}_i = \mathbf{h} + \mathbf{R}\mathbf{p}'_i \tag{12.1}$$

where, $\mathbf{p}_i = [\, p_{ix}, p_{iy}, p_{iz} \,]^T$ is the vector representing the position of the $i$th spherical joint in the global coordinate system O-$xyz$, $\mathbf{p}'_i$ is the vector representing the same point in the local coordinates C-$x'y'z'$, $\mathbf{h} = [\, x_c, y_c, z_c \,]^T$ is the vector representing the position of the moving platform, and $\mathbf{R}$ is the rotation matrix of the moving platform in terms of rotation angles $\theta_x$, $\theta_y$, and $\theta_z$ around $x$, $y$, and $z$ axes, respectively. Among the six motion components of the moving platform, it is known that $x_c$, $y_c$, and $\theta_z$ are dependent variables. Since the legs are constrained by the revolute joints, vector $\mathbf{p}_i$ is confined in the plane that is defined as

$$p_{iy} = \tan(\alpha_i)\, p_{ix} \tag{12.2}$$

where $\alpha_i$ is the angle of the $i$th guideway relative to the $x$ axis. By expressing the rotation matrix as

$$\mathbf{R} = \begin{bmatrix} r_{11} & r_{12} & r_{13} \\ r_{21} & r_{22} & r_{23} \\ r_{31} & r_{32} & r_{33} \end{bmatrix} \tag{12.3}$$

substituting Equation (12.3) into Equation (12.1), and then into Equation (12.2) yields

$$\left( y_c + r_{21}p'_{ix} + r_{22}p'_{iy} \right) = \tan\alpha_i \left( x_c + r_{11}p'_{ix} + r_{12}p'_{iy} \right) \tag{12.4}$$

In this particular tripod design, $\alpha_1$, $\alpha_2$, and $\alpha_3$ are $-30°$, $90°$, and $-150°$, respectively. After mathematic multiplication, this leads to

$$x_c = -\frac{\sqrt{3}}{3} L_p r_{12} \tag{12.5a}$$

$$y_c = -\frac{\sqrt{3}}{6} L_p r_{11} + \frac{\sqrt{3}}{6} L_p r_{22} \tag{12.5b}$$

$$r_{12} = r_{21} \tag{12.5c}$$

where $L_p$ is the side length of the triangular moving platform. In terms of rotation angles $\theta_x$, $\theta_y$, and $\theta_z$, rotation matrix $\mathbf{R}$ can be expressed as

$$\mathbf{R} = \begin{bmatrix} \cos\theta_y \cos\theta_z \\ \sin\theta_x \sin\theta_y \cos\theta_z + \cos\theta_x \sin\theta_z \\ -\cos\theta_x \sin\theta_y \cos\theta_z + \sin\theta_x \sin\theta_z \end{bmatrix}$$

$$\left. \begin{array}{cc} -\cos\theta_y \sin\theta_z & \sin\theta_y \\ -\sin\theta_x \sin\theta_y \sin\theta_z + \cos\theta_x \cos\theta_z & -\sin\theta_x \cos\theta_y \\ \cos\theta_x \sin\theta_y \sin\theta_z + \sin\theta_x \cos\theta_z & \cos\theta_x \cos\theta_y \end{array} \right] \tag{12.6}$$

Based on Equations (12.3) and (12.6), the constraint equations can be derived from Equations (12.5a)–(12.5c) as

$$x_c = -\frac{\sqrt{3}}{3} L_p \cos\theta_y \sin\theta_z \tag{12.7a}$$

$$y_c = \frac{\sqrt{3}}{6} L_p \left( -\cos\theta_y \cos\theta_z + \cos\theta_x \cos\theta_z - \sin\theta_x \sin\theta_y \sin\theta_z \right) \tag{12.7b}$$

$$\theta_z = \arctan\left( -\frac{\sin\theta_x \sin\theta_y}{\cos\theta_x + \cos\theta_y} \right) \tag{12.7c}$$

From Equations (12.7a)–(12.7c), it can be seen that $x_c$, $y_c$, and $\theta_z$ are three dependent variables depending on $\theta_x$ and $\theta_y$. Although $z_c$ does not appear in Equations (12.7a)–(12.7c), it does not mean that $x_c$, $y_c$, and $\theta_z$ are completely independent of $z_c$. In fact, the allowable angles of $\theta_x$ and $\theta_y$ are functions of $z_c$.

Taking the derivative of Equations (12.7a)–(12.7c) with respect to time yields expressions for the velocities as

$$\dot{x}_c = -\frac{\sqrt{3}L_p}{3}\left( -\sin\theta_y \sin\theta_z \dot{\theta}_y + \cos\theta_y \cos\theta_z \dot{\theta}_z \right) \tag{12.8a}$$

$$\dot{y}_c = \frac{\sqrt{3}L_p}{6}(\sin\theta_y \cos\theta_z \dot{\theta}_y + \cos\theta_y \sin\theta_z \dot{\theta}_z - \sin\theta_x \cos\theta_z \dot{\theta}_x - \cos\theta_x \sin\theta_z \dot{\theta}_z$$

$$- \cos\theta_x \sin\theta_y \sin\theta_z \dot{\theta}_x - \sin\theta_x \cos\theta_y \sin\theta_z \dot{\theta}_y - \sin\theta_x \sin\theta_y \cos\theta_z \dot{\theta}_z) \tag{12.8b}$$

$$\dot{\theta}_z = -\cos^2\theta_z \frac{\left(1 + \cos\theta_x \cos\theta_y \right)\left( \sin\theta_y \dot{\theta}_x + \sin\theta_x \dot{\theta}_y \right)}{\left( \cos\theta_x + \cos\theta_y \right)^2} \tag{12.8c}$$

Similarly, expressions for accelerations can be obtained from Equations (12.8a)–(12.8c), which are omitted here due to the page limitation.

### 12.6.2 Inverse Kinematic Model

While constrained kinematics of the tripod is used for monitoring, its inverse kinematics is needed for position control. Considering the $i$th sliding-leg/guideway system, the kinematic equation of the position of the $i$th spherical joint, $i.\,e.$ Equation (12.1), can be re-written as

$$\mathbf{p}_i = \mathbf{b}_i + \mathbf{s}_i + \mathbf{l}_i \tag{12.9}$$

where $\mathbf{b}_i$ is the vector representing the position of the lower end of the $i$th guideway attached to the base, $\mathbf{s}_i$ is the vector representing the displacement along the $i$th guideway, and $\mathbf{l}_i$ is the vector representing the $i$th sliding leg. Since $\mathbf{s}_i = s_i \mathbf{u}_i^s$ and $\mathbf{l}_i = l_i \mathbf{u}_i^l$, where $\mathbf{u}_i^s$ and $\mathbf{u}_i^l$ are the direction vectors of the $i$th guideway and the $i$th leg, respectively, the actuator displacement $s_i$ can be solved considering that the leg length is a constant

$$\left| \mathbf{p}_i - \mathbf{b}_i - s_i \mathbf{u}_i^s \right| = l_i \qquad (12.10)$$

where $l_i$ is the length of the $i$th sliding leg. For given $\theta_x$, $\theta_y$, and $z_c$, dependent variables $x_c$, $y_c$, and $\theta_z$ can be determined by Equations (12.7a)–(12.7c), then $\mathbf{h}$ and $\mathbf{R}$ are fully defined. With this, $\mathbf{p}_i$ can be determined by Equation (12.1), and subsequently $s_i$ can be solved using Equation (12.10). The true solution of Equation (12.10) should be the one closer to the previous value, that is

$$s_i = \left( s_i^{(k)}, \; \min_{k=1,2} \left| s_i^{(k)} - s_i(j-1) \right| \right) \qquad (12.11)$$

where $j$ stands for the $j$th step. In practice, the initial value of $s_i$ is provided by an encoder.

### 12.6.3 Java 3D Scene-graph Model

Java 3D is designed to be a mid to high-level fourth-generation 3D API [12.19]. What sets a fourth-generation API apart from its predecessors is the use of scene-graph architecture for organizing graphical objects in the virtual 3D world. Unlike the display lists used by the third-generation APIs (*e. g.* VRML, OpenInventor, and OpenGL), scene graphs mercifully hide a lot of rendering details from developers while offering opportunities for more flexible and efficient rendering. Enabled by the scene-graph architecture, Java 3D provides an abstract, interactive imaging model for behavior control of 3D objects. Because Java 3D is part of the Java pantheon, it assures users ready access to a wide array of applications and network support functionality [12.14], crucial to the *WISE-SHOPFLOOR* implementation. Figure 12.10 illustrates a Java 3D scene graph model of the tripod test bed. This test bed is a gantry system, which consists of an x-table and a tripod unit mounted on a y-table. The end effecter on the moving platform is driven by three sliding-legs that can move along three guideways, respectively.

Java 3D differs from other scene graph-based systems in that scene graphs may not contain cycles. Thus, a Java 3D scene graph is a directed acyclic graph. The individual connections between Java 3D nodes are always a direct relationship: parent to child, as shown in Figure 12.10. The scene graph contains a complete description of the entire scene, or *virtual universe*. This includes the geometry data, the attribute information, and the viewing information needed to render the scene from a particular point of view.

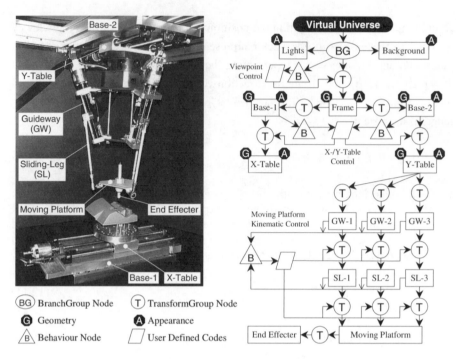

**Figure 12.10.** Java 3D scene-graph model of tripod

As shown in Figure 12.10, the *Virtual Universe* object provides grounding for the entire scene graph. All Java 3D scene graphs must connect to a *Virtual Universe* object to be displayed. A scene graph itself starts with the *BranchGroup* (BG) nodes (although only one BG node in this case). A *BranchGroup* node serves as the root of a sub-graph, or branch graph, of the scene graph. The *TransformGroup* nodes inside a branch graph specify the position, the orientation, and the scale of the geometric objects in the virtual universe. Each geometric object consists of a *Geometry* object, an *Appearance* object, or both. The *Geometry* object describes the geometric shape of a 3D object. The *Appearance* object describes the appearance of the geometry (color, texture, material reflection characteristics, *etc.*). The behavior of the tripod model is controlled by *Behavior* nodes, which contain the kinematic models, user-defined control codes and state variables. Sensor signal processing can be embedded into the codes as well for remote monitoring. Once applied to a *TransformGroup* node, the so-defined behavior control affects all the descending nodes. In this example, the movable objects (X-Table, Y-Table, and Moving Platform unit) can be controlled by using three control nodes, for both online monitoring/control and offline simulation. As the 3D model is connected with its physical counterpart through the control nodes by low-volume message passing (real-time sensor signals and control commands transmissions, *etc.*), it becomes possible to remotely manipulate the real tripod through its Java 3D model.

It is worthy of mention that although a Java 3D scene-graph model is defined in Java language as classes, a third-party tool can also be used to generate Java 3D models directly from a commercial CAD package, such as AutoCAD.

### 12.6.4 Remote Tripod Manipulation

As shown in Figure 12.2, our web-based remote device monitoring and control are conducted by using *StatusMonitor* and *CyberController*, which communicate indirectly with the device controller through its application server. In the case of tripod monitoring and control, they are further facilitated by the kinematic models derived in Sections 12.6.1 and 12.6.2, to reduce the amount of data traveling between web browsers and the tripod controller. Whereas all registered users can do monitoring simultaneously, only one user is authorized at any one time to manipulate the tripod. The required position $z_c$ and orientations $\theta_x$, $\theta_y$ of the moving platform are converted into the joint coordinates $s_i$ ($i = 1, 2, 3$) by the inverse kinematics for both Java 3D model rendering at client-side and device control at controller-side. With use of the embedded kinematic model, the actual data packet transmitted through the network is limited to only a few dominant parameters. Currently, the same data format is used in monitoring and control of the tripod test bed as shown in Figure 12.10. A typical data packet of eight floating numbers is listed below:

| $S_1$ | $S_2$ | $S_3$ | $x$ | $y$ | $z_c$ | $\theta_x$ | $\theta_y$ |
|-------|-------|-------|-----|-----|-------|-----------|-----------|

where $S_{i,i=1,2,3}$ is the displacement of $i$th sliding leg along its guideway; $x$ and $y$ are the coordinates of the x-y table; $\theta_x$ and $\theta_y$ are the pitch and roll angles of the moving platform; and $z_c$ is the center position of the moving platform along the z-axis. Figure 12.11 shows one screenshot of a *WISE-SHOPFLOOR* user interface for web-based tripod monitoring and control.

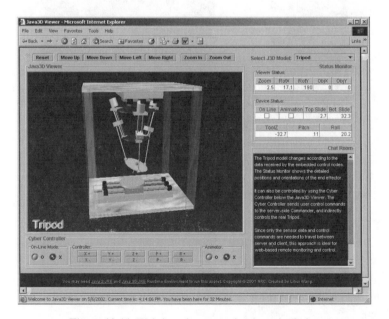

**Figure 12.11.** Web-based remote tripod manipulation

Regarding the physical setup, the three sliding-legs of the tripod are driven by three DC servomotors (24V) combined with three lead screws. Each of the actuators has a digital encoder (1.25 μm/count) for position feedback. The position data $s_i$ ($i = 1, 2, 3$) of the sliding-legs are multicasted to the clients, who subscribed to the sensor signal publishing service, for remote monitoring using the constrained kinematics. A sampling rate of 1 kHz is used for the case study.

The performance of the case study depends on both the PC configuration and the network throughput at the time of use. Our feasibility testing shows an average of 30 ms time delay in the lab environment, with instant data communication.

Other than the gantry system shown in Figure 12.10, the tripod unit and its web-based control approach can be applied to remote machining and inspection as demonstrated in Figure 12.12. In these cases, the tripod is used as a programmable add-on device for the purpose of enhancing the capability of the base machine by providing it with a more flexible rage of motion.

(a) A tripod-based spindle head attached to a low-end machine tool

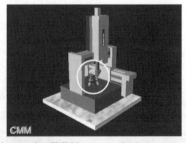

(b) Tripod as a work stage for CMM laser scanning

(c) Macro-micro robotic system with an industrial robot and a tripod

**Figure 12.12.** Other applications of web-based tripod monitoring and control

## 12.7 Case Study 2: Remote CNC Machining

This section introduces a *WISE-SHOPFLOOR* application on remote CNC machining, which is a constituent component of web-based manufacturing. Techniques of condition-based monitoring play important roles in remote CNC machining and can be integrated to our approach to guarantee high product quality and safe operations. However, this section mainly focuses on the enabling technology toward realization of remote CNC machining in web environment.

### 12.7.1 Test Bed Configuration

The test bed used as a case study for remote CNC machining is *Wahli-51* – a 5-axis horizontal milling machine. As shown in Figure 12.13, it is equipped with a PC-based OAC (open architecture controller) that serves as a gateway between Wahli and the application server. The Wahli is controlled through a user interface, implemented as a C++ application, running on the same OAC PC, which in turn, beyond a local operator, allows a remote user to send NC (numerical control) line commands to the OAC from a remote web browser and have them executed locally. Remote users can monitor and/or control the Wahli indirectly through the server-side modules (*SignalCollector*, *SignalPublisher*, and *Commander*, as shown in Figure 12.2), which perform data collection/distribution and facilitate remote CNC machining. While Java is used for implementing functions in applet and servlets, C++ is selected for local or remote execution control of the OAC. For security reasons, two different communication protocols are adopted for data transmissions at different levels: (1) high-level client–server HTTP Streaming Protocol that is firewall-transparent, and (2) low-level controller-server TCP (Transmission Control Protocol) for better hardware protection. Based on this configuration, it allows a remote user to monitor the absolute and relative motions of all axes as well as to control the spindle speed and feed rate for CNC machining.

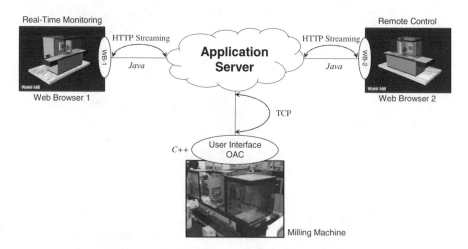

**Figure 12.13.** Test bed configuration for remote CNC machining

### 12.7.2 Java 3D Visualization

For the sake of network bandwidth conservation, Java 3D technology is again used for geometric modeling of the CNC machine, as an alternative to a camera-based solution for web-based visualization. Similar to Case Study 1, a 3D Wahli machine model is created in Java 3D scene graph, representing its physical counterpart in the *WISE-SHOPFLOOR* environment. Wrapped in an applet, the model only needs to be downloaded once from its application server, and remains alive through applet-servlet communications. The data transmitting through the network are limited to the data showing runtime status of the machine (position, orientation, feed rate, spindle speed, *etc.*). Other sensor data such as cutting force, workpiece thermography, and AE signals can be collected and processed by the application server. Only the results are sent to clients to guide remote machining operations.

Wahli is a 5-axis horizontal milling machine that requires linear motion control of X, Y, and Z axes, as well as rotary motion control of B and C axes (around Y and Z axes, respectively). A combined rotary stage having two rotary motions is mounted on top of an X-table, whereas the spindle head of the Wahli provides the other two linear motions along Y and Z axes. Figure 12.14 illustrates the Java 3D scene graph model of the Wahli machine.

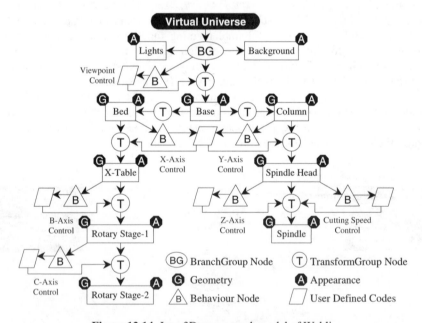

**Figure 12.14.** Java 3D scene graph model of Wahli

The scene graph model contains a complete description of the Wahli machine, including geometry data, attribute information, and viewing information needed for off-site machining monitoring. The behavior of the Wahli is controlled by *Behavior* nodes, which is subject to sensor data and is implementation-specific. The results of sensor data processing can be embedded into the codes for remote monitoring. In this case study, the 5-axis motions (X-Table, Rotary Stage-1, Rotary Stage-2, Spindle Head, and Spindle) are controlled by their corresponding behavior control nodes, for both online monitoring/control and offline simulation. As the Java 3D model is connected with its physical counterpart through the control nodes by low-volume message passing (real-time machining status and NC control commands), it becomes possible to remotely machine a part on the real Wahli machine through the *WISE-SHOPFLOOR* environment, where the physical security is addressed separately.

### 12.7.3 Data Communication

As mentioned in Section 12.4.2, Java security policy prevents an applet from talking directly with the Wahli controller through socket communications, which also limits the data communication within the firewall. As shown in Figure 12.7, the task of machining data collection and NC code distribution is handled by two server-side modules, namely, *SignalCollector* and *SignalPublisher*. These modules are developed based on the popular publish-subscribe design pattern. For every client registered with the *Registrar* (see Figure 12.2), there is one *Subscriber* who is knowledgeable about the user's request and responsible for maintaining an active channel between the client and the application server. Multi-threading techniques are applied to the module implementations to meet the data requirements from different end users.

Wahli data collection is accomplished over the TCP connection using a series of 12 floating numbers and one long integer that form one data packet. In the current implementation, a typical data packet is defined as follows,

| 1 | 2 | 3 | 4 | 5 | 6 | 7 | 8 | 9 | 10 | 11 | 12 | 13 |
|---|---|---|---|---|---|---|---|---|----|----|----|----|
| Relative position of 5 axes | | | | | Absolute position of 5 axes | | | | | FR | SS | CW |

where, FR, SS and CW denote feed rate, spindle speed and NC control word, respectively. A control word is a reserved long integer value indicating the status of the machine, including operation mode, such as *manual* (0x0001), *auto* (0x0002), or *jogging* (0x0040), coordinate system, axis status, *etc*. As with the real Wahli, controller the data packet provides both relative positions and absolute positions of the five motion axes for ease of off-site monitoring and control.

### 12.7.4 Remote Machine Control

Remote machine control of the Wahli is possible by sending proper movement commands (NC lines) through the applet–servlet–controller communication shown in Figure 12.13. In order to remotely control the Wahli, user authentication and control rights authorization must be accomplished for the client who requests this function. Control rights authorization is done by setting a bit in the control word in a data packet that is sent to the client. If the client has requested control rights and the bit is set, a message will appear on the screen notifying the user that he/she is now in control of the machine. Figure 12.15 shows a screenshot of the web user interface with Wahli selected for monitoring and NC control.

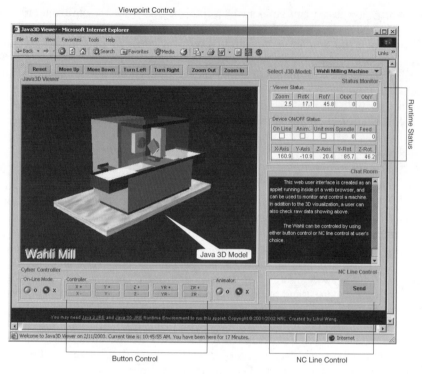

**Figure 12.15.** Web-based real-time monitoring and remote machine control

Motion control of the Wahli machine is accomplished in one of the two modes: jogging control or NC line control. The jogging control is with the use of the individual axis control buttons as labeled in Figure 12.15. Each button controls one axis along one direction. For instance, if we want to move the table of Wahli along X-axis in the positive direction, the following must occur.

1. Transmit a default control word from the remote web browser (applet) to the controller-side PC, with the following bit set.

```
0x04030040 = 0000 0100 0000 0011 0000 0000 0100 0000
                    A            BC                D
```

A – control request bit,
B – machine-on bit,
C – cycle start bit,
D – jogging mode bit (allowing machine control through jog buttons).

2. Transmit an active control word when the button X+ is pressed. This active control word remains the same until the button is released, at which time it returns to the default control word. The active control word has the following bit set for an X-axis positive button hit.

```
0x04430040 = 0000 0100 0100 0011 0000 0000 0100 0000
                A     E   BC                 D
```

E – X-axis positive bit.

Although the jogging button control is straightforward for users who may not be knowledgeable about NC codes, it cannot be used during the actual NC machining due to its limitations of single-axis control. Instead, the NC line control is recommended. This method sends individual NC lines directly to the Wahli for execution. An experienced remote user can type in one NC command line and send it as an NC block to the Wahli controller. For example, the following NC line tells the Wahli to proceed from the current position to the next, incrementally by (20, –30, 10) in linear rapid traverse mode. At the same time, the Wahli controller sets the spindle speed to 3000 rpm and turns the flood coolant on.

```
G0 X+20 Y-30 Z+10 S3000 M8
```

By using the NC line control, a remote user can send not only individual lines of NC code but an entire NC program line by line or in a batch. This allows the client to demonstrate full-scale NC machining and test new programs on the machine from a remote location [12.20]. However, for security reasons, NC code parsing must be accomplished by the controller interface to check the syntax and validity of any NC lines entered into the text box and sent to the OAC controller.

Most web-based systems rely on camera-based monitoring to properly guide remote control operations. Compared with an 8-bit VGA camera image of 640×480 (307,200 bytes), our packet size is only 52 bytes – a significant size reduction suitable for web-based real-time applications. In addition, a Java 3D model can provide users with a 360° full-range viewing angle, zooming function, and high resolution, useful for off-site trouble-shooting.

## 12.8 Toward Condition-based Monitoring

Although the *WISE-SHOPFLOOR* framework was initially designed as a web-based system for real-time monitoring and control, it imposes no limitation on data transmitting through the network. In addition to signals of position, orientation, feed rate, spindle speed, *etc.* for motion control, other sensor data such as cutting force, workpiece thermography, and AE signals can be collected and processed by the application server, using the techniques reported in other chapters. The results are then sent to clients to properly guide remote robot control or machining operations for product quality assurance. As part of our ongoing *iShopFloor* project [12.21], the required real-time manufacturing intelligence can also be collected by integrating the *WISE-SHOPFLOOR* with condition-based monitoring in the following area:

- *Cutting force-based tool condition monitoring and NC code compensation*
  The cutting forces will be collected and used to calculate proper feed rate and tool–workpiece relative deformation. The results can be directly used for NC code compensation and to meet tolerance requirements.

- *Thermal deformation compensation*
  The temperature distribution (collected by thermocouples or captured by thermography) over a machine tool and a workpiece can be used to calculate thermal deformations, especially the relative deformation between the cutting tool and the workpiece. The relative deformation is then used to offset G-code for proper compensation.

- *Chatter prevention*
  The same techniques reported in the other chapters can be implemented in the *WISE-SHOPFLOOR* to prevent chatter during NC machining.

- *Automatic robot path planning and setup calibration*
  Data retrieved through motion trackers, laser scanners, or position sensors can be integrated to a robot path planning or setup planning system in part assembly operations, which is helpful for remote robot calibration.

Empowered by the latest computing advancement and Java 3D technology, the above-mentioned force and temperature distributions can be easily displayed on Java 3D models in color maps or contour lines for remote real-time monitoring, which is otherwise impossible if using conventional monitoring systems.

## 12.9 Conclusions

This chapter presents a novel approach towards web-based real-time monitoring and control. It is implemented as a web-based system on top of the *WISE-SHOPFLOOR* framework, using a three-tier view-control-model architecture. The ultimate goal of the combined web-based and sensor-driven approach is to reduce network traffic by integrating real-time sensor data with 3D computer graphics for visualization, while still providing remote users with an well-informed intuitive environment. A publish-subscribe design pattern is implemented as Java servlets for efficient sensor data collection and distribution via a secure application server. As opposed to camera-based monitoring systems characterized by large numbers of bitmap images, our approach relies only on a narrow network bandwidth. The significantly reduced network traffic makes real-time applications practical over the web, which marks one step towards web-based collaborative manufacturing.

A platform-transparent prototype system is developed and validated through two case studies on remote robot control and CNC machining. The results demonstrate the promise of this approach in a distributed e-manufacturing environment. Using the same approach, other sensor data such as temperature, vibration and force can also be integrated and displayed in color maps or contour lines for machine condition monitoring, compensation and adaptive control. The real-time information is also found useful for distributed process planning, adaptive setup planning, tool path optimization, and dynamic scheduling.

As the decentralization of business grows, the application potential of this research is anticipated to grow withit. Examples include collaborative design, control simulation, virtual machining, operator training, facility touring, and off-site trouble-shooting, in addition to remote real-time monitoring and control.

# References

[12.1]   Caldwell, N. H. M. and Rodgers, P. A., 1998, "WebCADET: Facilitating Distributed Design Support," *IEE Colloquium on Web-based Knowledge Servers*, UK, pp. 9/1–9/4.

[12.2]   Smith, C. S. and Wright, P. K., 1996, "CyberCut: A World Wide Web Based Design-to-Fabrication Tool," *Journal of Manufacturing Systems*, **15**(6), pp. 432–442.

[12.3]   Adelson, B., 1999, "Developing Strategic Alliances: A Framework for Collaborative Negotiation in Design," *Research in Engineering Design*, **11**, pp. 133–144.

[12.4]   GE Fanuc, 2005, http://www.gefanuc.com/en/ProductServices/AutomationSoftware/Hmi_Scada/CIMPLICITY/index.html.

[12.5]   Waurzyniak, P., 2001, "Electronic Intelligence in Manufacturing," *Manufacturing Engineering*, (3), pp. 44–67.

[12.6]   MDSI, 2005, http://www.opencnc.com/Solutions/CNC_Controls/CNC_Controls.asp.

[12.7]   Flexlink, 2005, http://www.flexlink.com/.

[12.8]   Mazak, 2005, http://www.mazak.jp/english/cyber/outline/index.html.

[12.9]   Mori Seiki, 2005, http://www.moriseiki.co.jp/english/index.html.

[12.10]  Memex, 2005, http://www.e-manufacturing.com/.

[12.11]  Wang, L., Shen, W. and Lang, S., 2004, "Wise-ShopFloor: A Web-Based and Sensor-Driven e-Shop Floor," *ASME Journal of Computing and Information Science in Engineering*, **4**(1), pp. 56–60.

[12.12]  Wang, L., Wong, B., Shen, W. and Lang, S., 2002, "Java 3D Enabled Cyber Workspace," *Communications of the ACM*, **45**(11), pp. 45–49.

[12.13]  van den Broecke, J., 2000, "Pushlets, Part 1: Send Events from Servlets to DHTML Client Browsers," *JavaWorld*, March 2000.

[12.14]  Sowizral, H., Rushforth, K. and Deering, M., 2001, *The Java 3D API Specification*, 2nd Edition, Addison-Wesley, Boston, MA, USA.

[12.15]  NCMS, 2001, "Factory-Floor Internet: Promising New Technology or Looming Security Disaster," *Manufacturing in Depth*, National Center for Manufacturing Sciences, USA.

[12.16]  Xu, Y., Song, R., Korba, L., Wang, L., Shen, W. and Lang, S., 2005, "Distributed Device Networks with Security Constraints," *IEEE Transactions on Industrial Informatics*, **1**(4), pp. 217–225.

[12.17]  Zhang, D., Wang, L. and Lang, S., 2005, "Parallel Kinematic Machines: Design, Analysis and Simulation in an Integrated Virtual Environment," *ASME Journal of Mechanical Design*, **127**(4), pp. 580–588.

[12.18]  Wang, L., Sams, R., Verner, M. and Xi, F., 2003, "Integrating Java 3D Model and Sensor Data for Remote Monitoring and Control," *Robotics and Computer-Integrated Manufacturing*, **19**(1–2), pp. 13–19.

[12.19]  Barrilleaux, J., 2001, *3D User Interfaces with Java 3D*, Manning Publications Co.

[12.20]  Wang, L., Orban, P., Cunningham, A. and Lang, S., 2004, "Remote Real-Time CNC Machining for Web-Based Manufacturing," *Robotics and Computer-Integrated Manufacturing*, **20**(6), pp. 563–571.

[12.21]  Shen, W., Lang, S. and Wang, L., 2005, "iShopFloor: An Internet-Enabled Agent-Based Intelligent Shop Floor," *IEEE Transactions on Systems, Man, and Cybernetics*, Part C, **35**(3), pp. 371–381.

# 13

## An Intelligent Nanofabrication Probe for Surface Displacement/Profile Measurement

Wei Gao

Tohoku University
Sendai, 980-8579, Japan
Email: gaowei@cc.mech.tohoku.ac.jp

**Abstract**
This chapter describes a nanofabrication probe for surface displacement measurement and/or surface profile measurement. The probe is the combination of a fast-tool-control (FTC) cutting unit and a force sensor. The FTC cutting unit, which consists of a ring-type PZT actuator and a nanometer capacitance-type displacement sensor, is used for diamond turning of complex surface profiles. The force at the interface between the tip of the cutting tool and the surface can be detected by the force sensor with a sensitivity of 0.01 mN through employing an AC modulation technique. In the displacement/surface profile measurement mode, the surface is tracked by the tool through servo-control of the contact force by the FTC-unit, in which the displacement/surface profile can be obtained from the capacitance-type displacement sensor. Probe design and evaluation are presented.

## 13.1 Introduction

Diamond turning is an ultra-precision manufacuring process in which a single point diamond cutting tool with a sharp edge is used to cut ductile materials such as aluminum, copper and nickel, *etc.* [13.1]–[13.6]. Figure 13.1 shows a simple model of a diamond turning machine consisting of a spindle, an X-directional cross-slide, a Z-directional carriage and an NC controller (the carriage and NC controller are not shown in the Figure for clarity). The workpiece and tool are mounted on the spindle and the cross-slide, respectively. The spindle is mounted on the carriage with its motion axis parallel to that of the spindle. The tool is moved across the workpiece by the X-slide to cut the workpiece surface.

A flat surface can be generated by maintaining the depth of cut in the Z-direction, which corresponds to the position of the tool relative to the workpiece surface, at a certain value. It is also possible to generate a non-flat surface, such as a parabolic surface, through moving the workpiece/spindle along the Z-direction using the carriage. However, because the carriage has a very low bandwidth, on the order of

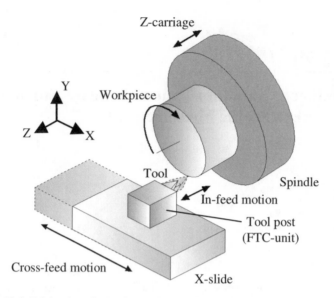

**Figure 13.1.** Fabrication of complex surface profiles on a diamond turning machine

10 Hz, it is dificult for the carriage to move the heavy spindle in a high speed, which is essential for generating complex surface profiles. Basically only rotationally symmetric surface profiles with limited spatial frequency bandwidth can be generated by a conventional diamonde turning machine.

The fast tool control (FTC) technique has been developed to overcome the shortcomings of conventional diamond turning machines when generating complex surfaces [13.7]–[13.13]. In the FTC, the depth of cut is controlled through changing the tool position by a PZT actuator instead of moving the spindle using the carriage. A PZT actuator has a high bandwidth (measured in kHz) and a high stiffness compared to that of the diamond turning machine. The single diamond cutting tool can be actuated by the PZT at high speed because the tool has a small mass, on the order of several grams. In addition, the PZT actuator also has the advantage of nanometer resolution. The advantages of the PZT actuator makes it possible to carry out nanofabrication of complex surfaces with high spatial frequency components on a diamond turning machine.

On the other hand, diamond turning is a complicated process, influenced by many factors, such as fabrication speed, feed rate, cooling fluid, tool edge geometry, temperature, *etc.* To assure the fabrication accuracy of complex surfaces, it is necessary to monitor the machining process. The cutting force is a good parameter for this purpose [13.14]–[13.16]. This chapter presents a FTC-unit integrated with a force sensor. Taking into consideration that cutting forces are small in the nanofabrication of complex surfaces, a piezoelectric force sensor with high sensitivity and high stiffness is employed. Measurement of the displacement and/or profile of the workpiece surface can also be carried out through detecting and controlling the contact force between the tool and the surface, based on the force sensor output.

## 13.2 Design of the Nanofabrication Probe

### 13.2.1 Concept of the Probe

Figure 13.2 shows a schematic of the nanofabrication probe. The probe is composed of a fast-tool-control unit (FTC-unit) and a force sensor. The FTC-unit consists of a diamond tool, a hollow ring-type PZT actuator and a displacement sensor. The diamond tool is mounted on the moving end of the PZT. The displacement sensor, which is the capacitance-type in most cases, is located inside the PZT. The displacement of the PZT as well as that of the diamond tool along the Z-direction can be measured by the displacement sensor. The hysteresis error and linearity error of the PZT can be compensated for with servo control based on measurements from the displacement sensor.

The FTC-unit is pushed against the force sensor mounted on the base by cap screws. Considering the sensitivity and stiffness necessary for nanofabrication and profile measurement, a piezoelectric force sensor is employed. The pre-load applied to the force sensor can be adjusted by the cap screws.

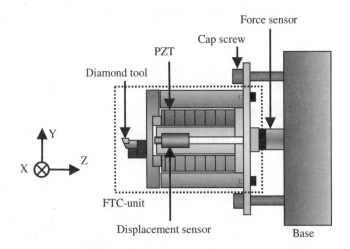

**Figure 13.2.** Conceptual design of the nanofabrication probe

Figure 13.3 shows a schematic for generating a complex surface profile using the nanofabrication probe. The workpiece is mounted on the spindle and the nanofabrication probe is placed on the X-slide of a diamond turning machine. The movement direction of the X-slide is from the outermost circle with a radius of $r_0$ to the center of the workpiece. Assume that the rotation motion of the spindle and the translational motion of the X-slide are synchronized. As can be seen in Figure 13.4, the polar coordinates of the tool tip position in the XY plane can be expressed by

$$\left(r_i, \theta_i\right) = \left(r_0 - \frac{Fi}{PT}, 2\pi \frac{i}{P}\right) \qquad i=0, 1, \dots, N-1 \qquad (13.1)$$

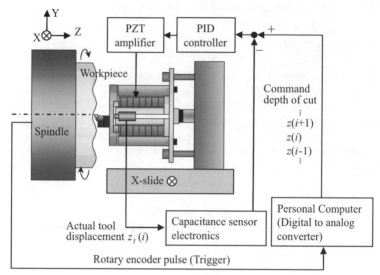

**Figure 13.3.** Schematic for generating complex surface profile by the probe

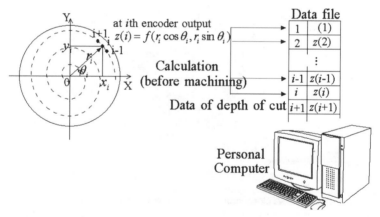

**Figure 13.4.** Depth of cut data for generation of complex surface profile

where F is the feed rate of the X-slide in units of mm/min, $P$ is the pulse number of the rotary encoder of the spindle in each revolution with a unit of pulse/rev, $T$ is the rotational speed of the spindle with units rpm (revolution per minute), $i$ is the $i$th rotary encoder pulse, and $N$ is the total pulse number of the rotary encoder for the X-slide reaches the center, respectively.

To generate a complex surface profile expressed by a function of $f(x, y)$, the depth of cut at each point is calculated as follows and stored in the memory of a personal computer before fabrication:

$$z(i) = f(x_i, y_i)$$

$$= f(r_i \cos \theta_i, r_i \sin \theta_i) \tag{13.2}$$

$$= f\left[\left(r_0 - \frac{Fi}{PT}\right)\cos\left(2\pi \frac{i}{P}\right), \left(r_0 - \frac{Fi}{PT}\right)\sin\left(2\pi \frac{i}{P}\right)\right]$$

When fabrication starts, the depth-of-cut data in the memory of the personal computer are output to the controller one by one through a digital-to-analog converter, responding to the trigger signals from the spindle rotary encoder. High-speed data transfer from the memory to the FTC controller can be carried out using direct memory access.

Figure 13.5 shows the concept for profile measurement of the workpiece surface by the nanofabrication probe. To use the same probe for measurement, the diamond tool is brought into contact with the machined surface. The contact force between the tool and the surface can be detected by the force sensor in the nanofabrication probe. To measure the surface profile, the diamond tool is scanned along the surface. During scanning, the Z-directional position of the diamond tool is servo controlled by the PZT based on the force sensor output in such a way that the tool tip is kept in contact with the surface at a constant force. The displacement of the PZT, which corresponds to the profile height of the machined surface, can thus be obtained from the capacitance displacement sensor. Assume that the rotation of the spindle and the movement of the X-slide are synchronized as shown in Equation (13.1), the X- and Y-coordinates of the each of the measurement points on the workpiece surface can be obtained from the encoder output as shown in Figure 13.5. If the output of the

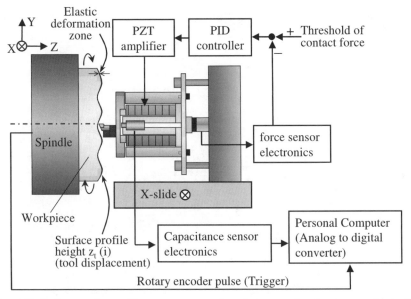

**Figure 13.5.** Concept for profile measurement of workpiece surface by the nanofabrication probe

displacement sensor is denoted $z_t(i)$, the profile of the workpiece surface can then be expressed as:

$$f(x_i, y_i) = f(r_i \cos\theta_i, r_i \sin\theta_i) = z_t(i) \qquad i=1, 2, ..., N \qquad (13.3)$$

where $i$ is the $i$th rotary encoder pulse and $N$ is the total number of sampling over the entire measurement area.

### 13.2.2 Design of the Probe

Figure 13.6 shows a schematic of the FTC-unit of the nanofabrication probe. Figure 13.7 shows a photograh of the FTC-unit. A hollow-type piezoelectric ring actuator with a stainless casing [13.17] is employed to actuate the diamond tool. The commercially available PZT is preloaded to achieve high stiffness. The PZT also has good characteristics including high frequency response and high displacement resolution. Table 13.1 shows the specification of the PZT. It can be seen that the PZT has extremely high stiffness up to 450 N/μm. The pre-load and casing assembly has a high resonant frequency, which is important for the fast tool control. The hollow structure of the actuator makes it easy to mount the capacitance-type displacement sensor inside the actuator.

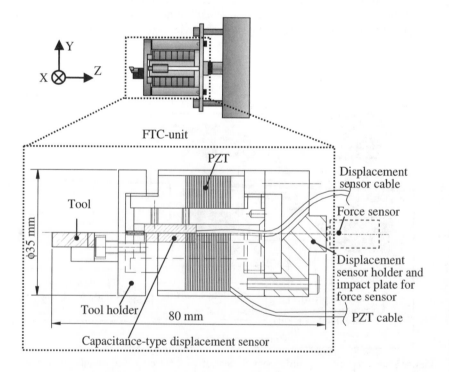

**Figure 13.6.** Design of the FTC-unit of the nanofabrication probe

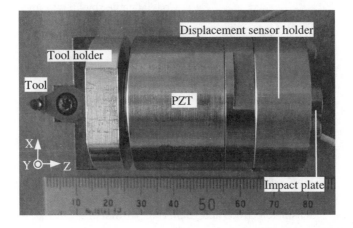

**Figure 13.7.** Photograph of the FTC-unit of the nanofabrication probe

**Table 13.1.** Specifications of the PZT ring actuator

| Dimension | $\phi$ 35 mm × L31 mm (inner diameter: 14 mm) |
|---|---|
| Displacement range | 12 μm/ (0 V thru 150 V) |
| Displacement resolution | 0.1 nm |
| Preload | 700 N |
| Stiffness | 450 N/μm |
| Maximum load | 11000 N |
| Maximum force generation | 8000 N |
| Electrical capacitance | 4.9 μF |
| Resonance frequency | 30 kHz |

Figure 13.8 shows a photograph of the capacitance-type displacement sensor [13.18]. Table 13.2 shows the specifications of the sensor. The sensor has a small electric electrode and guard ring with a diameter of 1 mm. The coaxial cable connecting the electrode to the sensor electronics is shielded from external electrical noise by a guard ring with a diameter of 4 mm. The length of the sensor is 10 mm. The sensor is small enough to be mounted inside the PZT actuator. The maximum bandwidth of the sensor is 20 kHz, and the bandwidth can be set lower for higher measurement resolution. In the experiment described in the following sections, the bandwidth is set at 5 kHz. The corresponding measurement resolution is 6 nm. The commercially available sensor is factory-calibrated using NIST traceable device. The nonlinearity is smaller than 0.1% of the measurement range. Figure 13. 9 shows a photograph of the sensor situated in the PZT. The center of the sensing electrode is aligned with the center of the PZT to avoid Abbe error. The sensor is also aligned co-axially with the PZT to avoid cosine error.

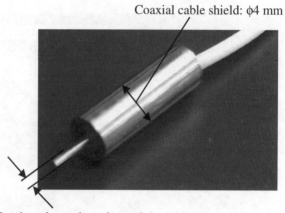

Coaxial cable shield: φ4 mm

Sensing electrode and guard ring: φ1 mm

**Figure 13.8.** Photograph of the capacitance-type displacement sensor

**Table 13.2.** Specification of the capacitance-type displacement sensor

| Dimension | Diameter of sensing electrode and guarding ring: 1 mm<br>Diameter of cable shield: 4 mm<br>Sensor length: 10 mm |
| --- | --- |
| Measurement range | 12.5 μm/ (0V thru 10 V) |
| Resolution | 6 nm@ 5 kHz |
| Bandwidth | DC–20 kHz |
| Nonlinearity | 0.1% measurement range |
| Temperature coefficient | 0.2% measurement range/◇C |

Sensing electrode (φ1 mm)

PZT

**Figure 13.9.** Photograph of the displacement sensor situated in the PZT

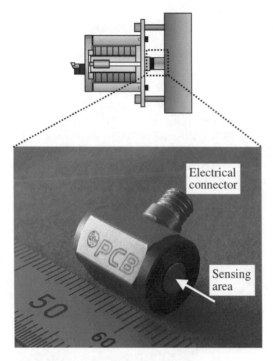

**Figure 13.10.** Photograph of the force sensor

**Table 13.3.** Specification of the force sensor

| Dimension | Sensor body: diameter 10 mm × length 15.2 mm<br>Sensing area: 4 mm |
|---|---|
| Measurement range | 9.8 N |
| Sensitivity | 524.2 mV/N |
| Stiffness | 380 N/μm |
| Maximum force | 48.9 N |
| Resonance frequency | 200 kHz |
| Bandwidth | 0.05 Hz–10 kHz |
| Rise time | 5 μs |
| Discharge time constant | > 10 s |
| Temperature coefficient | <0.09%/◇C |

Figure 13.10 and Table 13.3 show a photograph and specification of the force sensor [13.19]. The commercial sensor employs quartz as the sensing element. The pre-loaded quartz crystals are packaged in a rigid stainless steel casing. The electrostatic charge of the crystals corresponding to the input force is converted into a low impedance voltage output by an amplifier. The sensor is suited to measurement of dynamic force arising in the generation of complex surface profiles

using the nanofabrication probe. The maximum measurement range of force is 9.8 N and the resolution is on the order of 0.01 mN, which depends upon the amplifier and system electrical noise. The high force resolution is important for use in profile measurement of the workpiece surface by the nanofabrication probe. The high stiffness of the sensor also satisfies the requirement for nanofabrication on a diamond turning machine. The sensor has small dimensions, which make it easy to integrate in the nanofabrication probe. To maintain stiffness at the interface between force sensor and FTC-unit, a stiff impact plate with a stiffness of 587.8 N/μm is employed to couple the fabrication force to the force sensor.

Figure 13.11 shows a schematic of the nanofabrication probe structure. The force sensor is mounted on the probe base by the mounting screw of the force sensor. The

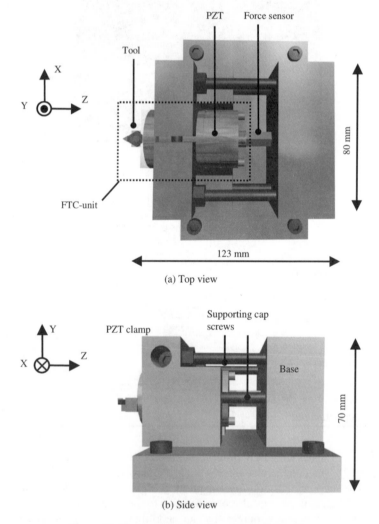

(a) Top view

(b) Side view

**Figure 13.11.** Structure of the nanofabrication probe

FTC-unit is clamped by a PZT clamp assembly. The clamp assembly is also mounted on the same probe base by four cap screws. The impact plate on the FTC-unit contacts with the sensing area of the force sensor. The fabrication force along the Z-direction is propagated to the force sensor through the impact plate. The pre-load against the force sensor can be adjusted by the cap screws supporting the PZT clamp assembly.

Figure 13.12 shows the model for analysing the stiffness of the nanofabrication probe along the Z-direction. The Z-directional stiffness $k_z$ can be expressed as

$$k_z = \frac{1}{\dfrac{1}{k_{pz}} + \left[ 4k_{sz} + \left( \dfrac{1}{k_{iz}} + \dfrac{1}{k_{fz}} \right)^{-1} \right]^{-1}} \tag{13.4}$$

Substituting the stiffness data in Table 13.4, $k_z$ is calculated to be

$$k_z = 315 \, \text{N/}\mu\text{m} \tag{13.5}$$

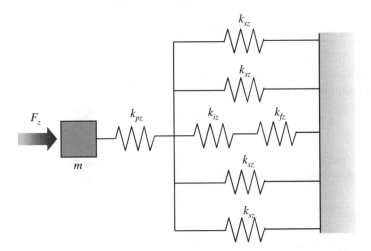

**Figure 13.12.** Model for stiffness analysis of the probe in the Z-direction

**Table 13.4.** Z-directional stiffness data of the elements used in the probe

| | |
|---|---|
| PZT actuator $k_{pz}$ | 450 N/μm |
| Force sensor $k_{fz}$ | 380 N/μm |
| Impact plate $k_{iz}$ | 587.8 N/μm |
| Each supporting cap screw $k_{sz}$ | 204.6 N/μm |

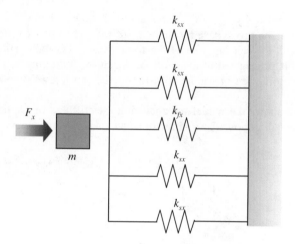

**Figure 13.13.** Model for stiffness analysis of the probe in the X-direction

**Table 13.5.** X-directional stiffness data of the elements used in the probe

| | |
|---|---|
| Force sensor $k_{fx}$ | 28.1 N/µm |
| Each of the supporting cap screws $k_{sx}$ | 4.6 N/µm |

The X-directional stiffness $k_x$, which is analysed by the simple model shown in Figure 13.13, can be expressed by

$$k_x = 4k_{sx} + k_{fx} \qquad (13.6)$$

The stiffness of the probe in the Y-direction is the same as that in the X-direction. Based on the stiffness data in Table 13.5, $k_x$ is calculated to be

$$k_x = 46.5 \text{ N/µm} \qquad (13.7)$$

The resonant frequencies in the two directions can then be obtained as follows:

$$f_z = \frac{1}{2\pi}\sqrt{\frac{k_z}{m}} = 2.8 \text{ kHz} \qquad (13.8)$$

$$f_x = \frac{1}{2\pi}\sqrt{\frac{k_x}{m}} = 1.1 \text{ kHz} \qquad (13.9)$$

where $m$ is the mass of the nanofabrication unit (the PZT, displacement sensor, tool, impact plate).

Figure 13.14 shows a photograph of the assembled nanofabrication probe.

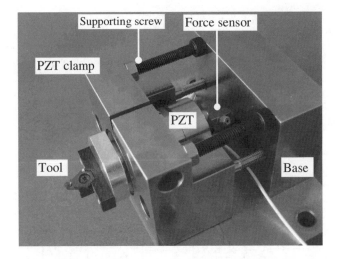

**Figure 13.14.** Photograph of the assembled nanofabrication probe

## 13.3 Evaluation of the Nanofabrication Probe

### 13.3.1 Evaluation of FTC Performance of the Probe

When the probe is employed for nanofabrication based on fast-tool-control (FTC), the static response and dynamic response are important for probe performances. Figure 13.15 shows the Z-directional static responses of the probe in open-loop actuation of the PZT. As can be seen in the figure, the displacement of the PZT shows a hysteresis response, resulting in a large linearity error of approximately

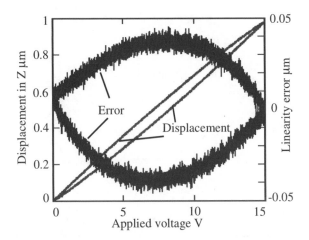

**Figure 13.15.** Static response of the probe in open-loop actuation of the PZT

0.1 μm over a range of 1 μm. A feedback control system for fast tool control is then constructed to remove the hysteresis motion based on the measurement result of the capacitance-type displacement sensor. Figure 13.16 shows the block diagram of the feedback control system. A PID controller is employed in the system. The PID gains are adjusted by trial and error. Figure 13.17 shows the static response of the PZT in closed-loop control. The hysteresis error is reduced from 100 nm (Figure 13.15) to 10 nm, which is close to the electronic noise level of the capacitance-type displacement sensor.

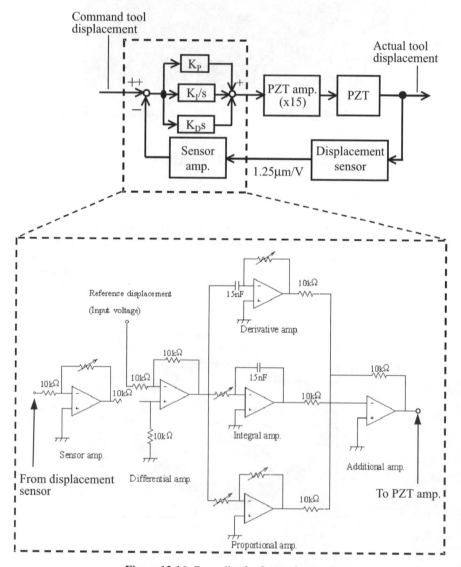

**Figure 13.16.** Controller for fast tool control

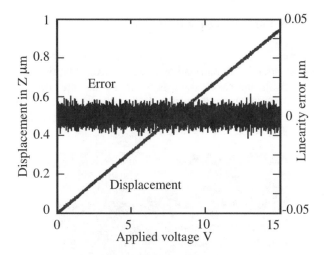

**Figure 13.17.** Static response of the probe in closed-loop feedback control of the PZT

Figure 13.18 shows the Z-directional dynamic response of the nanofabrication probe in open-loop actuation. As can be seen in the figure, the resonant frequency is approximately 2.5 kHz. This value is close to the theoretical value shown in Equation (13.8), indicating the feasibility of the stiffness analysis model shown in Figure 13.12. The –3dB bandwidth of the probe is evaluated to be approximately 1.5 kHz from Figure 13.18. This bandwidth is approximately the same as the FTC-unit without the force sensor, which has previously been developed for generation of large area microstructured surfaces [13.20].

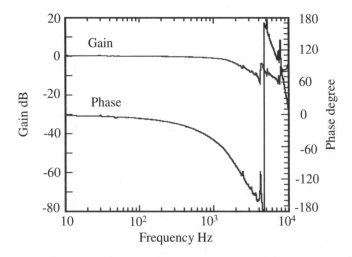

**Figure 13.18.** Z-directional dynamic response of the probe in open-loop actuation

### 13.3.2 Evaluation of Force Detection by the Probe

As can be seen from Figure 13.12, the force signal received by the force sensor is smaller than the Z-directional fabrication force $F_Z$. A calibration process is thus necessary to investigate the relationship between $F_Z$ and the force sensor output. Figures 13.19 and 13.20 show the calibration setup and the calibration result. A simple dynamic calibration method is employed [13.21]. As shown in

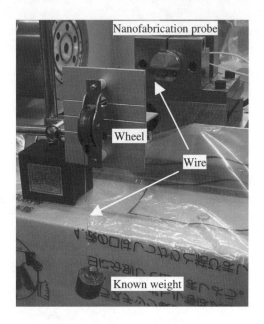

**Figure 13.19.** Setup for calibration of force detection

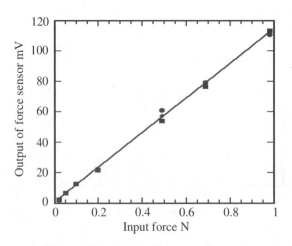

**Figure 13.20.** Calibration result of force detection

Figure 13.19, a known weight, which is used as the input force for the calibration, is hung on the nanofabrication probe via a wheel. The weight is quickly lifted up to release the force applied to the probe and the change in the force sensor output is monitored. The sensitivity of force measurement is evaluated to be 113.68 mV/N. This is approximately 21.3% of the sensitivity of the force sensor (see Table 13.3).

A modulation method is employed to increase the resolution of force detection [13.21]. As shown in Figure 13.21, the PZT actuator is oscillated at constant frequency and with small amplitude on the order of nanometers, by applying a sinusoidal voltage from a function generator. An AC force signal with the same frequency occurs when the tool contacts with the workpiece surface. The amplitude of the AC force signal can be detected sensitively with a lock-in amplifier in the presence of electronic noise.

When using the modulation technique, it is necessary to choose a suitable oscillation frequency for the PZT actuator. Figure 13.22 shows the result of testing the stability of the lock-in amplifier output with respect to the oscillation frequency. The oscillation amplitude is 1 nm and the oscillation frequency is variable from 10 Hz to 800 Hz. The time constant of the lock-in amplifier is set to 10 ms. As can be seen in the figure, the output is most stable when the oscillation frequency is approximately 340 Hz.

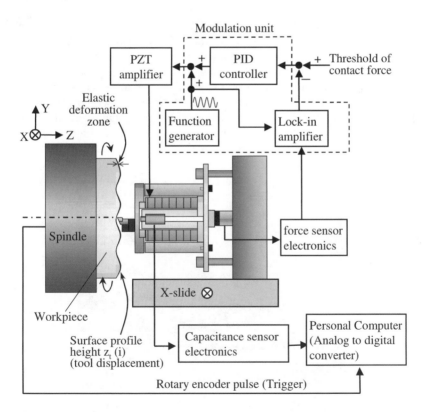

**Figure 13.21.** Modulation unit for improvement of the resolution of force detection

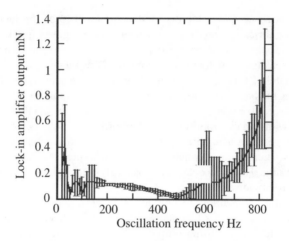

**Figure 13.22.** Stability of the lock-in amplifier with respect to oscillation frequency

Figure 13.23 shows the experimental setup for evaluating the detection of contact force between the workpiece surface and the diamond tool by the nanofabrication probe. The nanofabrication probe is shown on the right of the figure. A piece of optical glass (BK7) is used as the workpiece. Another force sensor (workpiece force sensor) is set against the backside of the glass. The pre-load is applied to the force sensor by the cap screws. A single crystal diamond tool with a nose radius of 1 mm is mounted on the tool holder. The diamond tool, which is oscillated with an amplitude of 1 nm and a frequency of 340 Hz, is moved to the sample with steps of 1 nm by the PZT actuator. The outputs of the force sensors with respect to the tool displacement are plotted in Figure 13.24. As can be seen in the figure, the outputs of the two force sensors show good correspondence in responding to the contact between the tool and the workpiece surface, indicating that it is possible for the

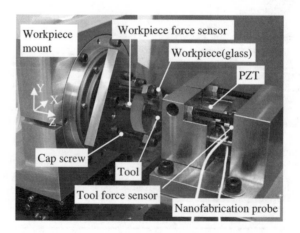

**Figure 13.23.** Setup for investigation of contact force

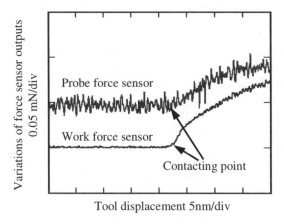

**Figure 13.24.** Detection of contact force

nanofabrication probe to detect the contact force on the order of 0.01 mN. The detectable minimum contact depth is evaluated to be on the order of 1 nm.

### 13.3.3 Evaluation of Displacement Detection by the Probe

Figure 13.25 shows the experimental setup for evaluation of workpiece displacement detection by the nanofabrication probe [13.22]. A laser interferometer [13.23] with a resolution of 0.6 nm is used as the reference. As can be seen in the figure, a piece of workpiece glass and the reflecting mirror of the interferometer are mounted on a Z-directional linear stage. Axes of the interferometer and the nanofabrication probe are aligned along the Z-direction. The diamond tool of the nanofabrication probe first makes contact with the workpiece glass surface at a force of 0.01 mN. Then the Z-stage moves the workpiece glass and the reflecting mirror in the Z-direction, while the diamond tool tracks the surface of the glass. For tracking the surface, the diamond tool is servo controlled by the PZT so that the tool tip is kept in

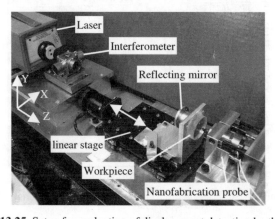

**Figure 13.25.** Setup for evaluation of displacement detection by the probe

contact with the surface at a constant threshold force, which is set to 0.01 mN. The servo control process is based on that shown in Figure 13.21. The displacement of the workpiece is obtained from that of the tool/PZT displacement, which is measured by the capacitance-type displacement sensor inside the PZT.

To evaluate the performance of displacement tracking/detection by the nanofabrication probe, the output of the capacitance sensor is compared with that of the interferometer. Figure 13.26 shows the results of stability tests of the interferometer and the capacitance-type displacement sensor in the nanofabrication probe. The test term is 5 minutes. As can be seen in the figure, the interferometer output has an instability of approximately 70 nm over the test term. The low frequency component of the instability is mainly caused by thermal drift of the interferometer. The high-frequency instability of the interferometer output, which is approximately 40 nm, is mainly caused by air disturbance in the optical path. In contrast, the output of the nanofabrication probe is stable within 10 nm, which is much more stable than the interferometer. The noise level of the instrument output is on the order of 3 nm, which is also much smaller than that of the interferometer.

Figure 13.27 shows the result of evaluating the performance of displacement tracking/detection by the nanofabrication probe. During the experiment, the linear stage moves the workpiece and the reflecting mirror with steps of 50 nm over a stroke of approximately 3.5 μm along the Z-direction. The displacement of the workpiece is tracked by the nanofabrication probe with a constant force of 0.01 mN. The horizontal axis of the Figure shows the output of the interferometer and the vertical axis shows that of the nanofabrication probe. The difference between the two outputs is also plotted in the figure. It can be seen that the maximum difference is approximately 70 nm over the 3.5 μm measurement range. Compared with the interferometer stability data shown in Figure 13.26, it is obvious that the difference is mainly caused by the instability of the interferometer. The experimental results indicate the capability of the nanofabrication probe for use as an *in situ* displacement probe for measurement of a machined surface.

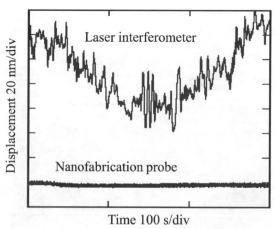

**Figure 13.26.** Stabilities of laser interferometer and nanofabrication probe used for displacement measurement

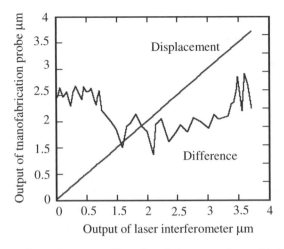

**Figure 13.27.** Comparison of outputs of laser interferometer and nanofabrication probe used for displacement measurement

## 13.4 Nanofabrication and Workpiece Surface Profile Measurement Using the Probe

Experiments of nanofabrication and workpiece surface profile measurement using the nanofabrication probe were carried out on an ultra-precision diamond turning machine. Figure 13.28 shows a schematic of the experimental system. Table 13.6 shows the specification of the diamond turning machine [13.24].

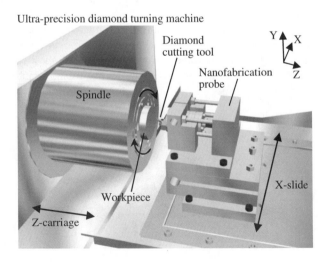

**Figure 13.28.** The experimental system for nanofabrication and surface profile measurement by the nanofabrication probe

**Table 13.6.** Specification of the diamond turning machine

| | |
|---|---|
| Spindle | static airbearing, rotary encoder feedback control<br>spindle diameter: 80 mm<br>rotational speed: 1–1500 rpm<br>spindle error: 50 nm (radial and axial motions)<br>rotary encoder resolution: 0.001◇<br>pulse number: 180,000 pulses/revolution |
| X-slide and Z-carriage | roller bearing, linear encoder feedback control<br>movement range: 220 mm (X), 150 mm (Z)<br>straightness: 500 nm/full movement range<br>linear encoder resolution: 10 nm<br>feed rate: 1mm–1260mm/min |

A fabrication experiment was conducted using a sinusoidal micro-structured surface. The surface was used as the reference for a surface encoder, which can measure multi-degree-of-freedom planar motions [13.25]–[13.30]. The surface profile, which is a superposition of sinusoidal waves in the X- and Y-directions, can be expressed by

$$f(x, y) = A_X \sin\left(\frac{2\pi}{\lambda_X} x\right) + A_Y \sin\left(\frac{2\pi}{\lambda_Y} y\right) \qquad (13.10)$$

where $A_X$, $A_Y$ are the amplitudes of the sine functions in the X-direction and the Y-direction, respectively. $\lambda_X$ and $\lambda_Y$ are the corresponding spatial wavelengths. In the fabrication experiment, $A_X$, $A_Y$ are set to 100 nm and $\lambda_X$, $\lambda_Y$ are set to 100 μm, respectively. Figure 13.29 shows a model of the surface profile. The area of the surface, which determines the measurement range of the surface encoder, is up to several hundreds of millimeters.

Figure 13.30 shows a fabrication result. The sinusoidal microstructures are successfully fabricated on an aluminium workpiece surface over an area of diameter 50 mm (maximum 150 mm). The fabrication conditions are shown in Table 13.7. The result indicates that the probe has the same ability for generation of complex profiles on soft material surfaces as the conventional FTC-unit without the force sensor [13.20].

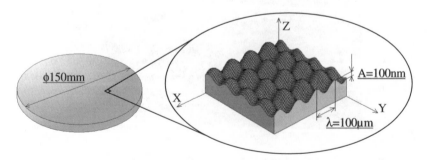

**Figure 13.29.** Schematic of the large area sinusoidal microstructured surface

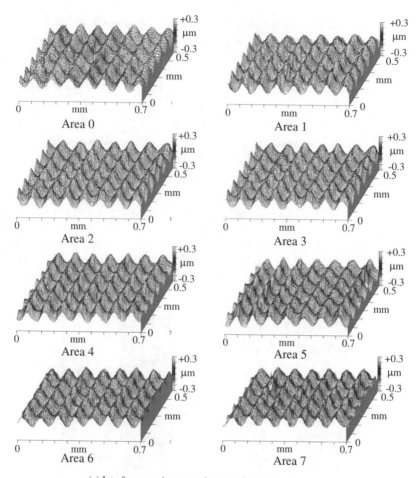

(a) Interference microscope images of the evaluation areas

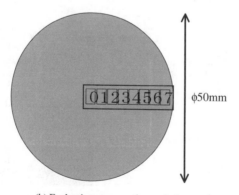

(b) Evaluation areas on the workpiece surface

**Figure 13.30.** Sinusoidal microstructured surface generated by the nanofabrication probe

**Table 13.7.** Parameters for generation of the surface shown in Figure 13.30

| Number of rotary encoder pulses | 300,000 per revolution |
|---|---|
| Rotation speed of the spindle | 20 rpm |
| Feed rate of the X-slide | 5 μm/revolution |
| Workpiece material | aluminum alloy |
| Fabrication area | φ50 mm |
| Single-crystal diamond cutting tool | nose radius: 1 mm, rake angle: 0 deg, clearance angle: 7 deg |
| Cutting fluid | electrical discharge machining oil |

On the other hand, compared with the conventional FTC-unit, it is much easier for the nanofabrication probe to generate complex profiles on brittle material surfaces. Figure 13.31 shows an example of generating a complex profile on a flat surface of brittle material by a conventional FTC. Although it is possible to make a brittle surface flat through a grinding process, there is always a tilt component

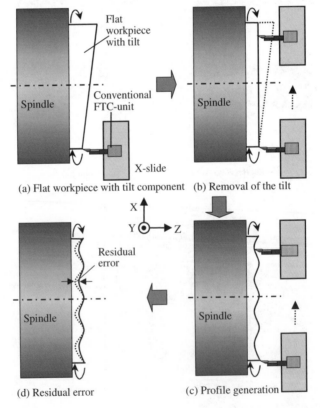

**Figure 13.31.** Nanofabrication of complex profile on a flat brittle surface by a conventional fast-tool-control (FTC) unit

occurring when mounting the workpiece on the spindle of a diamond turning machine as shown in Figure 13.31(a). Because it is difficult to accurately measure the tilt component relative to the tool position, it is necessary to remove the tilt component before generation of the complex profile (Figure 13.31(b), (c)). To make ductile cutting on a brittle surface, the depth-of-cut has to be limited within a certain value, which is on the order of 100 nm for most brittle materials [13.31]–[13.40]. This makes the tilt removal process difficult and time consuming. The tilt removal process also increases the tool wear. In addition, because the large elastic recovery ratio of the brittle material, the actual depth of cut is always smaller than the command depth of cut, resulting in a residual profile error in the fabricated surface (Figure 13.31(d)). Research that measures the machined surface profile with various types of displacement sensors on the machine have been conducted [13.41–13.54], but they have a problem that the measured points do not correspond to the machined points. Therefore the measurement information could not be fed back accurately to the machining process, making it difficult to realize precision nanofabrication of complex surface profiles in such a way.

Figure 13.32 shows the case of using the nanofabrication probe for surface displacement/profile measurement. As can be seen in Figure 13.32(a) and (b), the tilt

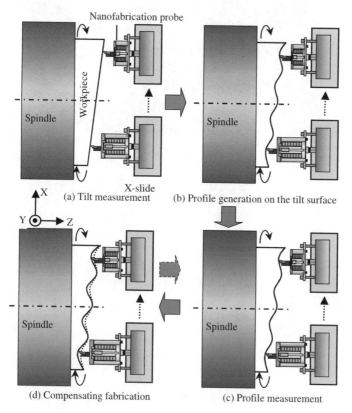

**Figure 13.32.** Nanofabrication of complex profile on a flat brittle surface using the nanofabrication probe for surface displacement/profile measurement

component is first measured by the probe so that the complex profile can be directly generated on the tilt surface without the process of tilt removal. The generated profile is then measured by the probe (Figure 13.32(c)) and a compensating fabrication process is carried out based on the measurement result (Figure 13.32(d)). Because the measurement probe and the fabrication probe have exactly the same coordinates, an accurate compensation process can thus be realized. The measurement and compensation process shown in Figure 13.32(c) and (d) can be repeated till the residual profile error is reduced to the required value.

Figure 13.33 shows the setup for measurement and fabrication of a complex profile on the surface of a glass workpiece. The glass material is BK7 and the diameter of the workpiece is 50 mm. Figure 13.34 shows the result of measurement

**Figure 13.33.** Setup for measurement and fabrication of complex profile on glass

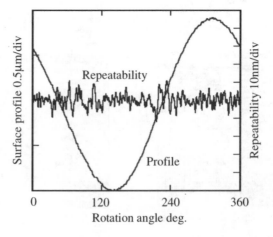

**Figure 13.34.** Measurement of surface profile made by the nanofabrication probe before fabrication

of the surface profile by the probe before fabrication. As can be seen in the figure, the surface has a tilt component of approximately 2 μm. The repeatability of the measurement is on the order of 10 nm. In the measurement, the threshold of the contact force is set to be 0.01 mN. Figure 13.35 shows an interference microscope image of the surface after the measurement. It can be seen that there is no damage caused to the surface by the measurement.

A nanofabrication experiment was carried out to add the profile component shown in Figure 13.36 to the original profile of Figure 13.34. The profile component has an amplitude of 14 nm (PV) and a spatial frequency of 4 per revolution. Figures 13.37 and 13.38 show spectrums of the original profile and the objective profile after adding the component shown in Figure 13.36, respectively. Table 13.8 shows the fabrication conditions.

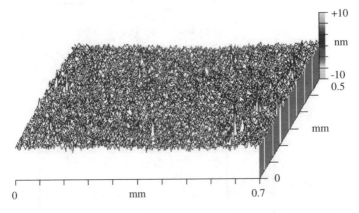

**Figure 13.35.** Image of the surface after measurement by the nanofabrication probe

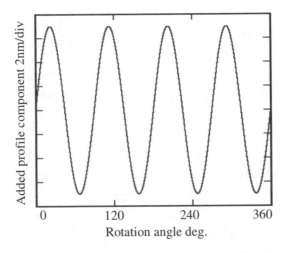

**Figure 13.36.** The profile component added to the original profile of the surface shown in Figure 13.31

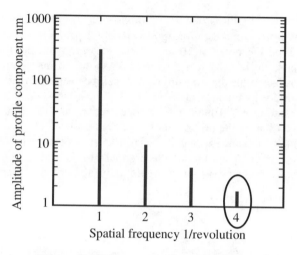

**Figure 13.37.** Spectrum of the original profile

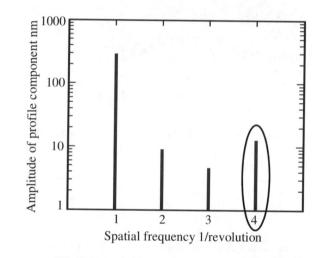

**Figure 13.38.** Spectrum of the objective profile

**Table 13.8.** Parameters for generating the surface shown in Figure 13.36

| | |
|---|---|
| Number of rotary encoder pulses | 15,000 per revolution |
| Rotation speed of the spindle | 20 rpm |
| Workpiece material | BK 7 glass |
| Single-crystal diamond cutting tool | nose radius: 1 mm, rake angle: 0 deg, clearance angle: 7 deg |
| Cutting fluid | ethanol |

After the first fabrication by the nanofabrication probe, the surface profile was measured by the same probe. Figure 13.39 shows the spectrum of the measured profile. It can be seen that the amplitude of the fourth-order component of interest is only approximately 2 nm, which is much smaller than the command value shown in Figure 13.36. A compensating fabrication is then carried out based on the measurement result. Figure 13.40 shows the result measured by the nanofabrication probe. It can be seen that the error in the amplitude of the fourth-order component is reduced to approximately 1 nm. The experimental results shown above indicate the feasibility of using the nanofabrication probe for accurate fabrication of complex surfaces on both ductile materials and brittle materials.

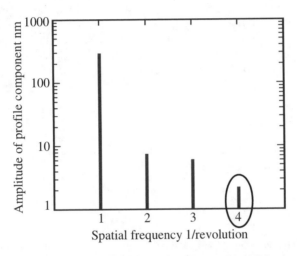

**Figure 13.39.** Spectrum of the profile after the first fabrication

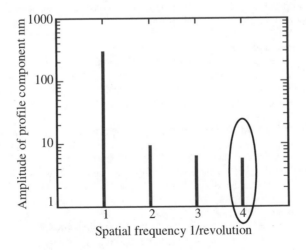

**Figure 13.40.** Spectrum of the profile after the compensating fabrication

## 13.5 Conclusions

The nanofabrication probe, which combines a fast-tool-control (FTC) cutting unit and a force sensor, cannot only provide in-process data of fabrication force but also carry out *in situ* displacement and/or profile measurement of the workpiece surface. A careful design has been conducted to avoid the influence of integrating the force sensor on the dynamic response of the FTC-unit.

The performance of the probe has been evaluated experimentally. It has been verified that the probe has a bandwidth of 2.5 kHz, which is close to a conventional FTC-unit without a force sensor. The force resolution of the probe is on the order of 0.01 mN. The ability of displacement measurement by the probe is comparable to a laser interferometer. Fabrication and measurement of complex profiles on both ductile and brittle material surfaces have been conducted. It has been confirmed that the probe is especially suited to use in ductile fabrication of brittle materials and the profile measurement function of the probe makes it possible to conduct accurate compensating nanofabrication of complex surface profiles.

## Acknowledgment

The author thanks Yasuto Kudo for his valuable assistance in the experiments and in preparing some of the figures used in this chapter.

## References

[13.1]   Taniguchi, N., 1994, "The state of the art of nanotechnology for processing of ultraprecision and ultrafine products", *Precision Engineering*, **16**(1), pp. 5–24.
[13.2]   Saito, T. T., 1978, "Diamond turning of optics: the past, the present, and the exciting future", *Optical Engineering*, **17**(6), pp. 570–573.
[13.3]   Krauskopf, B., 1984, "Diamond turning: reflecting demands for precision", *Manufacturing Engineering*, pp. 90–100.
[13.4]   Moriwaki, T. and Okuda, K., 1989, "Machinability of copper in ultra-precision micro diamond cutting", *Annals of the CIRP*, **38**(1), pp. 115–118.
[13.5]   Ikais, N., Shimada, S. and Tanaka, H., 1992, "Minimum thickness of cut in micromachining", *Nanotechnology*, 3, pp. 6–9.
[13.6]   Kim, S. D., Chang, I. C. and Kim, S. W., 2002, "Microscopic topographical analysis of tool vibration effects on diamond turned optical surfaces", *Precision Engineering*, **26**, pp. 168–174.
[13.7]   Kouno, K., 1984, "A fast response piezoelectric actuator for servo correction of systematic errors in precision machining", *Annals of CIRP*, **33**(1), pp. 369–372.
[13.8]   Patterson, S. R. and Magrab, E. B., 1985, "Design and testing of a fast tool servo for diamond turning", *Precision Engineering*, **7**(3), pp. 123–128.
[13.9]   Dow, T. A., Miller, M. H. and Falter, P. J., 1991, "Application of a fast tool servo for diamond turning of nonrotationally symmetric surfaces", *Precision Engineering*, **13**(4), pp. 233–250.
[13.10]  Fawcett, S. C. and Engelhaupt, D., 1995, "Development of Wolter I x-ray optics by diamond turning and electrochemical replication", *Precision Engineering*, **17**(4), pp. 290–297.

[13.11] Miller, M. H., Garrard, K. P., Dow, T. A. and Taylor, L. W., 1994, "A controller architecture for integrating a fast tool servo into a diamond turning machine", *Precision Engineering*, **16**(1), pp. 42–48.

[13.12] Okazaki, Y., 1998, "Fast tool servo system and its application to three dimensional fine figures", *In Proceedings of 13th ASPE Annual Meeting*, pp. 100–103.

[13.13] Ludwick, S. J., Chargin, D. A., Calzaretta, J. A. and Trumper, D. L., 1999, "Design of a rotary fast tool servo for ophthalmic lens fabrication", *Precision Engineering*, **23**(4), pp. 253–259.

[13.14] Drescher, J. D. and Dow, T. A., 1990, "Tool force development for diamond turning", *Precision Engineering*, **12**(1), pp. 29–35.

[13.15] Santochi, M., Dini, G., Tantussi, G. and Beghini, M., 1997, A" sensor-integrated tool for cutting force monitoring", *Annals of CIRP*, **46**(1), pp. 49–52.

[13.16] Gao, W., Hocken, R. J., Patten, J. A. and Lovingood, J., 2000, "Force measurement in nanomachining instruments", *Review of Scientific Instruments*, **71**(11), pp. 4325–4329 (2000).

[13.17] http://www.piezomechanik.com/.

[13.18] http://www.mtiinstruments.com/.

[13.19] http://www.pcb.com/.

[13.20] Gao, W., Araki, T., Kiyono, S., Okazaki, Y. and Yamanaka, M., 2003, "Precision nano-fabrication and evaluation of a large area sinusoidalgrid surface for a surface encoder", *Precision Engineering*, **27**(3), pp. 289–298.

[13.21] Gao, W., Hocken, R. J., Patten, J. A., Lovingood, J. and Lucca D., 2000, "Construction and testing of a nano-machining instrument", *Precision Engineering*, **24**(4), pp. 320–328.

[13.22] Gao, W., Kudo, Y., Kiyono, S. and Patten, J. A, 2004, "An instrument for nano-machining and nanometrology of free-form surface profiles with a diamond turning machine", *Journal of Chinese Society of Mechanical Engineers*, **25**(5), pp. 449–456.

[13.23] http://www.agilent.com/.

[13.24] http://www.toshiba-machine.co.jp/.

[13.25] Kiyono, S., Cai, P. and Gao, W., 1999, "An angle-based position detection method for precision machines", *JSME International Journal*, **42**(1), pp. 44–48.

[13.26] Gao, W., Dejima, S., Shimizu, Y. and Kiyono, S., 2003, "Precision measurement of two-axis positions and tilt motions using a surface encoder", *Annals of CIRP*, **52**(1), pp. 435–438.

[13.27] Watanabe, Y., Gao, W., Shimizu, H and Kiyono, S, 2003, "Analysis of a surface encoder in wave optics", *Key Engineering Materials*, **257–258**, pp. 219–224.

[13.28] Gao, W., Dejima, S., Yanai, H., Katakura, K., Kiyono, S. and Tomita, Y., 2004, "A surface motor-driven planar motion stage integrated with an $XY\theta_Z$ surface encoder for precision positioning", *Precision Engineering*, **28**(3), pp. 329–337.

[13.29] Gao, W., Dejima, S. and Kiyono, S., 2005, "A dual-mode surface encoder for position measurement", *Sensors and Actuators A*, **117**(1), pp. 95–102.

[13.30] Dejima, S., Gao, W., Katakura, K., Kiyono, S. and Tomita, Y., 2005, "Dynamic modeling, controller design and experimental validation of a planar motion stage for precision positioning", *Precision Engineering*, **29**(3), pp. 263–271.

[13.31] Brehm, R., Dun, K. V., Teunissen, J. C. G. and J. Haisma, 1979, "Transparent single-point turning of optical glasses, *Precision Engineering*, **1**, pp. 207–213.

[13.32] Krauskopf, B., 1984, "Reflecting demands for precision," *Manufacturing Engineering*, **92**(5), pp. 90–100.

[13.33] Blake, P. N. and Scattergood, R. O., 1990, "Ductile-regime machining of germanium and silicon", *Journal of American Ceramic Society*, **73**(4), pp. 949–957.

[13.34] Blackle W. S. and Scattergood, R. O., 1991, "Ductile-regime machining model for diamond turning of brittle materials", *Precision Engineering*, **13**(2), pp. 96–103.

[13.35] Morris, J. C., Callahan, D. L., Kulik, J., Patten, J. A., Scattergood, R. O., 1995, "Origins of the ductile regime in single-point diamond turning of semiconductors", *Journal of American Ceramic Society*, **78**(8), pp. 2015–2020.

[13.36] Shibata, T., Fujii, S., Makino, E. and Ikeda, M., 1996, "Ductile-regime turning mechanism of single-crystal silicon", *Precision Engineering*, **18**(2/3), pp. 129–137.

[13.37] Fang, F. Z. and Venkates, V. C., 1998, "Diamond cutting of silicon with nanometric finish", *Annals of CIRP*, **47**(1), pp. 45–49.

[13.38] Gao, W., Hocken, R. J., Patten, J. A., Lovingood, J. and Lucca D., 2000, "Experiments using a nano-machining instruments for nano-cutting brittle materials", *Annals of CIRP*, **49**(1), pp. 439–442.

[13.39] Patten, J. A. and Gao, W, "Extreme negative rake angle technique for single point diamond nano-cutting of silicon", *Precision Engineering*, **25**(2), pp. 165–167.

[13.40] Patten J. A., Gao, W. and Kudo, Y., 2005, "Ductile regime nano-machining of single crystal silicon carbide," *ASME Journal of Manufacturing Science and Engineering*, in press.

[13.41] Whitehouse, D. J., 1976, "Some theoretical aspects of error separation techniques in surface metrology," *Journal of Physics E: Scientific Instruments*, **9**, pp. 531–536.

[13.42] Shiraishi, M., 1989, "Scope of in-process measurement, monitoring and control techniques in machining processes", *Precision Engineering*, **11**(1), pp. 39–47.

[13.43] Fan, K. C. and Chao, Y. H., 1991, "In-process dimensional control of the workpiece during turning", *Precision Engineering*, **13**(1), pp. 27–32.

[13.44] Evans, C. J., Hocken, R. J. and Estler, W. T., 1996, "Self-calibration: reversal, redundancy, error separation and 'absolute testing'", *Annals of CIRP*, **45**(2), pp. 617–634.

[13.45] Gao, W., Kiyono, S. and Nomura, T., 1996, "A new multi-probe method of roundness measurements", *Precision Engineering*, **19**(1), pp. 37–45.

[13.46] Uda, Y., Kohno, T. and Yazawa,T., 1996, "In-process measurement and workpiece-referred form accuracy control system (WORFAC): Application to cylindrical turning using an ordinary lathe", *Precision Engineering*, **18**(1), pp. 50–55.

[13.47] Gao, W. and Kiyono, S., 1997, "On-machine measurement of machined surface using the combined three-point method", *JSME International Journal*, **40**(2), pp. 253–259.

[13.48] Gao, W. and Kiyono, S., 1997, "On-machine roundness measurement of cylindrical workpieces by the combined three-point method", *Measurement*, **21**(4), pp. 147–156.

[13.49] Gao, W. and Kiyono, S., 1997, "Development of an optical probe for profile measurement of mirror surfaces", *Optical Engineering*, **36**(12), pp. 3360–3366.

[13.50] Gao, W., Kiyono, S. and Sugawara, T., 1997, "High accuracy roundness measurement by a new error separation method", *Precision Engineering*, **21**(2/3), pp. 123–133.

[13.51] Nomura, T., Kamiya, K., Miyashiro, H., *et al.*, 1998, "Shape measurements of mirror surfaces with a lateral-shearing interferometer during machine running", *Precision Engineering*, **22**(4), pp. 185–189.

[13.52] Gao, W., Yokoyama, J., Kojima, H. and Kiyono, S., 2002, "Precision measurement of cylinder straightness using a scanning multi-probe system", *Precision Engineering*, **26**(3), pp. 279–288.

[13.53] Gao, W., Huang, P. S., Yamada, T. and Kiyono, S., 2002, "A compact and sensitive two-dimensional angle probe for flatness measurement of large silicon wafers", *Precision Engineering*, **26**(4), pp. 396–404.

[13.54] Gao, W., Kiyono, S. and Satoh, E., 2002, "Precision measurement of multi-degree-of-freedom spindle errors using two-dimensional slope sensors", *Annals of CIRP*, **51**(1), pp. 447–450.

# 14

# Smart Transducer Interface Standards for Condition Monitoring and Control of Machines

Kang B. Lee

National Institute of Standards and Technology
Gaithersburg, MD 20899-8220, USA
Email: kang.lee@nist.gov

**Abstract**
This chapter presents a summary of the distributed architecture-based IEEE 1451 suite of smart transducer interface standards for sensors and actuators. These standards specify communication protocols and transducer electronic data sheets (TEDS) for networked digital smart sensors and actuators, high-speed synchronized distributed multi-drop sensor systems, and wireless sensor interfaces. The concept of IEEE 1451 is based on a distributed architecture, which means intelligence is decentralized and is pushed down to the sensor module level. This arrangement is well suited for remote monitoring and control applications, such as condition-based maintenance (CBM). Machinery condition and health could affect machine performance, part quality, and productivity. Machinery Information Management Open Systems Alliance (MIMOSA) was organized to establish an open architecture and a set of protocols for exchanging complex sensor information between condition-based maintenance (CBM) systems. With the capability and wide availability of the Web, any sensor connected to a wired or wireless network can be accessed anywhere via the network or Internet. This will greatly enhance the effectiveness and application of machinery health monitoring and control in the manufacturing and production environments.

## 14.1 Introduction

When a machine is down, it is no longer productive in generating revenues and it could be costly to obtain repairs in order to restore it to a healthy running condition. Periodic maintenance of a machine is costly because any time the machine is taken down for maintenance, it produces no income. In today's extremely competitive market with the demand for higher productivity, we would like to minimize machine down time if at all possible. Therefore research on improving machine health aims to go beyond regular periodic maintenance. Monitoring a machine's condition during operation and using the data to predict its health and schedule maintenance is the most economical form of maintenance. Applying this predictive maintenance principle, the overall machinery condition at any time is known and thus accurate and efficient downtime planning is possible. Predictive maintenance uses many sensors, such as temperature, pressure, and vibration sensors. In the past, personnel have used portable instruments to measure a machine's condition periodically. Recently, continuous or real-time monitoring of machinery has been made possible by mounting sensors on the machine and using a data acquisition system and a host computer to collect the data. The application of permanently installed sensors on all the machines in a manufacturing plant is very costly. However, standardized smart sensor interfaces can potentially drive down equipment price and enhance its functionality and features through competition.

Sensor-to-network interfacing is a very important issue in the machine maintenance and asset management industry for mission-critical and safety-critical applications [14.1][14.2]. Unexpected machine failures are not only costly to correct but may even cost human lives. The high costs of machinery and maintenance software management systems call for a high degree of operational reliability and machine uptime. Despite the technological advancement made over the past decades, state-of-the-art machine condition monitoring and diagnostic systems have yet to be widely accepted and deployed on the factory floor. This is attributed mainly to the lack of standardized sensor interfaces for data acquisition, the great variety of ad-hoc algorithms for information processing, data interchange, fault diagnosis, and machine remaining life prognosis, and the lack of standards for a plug-and-play sensor for interoperability.

This chapter discusses the significance of the suite of smart transducer interface standards including IEEE 1451.1, 1451.2, 1451.3, and 1451.4 [14.3]–[14.8] and the proposed IEEE 1451.0 and 1451.5 [14.9][14.10] being drafted for networking smart transducers. It also briefly discusses the standardized digital interfaces and communication protocols as well as the networked smart transducer model. An application model of the IEEE 1451 standards in distributed measurement and control applications is also addressed. The IEEE 1451 standards provide for totally self-describing measurement and control devices, allowing one to choose best-in-class products from preferred vendors for true plug and play capability. The IEEE 1451 set of standards define a Transducer Electronic Data Sheet (TEDS), electronic interfaces, and wired/wireless connection of analog and digital communications – creating an ideal modular distributed architecture. Network Capable Application Processor (NCAP)s being developed by companies for the US Navy supporting these standard interfaces will make future naval systems instantly reconfigurable and adaptive [14.11]. These technologies will gradually migrate to industrial, commercial, and consumer markets for condition-based maintenance (CBM) of machines, heating, ventilation, and air conditioning (HVAC) systems, and home appliances, respectively, through the Department of Defense's dual-use technology and small business innovative research (SBIR) programs.

Machinery Information Management Open Systems Alliance (MIMOSA) [14.12], a non-profit industry trade association, aimed to establish an open architecture and a set of protocols for exchanging complex sensor information between CBM systems. It focused on the development of a Common Relational Information Schema (CRIS) protocol, which was a common language for transferring and exchanging machine condition monitoring and assessment data such as pressure and temperature between a client's proprietary database and another remotely located user, using common data communication conventions such as Java, eXtensible Markup Language (XML), and Distributed Component Object Model (DCOM). Another related effort, Open System Architecture for Condition Based Maintenance (OSA-CBM) program, developed an open architecture and standard for distributed CBM software components. MIMOSA/CRIS has been adopted by OSA-CBM as the core infrastructure for distributed machine maintenance information communication [14.13].

However, neither the MIMOSA nor the OSA-CBM specifications defines a sensor interface and communication protocols for acquiring sensor data. OSA-CBM recommended interfaces between its program structure and other published standards from the IEEE (AI-ESTATE and 1451) and International Organization for Standardization (ISO). Thus the IEEE 1451 smart transducer interface standard seemed to be able to play a key role in completing the process from the acquisition of data at the sensor level to the transfer of the sensor information to the enterprise level, where MIMOSA and OSA-CBM are set up to manage the information. The approach of integrating the IEEE 1451 standard with the MIMOSA and OSA-CBM specifications was investigated and the feasibility determined [14.14]. It was feasible to establish the proposed link between the IEEE 1451 standard and MIMOSA and OSA-CBM architecture. The OSA-CBM data model can be used as the entry points for interfacing with the IEEE 1451.1 Network Capable Application Processor (NCAP).

The IEEE 1451.1 standard defines a common object model for the abstract components of networked smart transducers, together with interface specifications to these components. The model was described using an Interface Definition Language (IDL). Researchers at NIST were interested in developing a Unified Modeling Language (UML) model of IEEE 1451.1 in order to further their research on IEEE 1451 for smart sensor integration and enhancing the interoperability of this standard with other industry standards such as MIMOSA and the OSA-CBM. Since the OSA-CBM data model can be readily used as the entry points for the interfacing requirements of an IEEE 1451.1 NCAP and the OSA-CBM was developed based on UML modeling, it would be much easier to develop an interface between the IEEE 1451.1 and OSA-CBM standards, if a UML model for IEEE 1451.1 is readily available. Thus, a UML model for IEEE 1451.1 was developed [14.15]. Now, let us review the IEEE 1451 standards.

## 14.2 IEEE 1451 Smart Transducer Interface Standards

The objective of IEEE 1451, Standard for A Smart Transducer Interface for Sensors and Actuators, is to define a set of common communication interfaces for connecting transducers to microprocessor-based systems such as instruments and control/field networks in a network-independent environment. The IEEE 1451 family of standards has grown from the original two basic standards, IEEE 1451.1 and 1451.2, to a total of seven standards [14.16]. It is basically divided into two parts: (1) defining a set of hardware interfaces or physical layers for connecting transducers to a microprocessor, a network, or an instrumentation system, and (2) defining a set of software interfaces for connecting transducers to different networks via a set of application programming interfaces (API) on the transducer side and network protocol logical interface on the network side. The main idea behind IEEE 1451 is vendor independent and network neutral. But there are no set requirements for the use of different analog-to-digital converters, microprocessors, network protocols, and transceivers. This in turn reduces the industry's effort needed to develop and migrate to networked smart transducers and allows manufacturers a flexibility to add value to their products. The ultimate goals of the standards are to provide the means for achieving transducers-to-network interchangeability and transducer-to-networks interoperability. Sensors and actuators are ubiquitous and used across industries. Thus the family of IEEE 1451 standards, as shown in Figure 14.1, is designed to meet the needs of various industries and user communities for applications that could connect sensors in a point-to-point, distributed multi-drop bus, mixed-mode (analog signal+digital TEDS), intrinsically-safe, or wireless fashion. The interfaces between the IEEE 1451 sensors and the digital network is unique to the user's network like a wireline-based Ethernet, Profibus, DeviceNet, or a wireless-based WiFi, Bluetooth, or ZigBee.

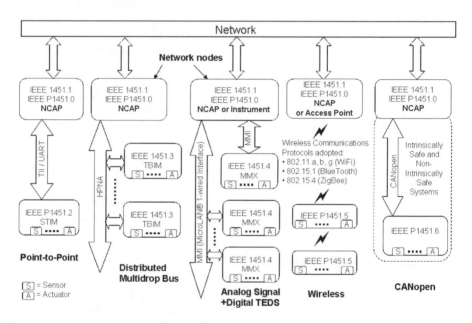

**Figure 14.1.** The family of IEEE 1451 standards

## 14.2.1 IEEE 1451.0 – Common Functions and Commands

The IEEE 1451.0 standard defines a set of commands, operations, communication protocols, and TEDS that serve as the basis for the family of IEEE 1451 smart transducer interface standards. The functionality is independent of the physical communications media between the transducers and NCAP. This makes possible to access any transducers in the system using the same command set. This also makes it easy to add other proposed IEEE 1451.X physical layers to the family. The interrelationship between IEEE 1451.0 and the application and physical layers is shown in Figures 14.1 and 14.2. The IEEE 1451.0 standard enables users to use the same command set to access any sensor and actuator in a system of networks where multiple wired and wireless sensor networks are connected, thus achieving data-level interoperability [14.17].

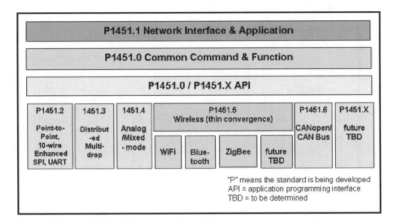

**Figure 14.2.** Logical layout of the family of IEEE 1451 standards

## 14.2.2 IEEE 1451.1 – Networked Smart Transducer Model

The IEEE 1451.1 standard defines a common object model for networked smart transducers (sensors or actuators) and the software interface specifications for each class representing the model [14.18]. Some of these classes form the blocks, components, and services of the conceptual transducer. The model of the networked smart transducer is shown in Figure 14.3. The networked smart transducer object model encapsulates the details of the transducer hardware implementation within a simple programming model. This makes programming the sensor or actuator hardware interface less complex by using an input/output (I/O)-driver paradigm. The network services interfaces encapsulate the details of the different network protocol implementations behind a small set of communications methods – client–server and publish–subscribe models.

In order to reduce the time for developing IEEE 1451.1 applications, an object-oriented framework for IEEE 1451.1 applications has been developed [14.19]. This framework consists of four layers: operating system layer, middleware and tools layer, IEEE 1451.1 layer, and IEEE 1451.1 application layer. The layer architecture of object-oriented application framework is shown in Figure 14.4.

The IEEE 1451.1 layer focuses on the class hierarchy of the IEEE 1451.1 standard. It consists of the IEEE 1451.1 Neutral Model and middleware-based IEEE 1451.1 Model. The IEEE 1451.1 Neutral Model is neutral to any networks and middleware. This model has been compiled and linked using I-Logix** Rhapsody UML tool to create a static and dynamic library. The C++ source code of the IEEE

1451.1 Neutral Model is available in the open source SourceForge web site [14.20]. IEEE 1451.1 application developers can extend this neutral model, and then implement their specific IEEE 1451.1 applications based on the C++ source code.

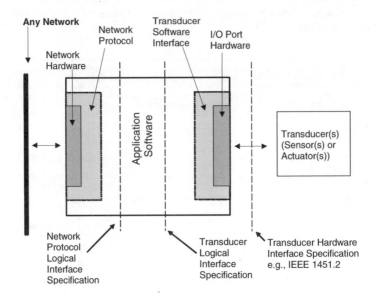

**Figure 14.3.** Networked smart transducer model

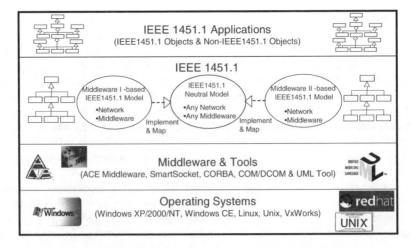

**Figure 14.4.** Layer architecture of object-oriented application framework

The IEEE 1451.1 application layer focuses on application system design that is the composition or aggregation among objects of IEEE 1451.1 application systems. The initial implementation of object-oriented framework for IEEE 1451.1

applications is provided based on the UML tool and ADAPTIVE Communication Environment (ACE) middleware. The ACE-based IEEE 1451.1 model uses ACE objects and design patterns to implement IEEE 1451.1 Neutral Model. Application developers can inherit the ACE-based IEEE 1451.1 model to implement their specific IEEE 1451.1 applications. They can customize their specific applications using this framework and thus dramatically reduce the time in IEEE 1451.1 application development.

### 14.2.3 IEEE 1451.2 – Transducer-to-Microprocessor Communication Interface

The IEEE 1451.2 standard defines (1) a Transducer Electronic Data Sheet, or TEDS, and its data format, and (2) a digital interface and communication protocol between transducers and a microprocessor, applicable to instruments and networks. The framework of the IEEE 1451.2 interface is shown in Figure 14.5. The IEEE 1451.2 interface defines the Smart Transducer Interface Module (STIM). It supports the concept of multi-variable transducers, that is, up to 255 sensors or actuators of various analog and digital mixes can be connected to a STIM. The STIM is connected to a network node called NCAP through the transducer independent interface using an enhanced Serial Peripheral Interface (SPI) protocol for data transfer.

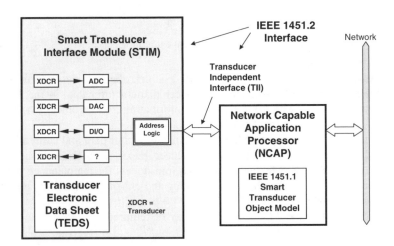

**Figure 14.5.** Framework of IEEE 1451.1 and 1451.2 interfaces

*14.2.3.1 TEDS*

The TEDS, stored in a nonvolatile memory, contains fields that describe the type, attributes, operation, and calibration of the transducer. An example of an IEEE 1451.2 TEDS for a single channel pressure sensor is described below. The pressure sensor has an analog output between 0 and 5 V dc corresponding to 0 to 20.68 MPa (3000 lb/in$^2$) pressure input. The sensor has a 10 ms response time, no warm-up requirement and the maximum non-linearity is 0.56% of supply voltage. The main components of the single channel STIM are: (1) a serial 12-bit analog-to-digital converter (ADC) for data conversion with a 75-μs conversion cycle and (2) an 8-MHz, 8-bit microprocessor with 4K by 12-bit on-chip EEPROM. Calibration is fixed. It is specified using five equal segments with non-zero offsets for each segment. This allows a first-order calibration functions to be used to reduce the non-linearity in the analog output. For the pressure sensor, non-linearity is reduced from 0.56% to 0.03%. A multi-segment calibration curve implemented in multiple linear segments is shown in Figure 14.6.

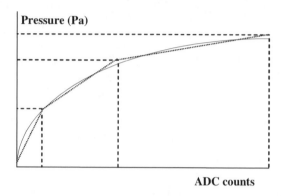

**Figure 14.6.** Multi-segment calibration curve

A transducer integrated with a TEDS enables the self-description of the transducer to the network. Since the transducer manufacturer data in the TEDS is always supplied with the transducer and this information is electronically transferred to an NCAP or host, human errors associated with entering sensor parameters manually is completely eliminated. Since the manufacturer data and calibration data are stored in the TEDS, losing the transducer paper data sheet is no longer a concern. With the TEDS features, replacing transducers for maintenance purposes and upgrading transducers with higher accuracy and enhanced capability are simply plug and play operations [14.21][14.22].

In the standard, some TEDS are mandatory or required, and some are optional. They are listed below:

1. TEDS that is mandatory
- Meta TEDS – contains the overall description of the TEDS data structure, worst case STIM timing parameters, and channel grouping information. Usually one Meta TEDS is required for each STIM.
- Channel TEDS – contains upper and lower range limits, physical units, warm-up time, presence of self-test, uncertainty, data model, calibration model, and triggering parameters. Usually one Channel TEDS is required for each STIM channel. Sample fields of the pressure sensor channel TEDS is shown in Figure 14.7. The fields U8, U16, and U32 represent unsigned integers in respective length.

2. TEDS that is optional

- Calibration TEDS – contains the last calibration date, calibration interval, and all the calibration parameters supporting the multi-segment model. Usually one Calibration TEDS is required for each STIM channel.

- Application Specific TEDS – for application specific use. Multiple TEDS are required per STIM.

- Extension TEDS – for use to implement future and industry extensions to IEEE 1451.2. Multiple TEDS are required per STIM.

| Field # | Description | Field Length (Bytes) | Field type | Field Contents |
|---|---|---|---|---|
| | *Data structure related information* | | | |
| 42 | Channel TEDS Length | 4 | U32 | 80 |
| 43 | Calibration Key | 1 | U8 | 1 (FIXED) |
| 44 | Industry Extension Key | 1 | U8 | 0 (NONE) |
| | *Transducer related information* | | | |
| 45 | Lower Range Limit | 4 | F32 | 0 |
| 46 | Upper Range Limit | 4 | F32 | 20684190 |
| 47 | Physical Units | 10 | UNITS | Pa (0,128,128,126,130, 124,128,128,128,128) |
| 48 | Unit Type Key | 1 | U8 | 0 (SENSOR) |
| 49 | Unit Warm Up Time | 4 | F32 | 1 |
| 50 | Self Test Key | 1 | U8 | 0 (NONE) |
| 51 | Uncertainty | 4 | F32 | 206842 |
| | *Data converter related information* | | | |
| 52 | Channel Data Model | 1 | U8 | 0 (N BYTE) |
| 53 | Channel Data Model Length | 1 | U8 | 2 |
| 54 | Channel Model Significant Bits | 2 | U16 | 12 |
| 55 | Channel Data Repetitions | 2 | U16 | 1 |
| 56 | Series Increment | 4 | F32 | 0 |
| 57 | Series Units | 10 | UNITS | 0 |
| 58 | Channel Update Time | 4 | F32 | 2.00E-05 |
| 59 | Channel Write Setup Time | 4 | F32 | 0 |
| 60 | Channel Read Setup Time | 4 | F32 | 8.00E-05 |
| 61 | Data Clock Frequency | 4 | U32 | 2.00E+05 |
| 62 | Channel Sampling Period | 4 | F32 | 2.00E-04 |
| 63 | Timing Correction | 4 | F32 | 0 |
| 64 | Trigger Accuracy | 4 | F32 | 5.00E-06 |
| | *Data integrity information* | | | |
| 65 | Checksum for Channel TEDS | 2 | U16 | 59968 |

**Figure 14.7.** Sample fields of the channel TEDS of the pressure sensor

## 14.2.4 IEEE 1451.3 – Distributed Multi-drop Systems for Interfacing Smart Transducers

During the course of the development of the IEEE 1451.1 and 1451.2 standards, some sensor manufacturers and users recognized the need for a standard interface for distributed multidrop smart sensor systems. In a distributed system, a large array of sensors, in the order of hundreds, needs to be read in a synchronized manner. The bandwidth requirements of these sensors might be relatively high, in the order of several hundred kilohertz. The IEEE 1451.3 standard defines specifications for meeting these requirements. The physical representation of the IEEE 1451.3 standard is shown in Figure 14.8. A 2-wire protocol has been selected by the working group as the physical medium for communications between the Transducer Bus Controller (TBC) and the Transducer Bus Interface Modules (TBIM). Power is delivered on the wires as well. The transducer bus is expected to have one bus controller and up to 255 TBIMs. A TBIM may contain one to 255 transducers. As a result, more than 65,000 transducers can be connected to a fully loaded distributed multi-drop system.

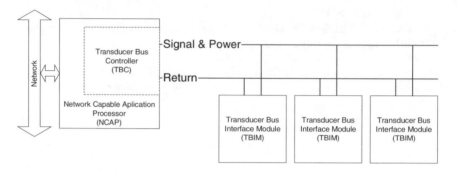

**Figure 14.8.** Physical representation of IEEE 1451.3 interfaces

### 14.2.5 IEEE 1451.4 – Mixed-mode Transducer Interface

In the condition-based monitoring and maintenance industry, analog transducers such as piezoelectric, piezoresistive, and accelerometer-based transducers are used widely with electronic instruments to measure the conditional state of machinery. Transducer measurements are collected by an instrument or computer for processing and analysis. The idea of having a small TEDS on the analog transducers brought transducer companies to work together defining the IEEE 1451.4 interface standard. An IEEE 1451.4 transducer, which could be a sensor or actuator with typically one addressable device, is referred to as a node containing a TEDS The IEEE 1451.4 transducer may be used to sense or control multiple physical phenomena. In order to reduce cabling and interfacing costs, models using different wiring configurations are chosen as a transducer connection interface. If a two-wire model is used, the analog transducer signal transmission and communication of the digital TEDS data to an instrument or a network are performed on the same wires, but at different times. This is called a mixed-mode operation and the communication is a mixed-mode interface. If a multi-wire model is used, communication of digital data and analog signals can be transmitted simultaneously, typically with 2 wires for sending analog signals and 2 wires for sending digital data. The digital communication scheme is used to read the TEDS information which is used to conFigure the IEEE 1451.4 transducers. Users can be confident because calibration data is directly acquired from the TEDS memories. Calibration due dates stored in TEDS can be used to generate calibration and maintenance notification. The mixed-mode transducer interface is shown in Figure 14.9.

In Figure 14.9, the IEEE 1451.4 transducer contains a mixed-mode interface (MMI) and a transducer electronic data sheet (TEDS). The MMI is a master-slave, multidrop, serial bus. A master device (an instrument or NCAP) initiates each transaction with each slave node (mixed-mode transducer) based on a digital communication protocol, known as the one-wire MicroLAN interface. The MMI may contain circuitry to detect and report a hot swap of transducers. The TEDS contains both fixed and dynamic data, stored in an electrically-erasable, programmable read only memory (EEPROM). An example accelerometer with a TEDS is shown in Figure 14.10. Since sensor manufacturers favored a very small TEDS for this standard, the TEDS information was encoded into 256 bits. A template, one for each class of transducers, is used to decode the TEDS data. The template is a piece of software describing the data structure of the TEDS Templates have been defined for the following transducers: accelerometers, bridge sensors, thermocouples, resistance temperature detectors (RTD), thermistor, voltage output sensors, current-loop output sensors, microphones, AC linear/rotary variable differential transformers, bridge sensors, and strain gages, as well as for calibration tables, calibration curves, and frequency response tables. The standard also allows manufacturers to submit new templates through the IEEE registration authority.

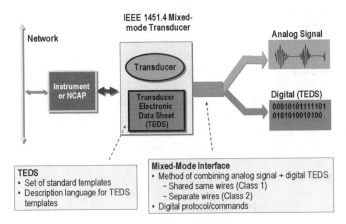

**Figure 14.9.** The IEEE 1451.4 mixed-mode transducer interface

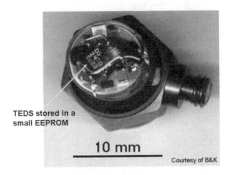

**Figure 14.10.** Accelerometer with a TEDS

The template is implemented in the Template Description Language (TDL) and is contained in a transducer block. The TDL is a scripted and tagged language providing a standard method to describe the functionality of the IEEE 1451.4 transducer. A transducer block, resides in the NCAP, is software object describing the IEEE 1451.4 transducer. The NCAP is a master device such as an instrument embedded processor or data acquisition system. The transducer block is used to access, decode, and encode TEDS using the TDL.

### 14.2.6 IEEE P1451.5 – Wireless Transducer Interface

The proposed IEEE 1451.5 standard defines a set of wireless communication interfaces between the transducer interface module (TIM) and NCAP. This standard specifies the sensor-to-NCAP connection, rather than the NCAP-to-network connection. The sensors-to-NCAP connection is conventional, implemented by sensor cable. It is replaced by a radio link. The standard adopts the three most popular wireless networking schemes, IEEE 802.11a/b/g (WiFi), IEEE 802.15.1 (Bluetooth), and IEEE 802.15.4 (ZigBee) as its communication protocols. An application programming interface (API) between the IEEE 1451.0 and 1451.5 layers as well as the physical or PHY TEDS have been defined. A conceptual block diagram of the IEEE 1451.5 standard is shown in Figure 14.11. In operation, a discovery command is executed from a host in the network. This command is broadcasted to all the NCAPs in the network. The IEEE 1451.1 resides in the NCAP decodes the command and passes it over to IEEE 1451.0. Through the IEEE 1451.0 to 1451.5 API, this command is communicated to the appropriate TIM(s) connected in the network via the NCAP. If a TIM is plugged into an NCAP after the discovery command, it can request a TIM registration service. Through the registration process, any TIM can basically announce itself to the network. The advantage of having the registration feature is that "plug and play" of TIM or sensors into the network is possible. Although three different wireless communication protocols are proposed to be adopted in the standard, "plug and play" of TIM by radio is not expected. That means a wireless Bluetooth TIM is not expected to be able to plug into a WiFi-based NCAP.

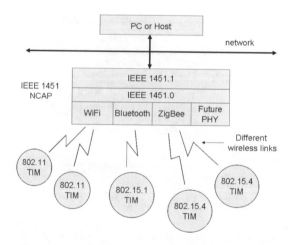

**Figure 14.11.** IEEE 1451.5 wireless sensor interfaces

The most important attribute of the IEEE 1451.5 wireless sensor standard is its common data format. In a system that contains multiple wireless networks running different protocols such as 802.11, 802.15.1 and 802.15.4, a host can access any sensor or actuator using the IEEE 1451.0 common command set. The sensor and TEDS data received would be the same using the common data format. At the time this manuscript was submitted, the proposed IEEE 1451.5 standard is still being developed in the IEEE Wireless Sensors Working Group.

## 14.3 Distributed Control Architecture

An industrial automation environment is shown in Figure 14.12, in which sensors information at the process level, the lowest level in the control hierarchy, is sent to the control network nodes where distributed intelligence can be executed. Sensor data received by the processors in the network nodes is then broadcast onto the network. Other nodes in the network can use the sensor information to make the appropriate control decision to manipulate the actuators or implement other algorithms. This distributed control scheme allows decisions to be made at the local level in a distributed manner based on simple commands from the upper level.

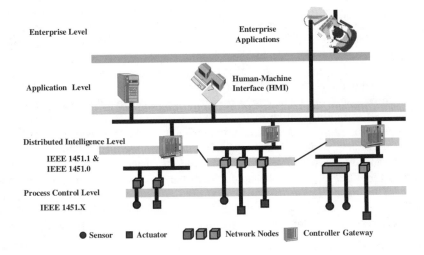

**Figure 14.12.** Distributed control architecture for an industrial automation environment

Operating in this distributed fashion, which is quite different from the traditional centralized control design, network traffic will be reduced and computing power will be distributed over the entire system. Higher-level, status-type information is reported to the upper application and enterprise level for system-level control, condition-based monitoring, or database archiving. In order to provide an environment that supports interoperability as shown in Figure 14.12, a framework for standardization must be included in the overall distributed measurement and control design.

An Internet-based distributed measurement control (DMC) system is a great platform for use in condition-based monitoring applications. The software, hardware, and standards used in the system are described below. First, the discussion focuses on the lowest level within the DMC framework; that is, what standards are applied for connecting transducers to the network nodes. Then discussion continues on using Ethernet as a control network medium and the software developed to support open network communications. Finally, the discussion converges to the software used to implement the distributed control applications on the network nodes.

### 14.3.1 Networked Smart Sensor Standards

At the lowest level of a distributed measurement and control system, sensors and actuators are used to sense environmental conditions and control physical parameters, respectively. Transducer interfacing refers to the process of physically and electrically connecting the transducer to the microprocessor of the network node. A key reason for standardizing the interface at the hardware interconnection level is due to the current compatibility problems that transducer manufacturers face when integrating their devices into multi-vendor networks [14.23][14.24]. Transducer interfacing also requires standardized software interfaces to provide application and network interoperability at the network-node level. Because the network and the transducer must expose their interfaces directly to transducer applications on each node, any attempt to migrate the application, the sensor hardware, or the network node to another platform requires a time-consuming and costly redesign of the application's interface to the new environment.

If standard interfaces such as IEEE 1451 are used in the DMC or any closed-loop manufacturing system, then sensor-to-sensor interchangeability and sensor-to-network interoperability can be realized. Two important features of the IEEE 1451 smart transducer interfaces are digital communication protocols between transducers and networks and self-identification of transducers via the TEDS This means that a sensor can self-describe itself to the network to make easy automatic system configuration. The TEDS can be uploaded to the system upon power up or upon request. The incorporation of IEEE 1451 TEDS into transducer hardware will simplify sensor maintenance, upgrade, and replacement by simple "plug and play". It also serves as self-documentation in the long run.

### 14.3.2 Network Communications using Ethernet

Network nodes must be connected to a control network in order to do useful work. Many current DMC-based control networks use proprietary hardware and software interfaces that limit the availability of data to higher-level networks and repositories. In addition, because DMC-based field networks are designed with specific application domains in mind, many companies possess several different types of device networks in order to solve different, specific problems. Supporting multiple types of control networks in order to target very specific domains is rapidly becoming impractical and costly. The need for distinct types of control networks to fulfill DMC requirements has been declining, moving towards open or *de facto* standards-based data communication networks based on widely accepted Ethernet technology.

The use of Ethernet as the preferred medium for DMC-based control networks is rapidly gaining in popularity because of its speed, cost-effectiveness, and the ability to leverage off-the-shelf application components to facilitate building distributed systems. Changes in the IEEE 802 Ethernet specification are making it quite formidable as a network for device-level control. For example, the IEEE 802.1p standard for message prioritization or quality of service (QoS) was initially developed for streaming live video, audio, and other multimedia content. Indeed, those in the control networking arena can leverage the real-time capabilities (due to its speed) in the Ethernet standard. The standard effectively guarantees that messages can be delivered and/or acknowledged in less than 4 milliseconds. This range is within most tolerances for providing adequate response times for process control applications.

### 14.3.3 Distributed Measurement and Control Model

Applications for control systems can take several forms. Most notable are segregated or individual, centralized, and distributed control [14.25]. Segregated control is not relevant to this discussion as it functions in a standalone non-networked environment. Centralized topologies typically use a master–slave application interaction scheme, where a master controller directly controls slave nodes on the network. Centralized control has by far been the most commonly implemented scheme in the past. Distributed control, however, is becoming the norm for designing control systems. Lower-cost microprocessors, higher-speed networks, open communications, and higher-level languages and development environments make them more cost-effective to implement. Implementing a standardized distributed control application framework is the key focus of this discussion. This standardized application framework is possible by using high-level languages, off-the-shelf development environments, *de facto* network interfaces, and IEEE 1451-based transducer integration techniques.

The control scheme discussed is based on the distributed model. In the model, two forms of distributed interaction are present, (1) autonomous low-level control algorithms implemented in C language on transducer-based network nodes, and (2) higher-level monitoring and control using a Java programming language and a Web browser. The application of standards to both of these areas is discussed. Using a distributed model for application development and system partitioning is rapidly becoming the norm.

The IEEE 1451.1 standard provides an abstract application programming interface (API) for networked transducer developers to write portable software for the nodes. The IEEE 1451.1 object-oriented framework for developing NCAP-based application objects lends itself to supporting highly sophisticated and distributed systems. This is a fundamentally different paradigm for application development than those found in traditional DMC system development partitioning.

Traditional centralized controller-based DMC system development makes use of the master–slave method for systems design. In the master–slave design, a node or device is designated as the master controller of all subordinate devices. Figure 14.13

shows the master executing system processing by polling slave devices under its control for information.

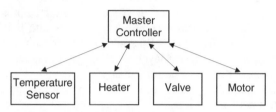

**Figure 14.13.** Using a master-slave control design for temperature control

As the master controller obtains information, it sends out control signals to each device needing attention on a one-to-one basis. The application shown represents a typical control layout for a temperature control system using this approach. The controller in the master–slave model becomes a bottleneck should information transfer to higher levels become necessary.

Since self-describing information and data access is not directly supported by the nodes, all data request, identification, and application interactions with the nodes must be going through the master controller. If the controller goes down or has other catastrophic problems, the entire system is at risk. Integrating this type of architecture with higher-level information models clearly becomes a problem. Efficient system designs as well as future standards integration issues demand a more scalable, distributed design to the master–slave model.

Distributed designs become more attractive because of a new breed of control systems category called soft control [14.26]. Soft control systems replace hardware controllers and interfaces with software-based control. This is in contrast to programmable logic controllers (PLC) that are based on specialized hardware, use inflexible architectures, and perform data exchange via hardware interfaces. Soft control-based systems use general purpose scalable hardware that facilitates data communication using higher level network interfaces. The software-based characteristics of soft control make easy the move to distributed measurement and control system architectures.

Using a distributed system design for temperature control as shown in Figure 14.14 illustrates the decentralization of processing to individual smart transducer-based networked nodes. In the distributed design, each node has autonomous control over the device it is controlling. Based on receipt of a variety of input, each node makes a decision and carries out the appropriate action. Using a messaging scheme, the nodes communicate requests/responses to their peers or perform publish–subscribe activities to transfer information on the network.

Adding distributed intelligence to nodes provides many benefits to system designers and integrators. Self-identifying features of the smart transducers and the nodes provide maintenance engineers with immediate feedback from device failures in the network when replacement parts are required. Failure and recovery schemes in distributed networks can be achieved gracefully. As the system learns of a node outage, it can respond by possibly starting a redundant node. On the other hand, master–slave systems typically require that all the intelligence be contained within the controller. The controller must provide for all contingencies due to the slave nodes' limited intelligence, preventing them from making autonomous decisions.

As the integration of microprocessors and network processors becomes more cost effective, more capable sensor network appliances will become more prevalent. Programming environments that support these systems will undoubtedly be object-oriented by design and support sophisticated modern operating systems. The resulting distributed applications will be extremely intelligent and highly integrated within the organization's information infrastructure and the manufacturing process.

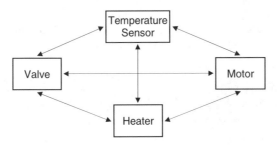

**Figure 14.14.** Using a distributed control design for temperature control

### 14.3.4 Web-based Access to Control Network

Web-based application has become more prevalent because of the flexibility it provides and the easy access to an application through the use of a common Web browser [14.27]. Hewlett-Packard (HP) was the first company to develop an NCAP that has a built-in micro Web server. This would enable one to access the transducers interfaced to the NCAP through the Internet using a Web browser.

Using the HP NCAP, NIST pioneered in developing an Internet-based DMC demonstration system. The topology is shown in Figure 14.15.

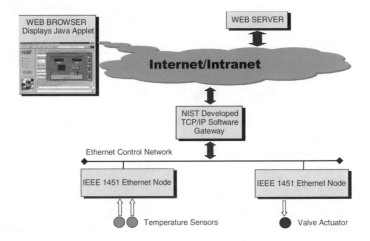

**Figure 14.15.** Internet-based DMC demonstration topology

Gateways convert a protocol from one network to the protocol of the other. Due to the differences in the HP communication mechanism, NIST had to develop a software gateway that provides transmission control protocol/Internet protocol (TCP/IP) clients with access to the HP NCAP nodes. In this case, the NIST gateway acts as a publish-subscribe intermediary between the HP nodes generating the data (using push technology) and the Java client applications consuming the data (using TCP/IP socket connections). The gateway extends the publish–subscribe HP communication services out into the realm of the Java clients by actually performing the HP communications functions on behalf of the client applications.

The NIST-developed gateway is a multithreaded Microsoft Windows-based program that extends the publish–subscribe messaging capabilities of the HP nodes to TCP/IP-based Java clients. To initiate a conversation with the gateway, a Java client must first connect to the gateway using a TCP/IP-based socket connection. After the connection phase is completed, the Java client issues topic subscription requests using the TCP/IP protocol. The gateway performs publish and subscribe functions on behalf of the Java client to complete the subscription process. Data received by the gateway from the HP nodes is then pre-pended with the topic name and forwarded directly to the appropriate Java client. The data received by the Java client is used to update all real-time data used by the applet. The applet then graphically displays the control loop with the updated values. The graphical simulation of the control loop provides the operator with an accurate picture of the control process in action. To provide even greater feedback to the operator, live real-time audio and video sequences of the system in operation are provided using a Microsoft NetMeeting video conferencing software client. The real-time graphical simulation of the system in operation combined with the audio and video provides a unique diagnostic and control mechanism on the Internet.

Because these operations can be monitored and controlled remotely, operators no longer need to be in front of the machine to monitor the temperature or change system parameters in the control loop. In fact, from any standard Web browser, the operator can log on and get up-to-the-second information about the process being controlled. Although the application domain in this case is concerned with remote access and control, this architecture is directly extensible to a variety of other domains. Figure 14.16 illustrates the NIST-developed Java applet for remote monitoring and control using the Microsoft Internet Explorer or other browser.

### 14.3.5 Internet-based Condition Monitoring

An implementation was developed to illustrate Internet-based condition monitoring applying the emerging IEEE 1451 smart transducer interface standards and the *de facto* networking standards from the Internet community by remotely monitoring and controlling an industrial process via the Internet. State-of-the-art hardware and software technology for DMC deployment is described in terms of a real-world Internet-based application. A high-speed, precision milling machine in the machine shop was used as a testbed. Large temperature differentials in an open machine shop environment can result in dimensional variation of the finished products. During the milling process, a means of maintaining a relatively constant temperature of the material is required. Coolant is used as lubricant during milling and also as a means to bath the part at constant temperature. A coolant tank, shown in Figure 14.17, serves as the coolant source to the part being milled.

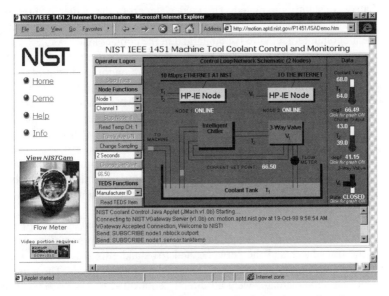

**Figure 14.16.** The Java-based remote monitoring and control applet

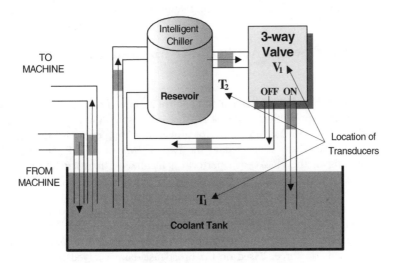

**Figure 14.17.** Maintaining constant coolant temperature in a tank

An Internet-based DMC closed-loop control system was designed to regulate the temperature of the coolant in the tank. Figure 14.17 illustrates the components of the DMC-based control system. The goal was to keep the coolant in the tank to within a specified temperature range. During the machining process, the coolant temperature tends to rise due to the heat generated in the cutting process. The increase in ambient air temperature around the machine also contributes to the rise in coolant temperature as well. An intelligent chiller maintains a reservoir of cold water that is constantly circulated. The cold water from the chiller is directed into the coolant tank when demanded using a 3-way valve. Through this control process, a consistent coolant temperature is maintained in the coolant tank.

The goal of the NIST Internet-based DMC demonstration was to provide a means via standardized interfaces to access, view, monitor, and control this DMC application in real-time. In order to provide this capability, several key hardware and software technologies were developed. The overall DMC demonstration topology is shown in Figure 14.16. The Java applet executes from within a Web browser. It provides an information delivery mechanism to the shop floor or operation manager's desk across the country. Any Web browser that supports Java can be used to view this application. A Web server located in NIST provided the Java and hypertext markup language (HTML) data file repository capabilities for the browser in this demonstration. The actual DMC system was developed using then two state-of-the-art HP Ethernet-based NCAP nodes.

These NCAP nodes support a reference implementation of the IEEE 1451 specifications that provide (a) standardized application programming interfaces (API) to the transducer and the network, and (b) standardized hardware interfaces for connecting transducers at the digital level to the Ethernet nodes. The temperature sensors and the valve actuator were all directly controlled in this distributed setting using C language applications for the embedded Ethernet-based IEEE 1451-compatible HP NCAP. Higher-level distributed control and monitoring capabilities were implemented in the Java programming language.

## 14.4 Networked Sensor Application – Machine Tool Condition Monitoring

Based on the distributed measurement and control architecture discussed in previous sections, a general application model of IEEE 1451 is shown in Figure 14.18. In the model, three NCAP/STIMs are used for illustration purposes. In scenario one, sensors and actuators are connected to the two STIMs of NCAP #1, application software running in the NCAP can perform a localized control function, such as maintaining a constant spindle speed. The NCAP reports process information and control status to a remote monitoring station or host. It frees the host from the processor-intensive, closed-loop control operation. In scenario two, NCAP #2 is connected to a STIM with sensors only. It can perform remote process or condition monitoring function; for example, monitor the vibration level of a set of bearings in a turbine. In scenario three, based on the broadcast data received from NCAP #2, NCAP #3 activates an alarm when the vibration level of the bearings exceeds a critical setpoint.

Based on the IEEE 1451 application model, a specific application of the IEEE 1451 networked sensor standards and technology is implemented for remote condition monitoring of a machine tool – a three-axis vertical machining center. The machine and sensor network components are shown in Figure 14.19.

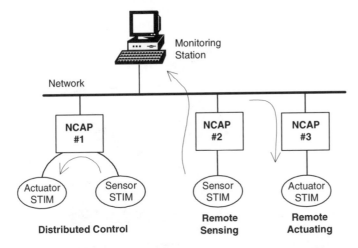

**Figure 14.18.** Application model of IEEE 1451-based sensor network

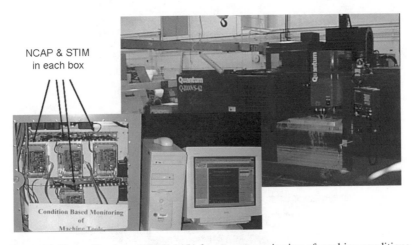

**Figure 14.19.** Application of IEEE 1451 for remote monitoring of machine conditions

Similar to the IEEE 1451 application model, the machine condition monitoring system consists of four sets of IEEE 1451-based Ethernet network nodes, temperature sensors, accelerometers, Ethernet hub, and power supplies. The network nodes, consisting of NCAP and STIM modules and related hardware, were integrated into a protective cabinet for day-to-day operation in a production shop environment. Sensors were strategically placed on some key components of a vertical machining center such as spindle bearings and axis motors as illustrated in Figure 14.20. Web capability was developed for remote monitoring of the sensors via the Ethernet and Internet using a common Web browser. A Web page, shown in Figure 14.21, acquires and displays a trend chart of the temperatures of the vertical spindle and Z-axis drive motors.

Networked Sensor System Configuration

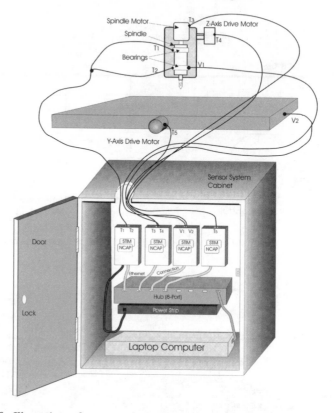

**Figure 14.20.** Illustration of a networked sensor system configuration for machine tool condition monitoring

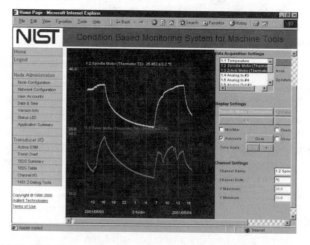

**Figure 14.21.** Web display of machine spindle and axis drive motors temperature

### 14.4.1 Design Approach

Like personal computer networking, sensor networking is finding favor in industry because of its simplicity in form, anticipated low cost to implement, and ease of deployment, maintenance, and upgrade. The IEEE 1451 smart transducer interface is an emerging sensor connection standard which provides a set of common interfaces for connecting sensors to networks. The approach used here applies the IEEE 1451 interface standards toward the implementation of the sensor-based CBM system for machine tool. Modularity was considered in the design for the ease of interchanging components. Commercially available sensors and networking products were used to build the system.

### 14.4.2 System Implementation

The machine tool chosen for this integration is a multi-axis vertical machining center. This computerized numerical control (CNC) machining center is used for production of metal parts in the main fabrication shop. Certain components in the vertical machining center can contribute to a machine tool failure. The components such as spindle bearings, spindle motors, and axes drive motors are subjected to heavy cutting loads, which can cause overheating and excessive vibration. Monitoring the temperature and vibration level of these components can help to identify their health and condition. The temperature and vibration signals from the spindle bearings can provide valuable information on the condition of the bearings. Likewise, monitoring the vibration signal on the table near the part being machined can inform the operator whether excessive or disproportionate cutting force is applied to the cutting tool. Excessive cutting tool vibration can cause tool chatter and can result in abrupt tool breakage or precipitate wear on the bearing. By monitoring the temperature of the spindle bearings, spindle motor, and axis drive motors, the thermal state of these units and their relative performance levels can be tracked.

### 14.4.3 Hardware System Layout

Thermistors and accelerometers were chosen to measure the temperature and vibration parameters, respectively. A schematic layout illustrating the sensor monitoring system is shown in Figure 14.18. The various temperature and vibration sensors mounted on the machine tool are connected to an Ethernet network via the IEEE 1451-based network nodes. Each IEEE 1451-based node consists of a STIM and NCAP pair. Sensors are connected to each STIM module, which contains TEDS for the sensors. The STIM also conditions the analog sensor signal and converts it into digital form, which can easily be processed by the NCAP. The Ethernet-based NCAP coordinates network communications and passes the digital sensor signal onto the network. The NCAP, integrated with a micro Web server, can facilitate Internet operation. A common Web browser such as Netscape or Internet Explorer can be used to access the sensor and TEDS data, and plot a trend chart of the characteristic being monitored. A temperature trend chart is shown in Figure 14.21.

There is more information in the trend chart that could be useful for tracking the usage of the machine tool. The two dips in temperature on the Z-axis motor temperature trend chart signifies the mid-morning coffee and lunch breaks of the machine operator.

## 14.5 Conclusions

The distributed measurement control concept with networked sensors and actuators would work quite well with distributed control, remote condition monitoring, and conditioned-based maintenance of machinery and machining systems. Taking this approach in advanced system design, it will reduce the total-life-cycle cost of the system in terms of modular sensor system design, use of commercially off-the-shelf (COTS) sensors and actuators, ease of maintenance and upgrade by simply "plug and play" of components. Also, using standardized components and interfaces, interoperability is no longer an issue.

The IEEE 1451 smart transducer interface standards are defined to allow transducer manufacturers to build transducers of various performance capabilities that are interoperable within a networking system. The IEEE 1451 standards have provided the common interface and enabling technology for the connectivity of transducers to microprocessors, control and field networks, and data acquisition and instrumentation systems. The standardized TEDS specified by the standards allow the self-description of sensors. It also provides a standardized mechanism to facilitate the "plug and play" of sensors to networks. The network-independent smart transducer model defined by IEEE 1451.1 and common command, TEDS, and functionality defined by IEEE 1451.0 will allow sensor manufacturers to support multiple networks and protocols; therefore transducer-to-network interoperability is on the horizon. The expanding Internet market has created a good opportunity for sensor and network manufacturers to exploit the web-based and smart sensor technologies.

The advancement of wireless technology will revolutionize the use of sensors and actuators. The emerging ZigBee, Bluetooth, Ultra-Wide Band, WiMax, and WiFi wireless technologies have promised to free the wires from the sensors and put billions of wireless devices into operation, particularly in the field of condition monitoring and conditioned-based maintenance of machinery and equipment. The sensor community is carefully evaluating these wireless connectivity schemes in an effort to continue pursuing a better way to ease the connection and integration of sensors into the application domain. Wireless sensor connectivity will continue to be explored until it is proven that wireless connection is as good as and as secure as a wired connection.

Advanced and high performance microprocessor technology will enhance the performance of NCAP/STIM modules and reduce their size and cost. The next generation of NCAP, which is a networkable, integrated sensing device, will fuse sensors, STIM, NCAP, and wireless technology into a single module. The modular wireless NCAP will be embedded with various algorithms (Fast Fourier Transform, Wavelet Transform, close-loop control, filtering, *etc.*), which will be selectable remotely by the host to perform specific operation. For example, a bearing NCAP is used to monitor the health of and predict the remaining life of a high-speed bearing, whereas a temperature NCAP is used to monitor the temperature of a critical machine tool component and it will alert the operator if the measured temperature exceeds the setpoint based on a complex temperature profile stored in the TEDS These low cost, wireless, modular, sophisticated, networkable sensing devices based on the IEEE 1451 standards can be deployed easily by simply place and play. As a result, users will greatly benefit from many innovations and new applications.

## Acknowledgment

This chapter is a contribution of the National Institute of Standards and Technology. The author also thanks the IEEE 1451 working groups for the use of some materials for this chapter.

## Disclaimer

** Commercial equipment and materials are identified in order to adequately specify certain procedures. In no case does such identification imply recommendation or endorsement by the National Institute of Standards and Technology, nor does it imply that the materials or equipment identified are necessarily the best available for the purpose.

## References

[14.1]    Talbott, C., January 24, 1998, "A MIMOSA White Paper on Prognosis of Remaining Machine Life Based on Condition", Talbott & Associates, West Chicago, IL.

[14.2]    Kostyukov, V., 1999, "Safe Maintenance of the Equipment is the Technology of the 21st Century – Russian Experience", *SPC Dynamics*, Omsk, Russia.

[14.3]    IEEE Std 1451.1-1999, *IEEE Standard for a Smart Transducer Interface for Sensors and Actuators–Network Capable Application Processor Information Model*, Instrumentation and Measurement Society, TC-9, Institute of Electrical and Electronic Engineers, New York, N.Y. 10016-5997, SH94767, April 18, 2000.

[14.4]    IEEE Std 1451.2-1997, *IEEE Standard for a Smart Transducer Interface for Sensors and Actuators–Transducer to Microprocessor Communication Protocols and Transducer Electronic Data Sheet (TEDS) Formats*, Instrumentation and Measurement Society, TC-9, Institute of Electrical and Electronic Engineers, New York, N.Y. 10016-5997, SH94566, September 25, 1998.

[14.5]    IEEE Std 1451.3-2003, *IEEE Standard for a Smart Transducer Interface for Sensors and Actuators–Digital Communication Protocols and Transducer Electronic Data Sheet (TEDS) Formats for Distributed Multidrop Systems*, Instrumentation and Measurement Society, TC-9, Institute of Electrical and Electronic Engineers, New York, N.Y. 10016-5997, SH95174, March 31, 2004.

[14.6]    IEEE Std 1451.4-2004, *IEEE Standard for a Smart Transducer Interface for Sensors and Actuators–Mixed-mode Communication Protocols and Transducer Electronic Data Sheet (TEDS) Formats*, Instrumentation and Measurement Society, TC-9, Institute of Electrical and Electronic Engineers, New York, N.Y. 10016-5997, SH95225, December 15, 2004.

[14.7]   Lee, K., 2000, "IEEE 1451: A Standard in Support of Smart Transducer Networking," *Proceedings of the IEEE Instrumentation and Measurement Technology Conference (IMTC)*, Baltimore, MD, **2**, pp. 525–528.

[14.8]   http://ieee1451.nist.gov

[14.9]   http://grouper.ieee.org/groups/1451/0/

[14.10]  http://grouper.ieee.org/groups/1451/5/

[14.11]  Brooks, T., Lee, K. B., Chen, S., "IEEE 1451 Smart Wireless Machinery Monitoring and Control for Naval Vessels," *Thirteenth International Ship Control Systems Symposium (SCSS)*, Orlando, Florida, April 7–9, 2003.

[14.12]  Mitchell, J., 2000, "MIMOSA: the Golden Opportunity," MIMOSA, Inc.

[14.13]  Michael Thurston, 2001, "OSA-CBM and MIMOSA: Overview for Sandia Labs", Applied Research Lab (ARL), Penn State University.

[14.14]  Lee, K., Schneeman, R., Gao, R., 2002, "Sensor Network and Information Interoperability – Integrating IEEE 1451 with MIMOSA and OSA-CBM," *Proceedings of the Instrumentation and Measurement Conference (IMTC)*, Anchorage, Alaska, pp. 1301–1305.

[14.15]  Lee, K. B., Song, E. Y., 2003, "UML Model for the IEEE 1451.1 Standard," *Proceedings of the Instrumentation and Measurement Conference (IMTC)*, Vail, Colorado, pp. 1587–1592.

[14.16]  Lee, K. B., 2005, A chapter on "A Smart Transducer Interface Standard for Sensors and Actuators," *The Industrial Information Technology Handbook*, CRC Press, ISBN 0-8493-1985-4, pp. 70-1 to 70-15.

[14.17]  Wiczer J., Lee K., 2005, "A Unifying Standard for Interfacing Transducers to Networks – IEEE 1451.0," *Proceedings of ISA Conference*, Chicago, Illinois, to be published.

[14.18]  Warrior, J., 1996, "IEEE-P1451 Network Capable Application Processor Information Model", *Proceedings of Sensors Expo*, Anaheim, Helmers Publishing, pp. 15–21.

[14.19]  Lee, K., Song, Y., 2005, "Object-oriented Application Framework for IEEE 1451.1 Standard," *IEEE Transactions on Instrumentation and Measurement*, **54**(4), pp. 1527–1533.

[14.20]  https://sourceforge.net/projects/open1451

[14.21]  Woods, S.; Lee, K.; Bryzek, J., 1997, "An Overview of the IEEE-P1451.2 Smart Transducer Interface Module", *An International Journal on Analog Integrated Circuits and Signal Processing, Special Issue: Smart Sensor Interfaces*, **14**(3), Kluwer Academic Publishers.

[14.22]  Johnson, R., 1997, "Building Plug-and-Play Networked Smart Transducers," *Sensors Magazine*, Helmers Publishing, pp. 40–46.

[14.23]  Lee, K., Schneeman, R., 1996, "A Standardized Approach for Transducer Interfacing: Implementing IEEE P1451 Smart Transducer Interface Standards." *Proceedings of the SENSORS Conference*, Philadelphia, PA.

[14.24]  Lee, K., Schneeman, R., 1997, "Multi-Network Access to IEEE P1451 Smart Sensor Information Using World Wide Web Technology," *Proceedings of SENSORS Conference/Expo 1997*, Boston, MA, pp. 15–34.

[14.25]  Bryan, L. A., Bryan, E. A., 1988, *Programmable Controllers: Theory and Implementation*, Industrial Text Company, Atlanta, GA.

[14.26]  Dieraur, P., 1998, "PLCs giving way to smart control," *InTech Magazine*. ISA Services Inc. Research Triangle Park, NC.

[14.27]  Lee, K., Schneeman, R., 1997, "Multi-Network Access to IEEE P1451 Smart Sensor Information Using World Wide Web Technology," *Proceedings of SENSORS Conference/Expo 1997*, Boston, MA, pp. 15–34.

15

# Rocket Testing and Integrated System Health Management

Fernando Figueroa[1] and John Schmalzel[2]

[1] NASA Stennis Space Center, Stennis Space Center
MS, 39529, USA
Email: Fernando.Figueroa-1@nasa.gov

[2] Rowan University
Glassboro, NJ 08028, USA
Email: j.schmalzel@ieee.org

**Abstract**
Integrated System Health Management (ISHM) describes a set of system capabilities that in aggregate perform: determination of condition for each system element, detection of anomalies, diagnosis of causes for anomalies, and prognostics for future anomalies and system behavior. The ISHM should also provide operators with situational awareness of the system by integrating contextual and timely data, information, and knowledge (DIaK) as needed. ISHM capabilities can be implemented using a variety of technologies and tools. This chapter provides an overview of ISHM contributing technologies and describes in further detail a novel implementation architecture along with associated taxonomy, ontology, and standards. The operational ISHM testbed is based on a subsystem of a rocket engine test stand. Such test stands contain many elements that are common to manufacturing systems, and thereby serve to illustrate the potential benefits and methodologies of the ISHM approach for intelligent manufacturing.

## 15.1 Introduction

Integrated system health management (ISHM) addresses health management of systems, particularly of high complexity encompassing large numbers of items such as actuators, pumps, pipes, tanks, instruments, sensors, and functional processes. These are collectively termed system "elements". Determining the impact of element degradation and anomalies is the fundamental problem that ISHM addresses. ISHM should detect impending failures, identify anomalies and failures when they occur, diagnose root cause, predict future element failures, and provide historical records of operation for each element in a system including data and the associated events that contribute to the determination of condition. However, the most important role of an ISHM is to provide insight into the state of a system to answer key questions including: "What can the system do now? How safe is it? What are ways to work around degraded elements and failures?"

Failures of system elements normally occur over time as the result of progressive degradation; for example, stress fractures resulting from cyclic loading. ISHM is concerned with managing the interplay between data, information, and knowledge (DIaK) in order to make possible detection, prediction, and response to failures. Data flows into an ISHM from a variety of sensors, which can include real-time data and historical data from collections of archival system files. Similarly, the ISHM needs access to knowledge bases that contain descriptions of elements, models for behaviors of interest, and examples of nominal and off-nominal behaviors. Sensor data combined with system-wide knowledge is operated on to develop information about the system. Further processing based on methods for information fusion, inference, and decision making, provides insight into system state and detects anomalous conditions that may signify future failure. When failures do occur, the ISHM shifts to determination of system impacts and recommendations for possible courses of action.

The authors contend the Integrated System Health Management philosophy to be predicated on:

- Lives and missions depend on vigilant systems.
- Data is valuable: No data should be left uninterpreted.
- Information is hidden: Intelligence is required to extract information.
- Interpretation, reasoning, and decision making require knowledge of how elements interact as part of an integrated system.

ISHM is a set of capabilities that indicates how well a system can perform the following suite of functions:

- Evaluate condition of system elements.
- Detect anomalies and their causes.
- Identify overall system state.
- Predict system impacts.
- Recommend responses to mitigate anomaly and failure effects.
- Communicate contextual and timely DIaK and situational awareness to system elements and system operators.

Increasing complexity and automation is required to achieve more demanding civilian and military technology objectives. NASA must push technology limits to deploy advanced systems-of-systems (SoS) to achieve its goals. The Space Shuttle, the International Space Station (ISS), and Rocket Engine Test Stands (RETS), are current examples. These complex SoS call for large life-cycle investments – especially for crew-rated spacecraft – and require large teams of human experts and support resources for monitoring, assessing, maintaining, upgrading, and operating.

NASA's Exploration Systems Mission Directorate (ESMD) [15.1] is currently focused on development of technologies for products to fulfill President Bush's New Vision for Exploration. The Exploration Systems Research and Technology (ESR&T) Program has defined specific capabilities needed; one of them is Integrated System Health Management (ISHM). ISHM approaches need to be matured to enable NASA's new Exploration Mission, which includes requirements for long-duration robotic-assisted human space travel with increased safety and decreased life cycle costs. ISHM is essential for dealing with SoS. ISHM is about knowing the condition of every element of the SoS at all times. It is about embedding knowledge and information so that systems can apply human-inspired strategies to monitor, capture anomalies, diagnose sources of anomalies, and predict future status. It is also about providing users with an integrated situational awareness of the SoS, and of every element in the context of its function within the SoS. It is appropriate to embed intelligence throughout SoS elements to achieve an overall ISHM capability. Use of a distributed intelligence approach also embodies other important attributes such as modularity, flexibility, and obtaining affordable life-cycle costs.

Although the focus of this book is on intelligent manufacturing systems, this chapter refers to systems most often associated with the aerospace industry when describing the background, needs, and benefits of ISHM. In particular, the application environment of rocket engine tests consists of complex pipe, valve, and control networks for the delivery of gases and liquids for a spectrum of pressures, temperatures, and mass flow rate requirements. Since many manufacturing systems share similar elements, ISHM for rocket testing is directly applicable to condition-based monitoring and control for intelligent manufacturing.

## 15.2 Background

The broad field of condition based maintenance has been extended and modified to support unique requirements for specialized applications. As one example, work on integrated vehicle health management (IVHM) systems for space applications started in the 1980s and continues today [15.2]. Similarly, the core ISHM activity of identifying unique features by monitoring and processing sensor data to detect anomalies is widely performed using a variety of standard algorithms, e. g. use of statistical measures, trend analysis, identification of time constants, and measurement of noise, are but a few of the many techniques available [15.3][15.4]. Integration of data from multiple sources to make better-informed decision is broadly termed sensor fusion; there is a large body of publications in this area [15.5]–[15.7]. Extracted features are input to multiple classifier systems [15.8] and other structures for purposes of reasoning about anomaly sources (diagnosis) and future implications (prognosis) and may include guidance on the actions required to maintain the system as close as possible to the desired performance goals.

These and related methodologies have generally been applied to detect anomalies on individual elements of a system (e. g. pumps, valves, airfoils). The challenge is to integrate these methods with embedded information and knowledge to implement an ISHM that can determine the health of every element and perform system-wide diagnosis and prognosis. Because of the complexity of the systems-of-systems approach required, ISHM remains an emerging area of research and development with relatively few operational implementations available. These examples provide basic capability by monitoring a subset of signals; malfunctions are indicated when certain key variables exceed fixed thresholds. Reaction responses are typically simplistic including process shutdown and event logging.

Among the most sophisticated of current ISHM implementations are those designed for the Boeing 777 [15.9][15.10], Northrop Grumman's B2 [15.11], and the Boeing Rocketdyne Space Shuttle Main Engine (SSME) [15.12]–[15.14]. The goals of the airplane health management (AHM) system include an ambitious reduction in unplanned maintenance from 75% to 25% of current levels. The AHM system operates on the onboard data collected by the central maintenance computer from the distributed built-in test components. First-generation threshold detection approaches have evolved to data mining that operates on real-time data transmitted from an in-flight aircraft in conjunction with the large database collected from the entire fleet. Achieving the 75% to 25% goal requires integration across subsystems. For example, a power bus shared between two systems means that anomalies reported from one subsystem need to be linked to analysis of anomalies in the other subsystem. Extension of the 777's AHM approach to other complex systems such as spacecraft or advanced manufacturing is likely to be difficult because there is no reported open architecture that supports scalability and flexibility.

Development of an ISHM for the SSME was defined as part of long-term goals for enhancing reliability and extending the operational life of the Space Shuttle. The work was phased to achieve near-term reliability enhancements while making future ISHM upgrades possible through an additional communication interface to the engine controller. However, it is unlikely that further development work will continue due to the planned phase out of shuttle operations by 2010. A new initiative to develop a replacement Crew Exploration Vehicle (CEV) will undoubtedly incorporate many of the lessons learned during the SSME ISHM development.

NASA's long interest in health management provides several other examples of ISHM-related development efforts. During the X33 program, NASA tested potential ISHM technologies that could be used to assess structural health of wings. Experiments produced equipment to collect data from distributed fiber optic strain gages thereby flight testing sensors, interconnects, signal processing electronics, and supporting computer system [15.15]. In another effort, the Propulsion IVHM Technology Experiment (PITEX) program implemented an integrated health management architecture that represented system elements with state models in order to detect anomalies and their sources. The prototype implementation incorporated Livingstone as the software environment to address a propulsion system composed of tanks, valves, and other basic elements; it was tested using simulated data [15.16].

NASA, joint with Australia's Commonwealth Scientific and Industrial Research Organization (CSIRO), investigated detection of structural damage on aircraft skins [15.17]. Consisting of distributed piezoelectric sensors mounted on the inside of aluminium plates, groups of four sensors shared data acquisition and processing, and communication with neighboring boxes. This work explored novel sensors and the means to embed sensor intelligence. More importantly, the collaboration considered network technologies for management of sensor data, information, and knowledge, while focusing on detection of surface damage.

A significant pattern in the latest ISHM work is a shift toward distributed intelligence. For example, MIT's Draper Laboratory proposed intelligent spacecraft encompassing networked intelligent components [15.18]. This is a fundamental departure from current spacecraft, such as the Space Shuttle, which are highly centralized architectures with individual sensors connected to central analog and digital processing equipment interconnected with MIL-STD-1553 data buses. The advance proposed by MIT is an architecture composed of intelligent nodes to support distributed processing at the sensor. Considerations included sensors and actuators with embedded processing, and digital communication among sensors, actuators, and computers. Key components of the novel architecture included two-wire power and data networks; fault-tolerant organization; smart sensors based on IEEE-1451 standards [15.19][15.20]; and packaging considerations. A laboratory prototype network with intelligent nodes was developed and tested to substantiate the benefits of the approach and found reduced wiring, weight, and complexity; increased reliability and safety; improved health management as measured by reduced operational costs; reduced software complexity and cost; and simplified vehicle development and integration. Other anticipated long-term benefits to an open architecture designed to support evolution is the flexibility to add new capabilities. More recently, the National Science Foundation (NSF) has begun promoting research on novel architectures that include intelligent sensors [15.21].

The work presented in this chapter builds on our earlier work [15.22] and is closely aligned with the architectures suggested by [15.18] and [15.21]. Figure 15.1 provides a graphical summary of the hierarchical distributed ISHM approach.

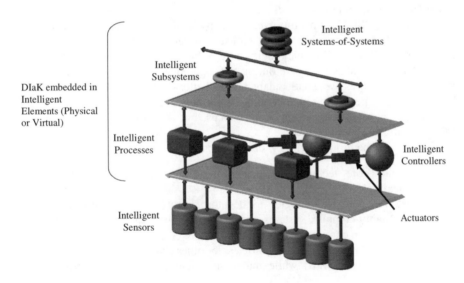

**Figure 15.1.** ISHM organized as a hierarchical network of distributed intelligent elements

Developing a state-of-the-art ISHM capability requires recruitment of technologies that enable (1) integration of overall ISHM components in a coherent architecture/framework; (2) embedding of intelligence; (3) algorithms and strategies involving fusion of DIaK for anomaly detection (model-based, numerical, statistical, empirical, expert-based, qualitative, *etc.*); (4) diagnostics/prognostics strategies and methods; (5) advanced control and decision strategies; and (6) user interface. While some of these technology areas are relatively mature, others require further development and theoretical underpinning. In particular, integration of an appropriate architecture and embedding of intelligence are key challenges to achieving credible ISHM capabilities. These and other issues are addressed in the embodiment described.

## 15.3 ISHM for Rocket Test

### 15.3.1 Implementation Strategy

This work began as a project to develop ISHM support for rocket engine testing that included smart sensors in the architecture. The Integrated Health with Networked Intelligent Elements (IHNIE) system is based on a generic taxonomy, framework, and methodologies to implement an ISHM. The proof of concept focuses on – among others – the hydraulic system for a test stand (Stennis Space Center), which provides facility fluid motive power to actuate valves that control propellant, oxidizer, and purge gases and/or liquids. The IHNIE system is organized in the hierarchical format of Figure 15.1. The framework and associated methodologies make possible the flow of information and knowledge among the elements of the system to determine element condition.

Developing ISHM capability shares many features in common with other complex system designs, but differs radically because of the need to evolve the DIaK content. Core ISHM technologies need to be integrated using available DIaK, yet the DIaK continues to evolve during SoS operation. Achieving optimum ISHM performance requires systematic adaptation to better exploit new knowledge gained by continuing experience with a system. Developing such an evolutionary system is beyond the scope of the current project focus, but is a long-term vision of what ISHM capability should include.

To this end, the prototype ISHM development seeks to

(1) Mature core technologies that provide the infrastructure required for a flexible/evolutionary ISHM system.
(2) Use the core technologies to develop prototypes for well-characterized, reasonably-complex systems with historical DIaK, available expertise, and which are in continuous operation.
(3) Port validated ISHM technologies to other systems (*e. g.* manufacturing facilities, spacecraft) while maintaining and evolving the prototypes and testbeds

### 15.3.2 DIaK Architecture

The IHNIE prototype consists of object models that encapsulate the system elements, their associated knowledge bases, generic methods and procedures, and communications processes. Figure 15.2 shows the overall data, information, and knowledge model for the RETS ISHM. *Data* flows within major components are considered to be raw and unprocessed. Modifying the data according to algorithms transforms the data to *information*, which is the desired exchange commodity, for example, conversion of raw sensor voltage readings to meaningful engineering units. Another significant factor is the availability of *knowledge* and its use. For example, armed with new information about parameters of the SoS, compilations of expert interpretations with system design files would provide the context to determine the actual SoS state. The explicit flow of data and health information with counterflow of evaluation and guidance is an important facet of the architecture. Higher architecture layers (right side of Figure 15.2) are in a better position to evaluate elements to determine condition and develop measures of trust for lower levels.

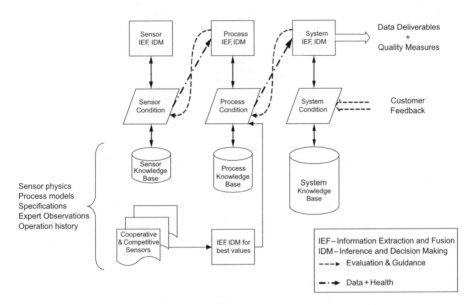

**Figure 15.2.** DIaK model for the rocket engine test stand ISHM

Implementation of the Figure 15.2 approach is accomplished using the software environment G2 [15.23] as the central organizing agent as shown in Figure 15.3. The major DIaK sources are provided with interfaces, termed "gateways," to the G2 knowledge management engine:

- Sensor gateways. Developing an ISHM in and for an operational RETS environment requires maximum use of existing resources including available sensors. However, unlike the future vision of ISHM systems populated by smart sensors, current test stands consist of traditional sensors hard-wired to centralized data acquisition portals. The prototype accommodates both present and future sensor paradigms through inclusion of a smart-sensor gateway, which operates on incoming data streams to produce virtual smart-sensor outputs visible to G2. This is discussed in greater detail in a later section.

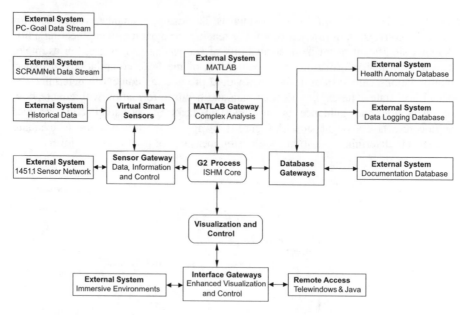

**Figure 15.3.** Gensym's G2 software environment provides the core ISHM framework

- Historical data available from legacy sensor data acquisition archives can be "played back" through the ISHM.
- Data available from real-time data acquisition feeds during idle (facilities maintenance) and article test at low (PC-Goal) and high (SRCAMNet) speeds can be presented to the ISHM.
- Data and information are available from prototype smart sensors developed to help evaluate their role in advanced ISHM.

- Analysis gateways. Implementation of algorithms for health evaluation and other analytic requirements can be performed either centrally or distributed. In keeping with the distributed processing philosophy, the prototype ISHM includes both centralized and dispersed components.
  - MATLAB® interface for exchange of data and analysis results with G2.

- Interface gateways. Effective interfaces are required between the ISHM and the broader SoS and between the ISHM and human operators (astronauts in flight or ground-based controllers).
  - Graphical User Interface (GUI). The ISHM needs to provide outputs that can be used to understand system state. Typical 2-d console displays are used.
  - Virtual Reality (VR). The immersive environment of advanced visualization provides enhanced communication between users and the ISHM.
  - Control interface. Some amount of interaction is needed between the ISHM and the actuator control points of the SoS. Strategies for

determining how to respond to state assessment must consider available response time. When there is too little time for human-in-the-loop evaluation, the ISHM will need to take action.

- Database gateways. ISHM requires access to an appropriate universe of knowledge encapsulated in a variety of databases.
  - Health anomaly database. Developing and distributing anomaly detection is facilitated with a database that includes anomalies (names), exemplar data, and detection algorithms. This can be used to distribute health algorithms and can serve as the source of training data sets.
  - Data logging. Historical data records are available through a database.
  - Documentation database. Access to SoS design documentation is made available through database access.

### 15.3.3 Object Framework

Systems are typically described (modeled) in terms of objects, thus it was natural to base the framework in the object-oriented programming paradigm. In this manner, DIaK and methods are encapsulated. Programming effort is decreased through the use of inheritance (objects in a subclass inheriting attributes from a parent class) and polymorphism (methods or functions for objects in a subclass are defined with variations with respect to their parent class).

Figure 15.4 shows the general model describing how ISHM objects (structure, parameters, methods, *etc.*) are implicitly linked through multiple inheritance and expressions of relationships such as "connected to" or "running in." For example,

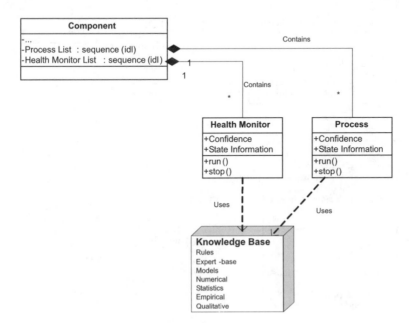

**Figure 15.4.** ISHM object linkage

*tank_pressurization_process running in tank, LOX1.*

Objects are also suited to encapsulate rule-based reasoning. For example, it is possible to define a rule that uses a qualitative process describing a step change in the tank parameters as follows:

*If the behavior of a temperature-sensor connected to tank LOX1 is step-change-up then the behavior of any pressure-sensor connected to tank LOX1 is step-change-up.*

The rule refers to three objects: a temperature sensor, pressure sensor, and a tank. The relationship between the temperature and pressure measurements is inferred from a process that is taking place in the tank, LOX1.

The diagram in Figure 15.5 depicts the object-oriented software environment based on the capabilities of G2. The main class is called an "Element". Subclasses are defined to represent sensors, components, and processes. In turn, tank, pump, etc. are subclasses of "Component". Similar subclasses are defined under "Sensor" and "Process". In this framework, a process is defined as an object associated with an environment (another object) where the process occurs.

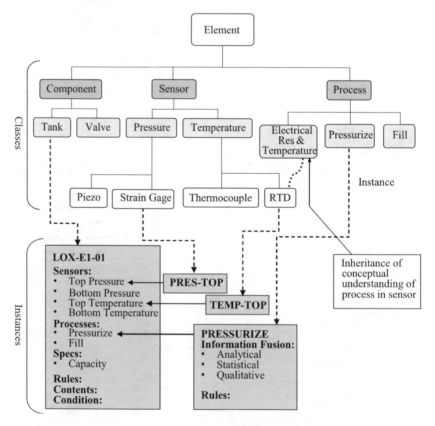

**Figure 15.5.** Object-oriented approach to ISHM architecture implementation

This overall approach also provides a basis to implement learning. For example, a tank_fill_process object defines a process model for filling tanks, thus it is a process associated with the object tank. Other processes may be tank_pressurization_process, or tank_flow_process. A pipe object will have an associated pipe_flow_process object. A sub-class of this object describing laminar flow may be pipe_flow_process_laminar object. Processes that take place in sensors, but not exclusively, may also be defined under the general Process Class. For example, processes associated with the behavior of electrical resistance and temperature provide a conceptual background to reason about sensors based in the underlying principle of operation (*e. g.* strain gages, RTDs). In all cases, the basic framework can be updated with progressively finer models as experience is gained with the overall system.

The SoS is defined as a hierarchical distributed network of intelligent elements. The architecture/framework is shown in Figure 15.6. The SoS has intelligent sensors at the bottom and the SoS at the top. Sensors feed information (= processed data) and measures of information quality and overall sensor health to the intelligent process layer. This can be a two-way exchange with the process layer informing the sensor layer to improve functional capability and refine quality measures. Rather than focusing on physical components as elements of a system, the framework focuses on processes that take place within elements. Intelligent processes can be defined by models; for example, pressurization, depressurization, flow, venting, leakage, fracture, corrosion, detachment, drift, wear, noise, *etc.* (see Figure 15.3). Other processes describe anomalies and their ontogenesis.

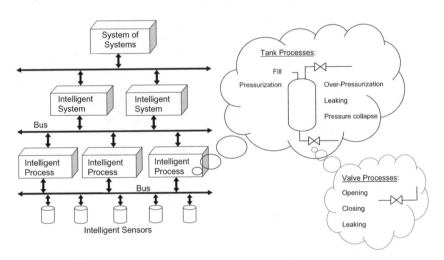

**Figure 15.6.** IHNIE process-centered fusion of data, information, and knowledge

The process-centric focus of Figure 15.7 is conducive to development and use of generic models for components such as tanks or pipes, which provide specific constraints or boundary conditions where processes are to take place. Hence, a generic pressurization process model applies to a pipe, a tank, a reservoir, or any containment element. Generalization improves modularity, reusability, maintainability, and evolution of the ISHM model to support scalability without losing the initial investment. This architecture also supports evolution as new information and knowledge are gained through experience and as new technologies come on line.

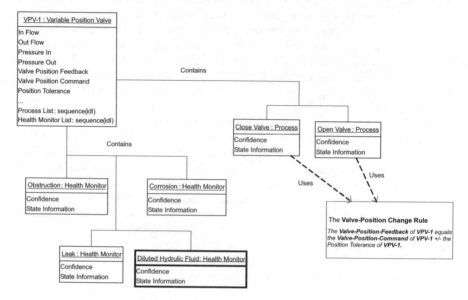

**Figure 15.7.** Specific view for a variable position valve (VPV), showing a new anomaly (in bold) that was added after the system was started. The two processes are shown using the same rule that governs how the valve position should compare to the valve command.

Another strength of the ISHM architecture is that intelligent sensors and certain other elements are not required to be physical entities. That is, legacy elements can be updated to participate in the ISHM using appropriate translation and interface layers that impart intelligence. Elements (or their intelligent attributes) are defined virtually; one example is the virtual smart sensor (VSS).

## 15.4 ISHM Implementation

### 15.4.1 Overall System

The prototype ISHM is based on a server where the G2 main program resides. Ethernet network infrastructure connects the G2 server with intelligent sensors and other computer nodes providing access to databases, historical data, and interfaces. Because of the limited number of protoype physical smart sensors available, the implementation also includes another computer that hosts a virtual smart sensor environment to make smart sensor behavior available to classical sensors (Figure 15.8). In the future, it will provide information and knowledge as well, and will be compatible with the IEEE 1451 standards.

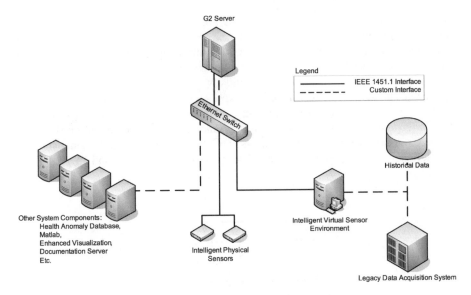

**Figure 15.8.** Physical ISHM implementation

## 15.4.2 Intelligent Sensors

Sensors are integral to all complex systems; they make possible implementation of automatic systems with feedback control. Classical control system design assumes that sensors provide data with an acceptable degree of accuracy. This is perhaps acceptable for relatively low complexity, low criticality systems, where consequences of sensor anomalies on the performance of the system may be tolerated. However, increasing complexity and degree of automation for current and future systems, combined with requirements for safety, availability, and life-cycle costs, drive up the numbers of sensors and tighten requirements for sensor reliability and availability.

Sensors are crucial to ISHM since intelligent sensors feed both data and certain types of information to the G2 environment.

### 15.4.2.1 Physical Smart Sensors

Prototype physical intelligent sensors have been built. The protocol to interface the IS with the G2 Main Program will be based on the IEEE1451 standards (this has been partially implemented in one version of the prototype IS).

Spurred by the computer networking community, the speed and reliability of networking technologies have significantly improved due to the demands from the computing community. Industry and technology developers have embraced the paradigm of sensor networks as a way to simplify complexity (significant decrease in number of wires) and increase the availability of sensor data to multiple processing elements (*e. g.* controllers connected to the network as nodes). To further proliferation of sensor networks, standard protocols have been defined to allow sensors from multiple manufacturers to be freely mixed in the network. One drawback of earlier protocols was speed. The throughput was usually less than 10 data updates per second, or even less for more complex systems. Specifically, speed and reliability have increased dramatically. These technologies are directly applicable to sensor networks and networks composed of other elements that makeup an SoS (controllers, processing computers, *etc.*). The National Institute of Standards and Technology (NIST) has promoted the development of the IEEE 1451.X standard series for smart sensors. These standards establish a foundation for development of "intelligent" networked sensors (see Figures 15.9 and 15.10).

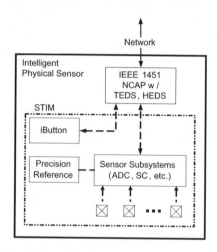

**Figure 15.9.** General model of an IEEE-1451 compliant smart sensor

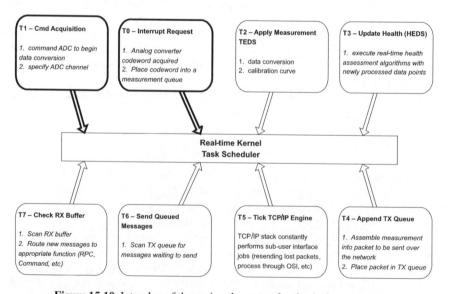

**Figure 15.10.** Interplay of the major elements of a physical smart sensor

The combination of faster network protocols and "Smart" sensor standards make possible the maturation of technologies to implement networks of intelligent sensors, or for that matter, of intelligent elements. Implementation of such networks may take different forms. In the case of intelligent sensors, "intelligence" can reside physically on (or near) the sensing element itself. In this case we have a physical smart sensor (PSS). However, intelligence in the form of a virtual smart sensor (VSS) can also reside on a processor (computer) attached remotely to the network. This latter option is suitable to retrofit existing systems that use classical sensors (Figure 15.11). There are two areas where technology needs to be matured to define Systems-of-Systems (SoS) as hierarchical networks of intelligent elements: (1) define what intelligence must go in the sensors (or other elements), and (2) develop an effective integration framework that focuses on management of data, information, and knowledge (DIaK) and not just data. This latter area includes developing appropriate software tools to manage DIaK. DIaK include information typically provided by product data sheets and manuals, algorithms for DIaK fusion, DIaK-based methods and strategies to extract features, diagnose, and predict. Furthermore, managing DIaK implies (a) storage, (b) sharing, (c) maintenance, (d) modification, and (e) evolution.

Intelligence that resides in a sensor must be defined according to the functional capability needed at the sensor. This capability is levied by higher level capability requirements from the SoS.

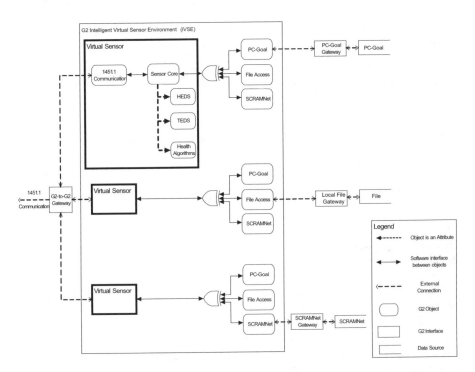

**Figure 15.11.** G2 Intelligent Virtual Sensor Environment. Each IVS receives "raw" data and processes it to produce an output identical to a physical smart sensor

To define intelligence resident in an intelligent sensor calls for sensors to provide data, qualification of the condition of the data (*e. g.* error estimate, useful or not), and the health of the sensor. This package of DIaK must be provided to the process (measurand) with which the sensor is associated (*e. g.* a pressure sensor measuring a pressurization process in a tank). In turn, the process should provide back to the sensor, information that may be used to update the sensor health. DIaK within each intelligent sensor should be sufficient so that it may use its own historical data, information about the sensor, knowledge of the physical phenomena (empirical, analytical, *etc.*) governing the operation of the sensor, and perhaps very basic information about the process that the sensor is serving (*e. g.* an estimate of the time-constant of the parameter being measured), in order to continuously assess its health and the quality of the data. In addition, measurements about the operating environment of the sensor (*e. g.* temperature, humidity) should also be used as they can help determine the sensor health.

### 15.4.3 Process Models

Process models may be analytic, qualitative, numerical, *etc.* It is expected that most models will be in the form of programs in C or C++ Language. Many algorithms (representing process models) to detect anomalies have been adapted to be run from the G2 VS program, the G2 Main Program, and from within PIS. For example, a pressurization process has been defined as a class object. A subclass is pressurization of gasses. An instance of that process object (pressurization of gasses) is defined as a rule:

> *For any temperature sensor and any pressure sensor attached to a tank, the slope of the temperature sensor has the same sign as the slope of the pressure sensor.*

There can be many types of models describing a pressurization process in gasses, and the architecture/taxonomy/ontology is defined to use all relevant models (or model-less) algorithms.

## 15.5 Implementation/Validation: Rocket Engine Test Stand

The prototype IHNIE is validated by implementing an ISHM system for the Hydraulic Subsystem of a Rocket Engine Test Facility at SSC. A model representing the hydraulic system is being created in the G2 Main Program. Instances of intelligent elements representing legacy elements in the hydraulic system are being defined. Models of virtual intelligent sensors representing legacy sensors in the hydraulic system are instantiated in the G2 Virtual Intelligent Sensor environment. The implementation uses historical and real-time data to test and validate ISHM capability. The final objective is to insert an ISHM computer/monitor in the Test Control Room to expose operators to ISHM technology capabilities and enlist their help to fine tune and evolve the ISHM system. When physical intelligent sensor prototypes become available for installation in the test stands (two laboratory prototypes with communications capabilities are now available), they will be fully integrated into the ISHM system.

## 15.6 Conclusions and Future Work

The prototype core elements will also be used to implement ISHM for a subsystem of the International Space Station (ISS). The ISHM Testbeds & Prototypes (ITP) project expands the use of the ISHM technology development resources (data, models, standard interfaces, *etc.*) and is based on the IHNIE core system. In effect the IHNIE is now a client to the ITP ISS Testbed.

The ITP project is a multi-center NASA effort. Key components include (1) development of advanced smart sensors (KSC), (2) application of the ISHM to a portable, low-thrust rocket engine test skid (SSC), (3) interface to advanced model-based reasoners (ARC), (4) interface to ISS (JSC), and (5) incorporation of advanced health-detection algorithms (GRC, JPL, MSFC, and others).

## Acknowledgments

This work has been supported by NASA: Exploration Systems Mission Directorate (Human and Robotic Technology Program) Project "ISHM Testbeds and Prototypes", Space Operations Mission Directorate (IR&D Program) Project "Intelligent Integrated Health Management System". The technical support of the entire project team is gratefully acknowledged.

## Acronyms

| | |
|---|---|
| AHM | Airplane Health Management |
| ARC | Ames Research Center |
| CEV | Crew Exploration Vehicle |
| CSIRO | Commonwealth Scientific and Industrial Research Organization |
| DIaK | Data, Information, and Knowledge |
| GUI | Graphical User Interface |
| GRC | Glenn Research Center |
| IHNIE | Integrated Health with Networked Intelligent Elements |
| ISHM | Integrated System Health Management |
| ISS | International Space Station |
| ITP | ISHM Testbeds & Prototypes |
| IVHM | Integrated Vehicle Health Management |
| JPL | Jet Propulsion Lab |
| JSC | Johnson Space Center |
| KSC | Kennedy Space Center |
| MSFC | Marshall Space Flight Center |
| NIST | National Institute of Standards and Technology |
| PITEX | Propulsion IVHM Technology Experiment |
| PSS | Physical Smart Sensor |
| RETS | Rocket Engine Test Stand |

RTD      Resistance Temperature Device
SoS      System of Systems
SSC      Stennis Space Center
SSME     Space Shuttle Main Engine
VR       Virtual Reality
VSS      Virtual Smart Sensor

# References

[15.1]   http:// www.exploration.nasa.gov.
[15.2]   Aaseng, G.B., 2001, "Blueprint for an Integrated Vehicle Health Management System," *Proceedings of 20th Conference on Digital Avionics Systems*, pp. 3C1/1-3C1/11.
[15.3]   Figueroa, F. and Mahajan, A., 1993, "Generic Model of an Autonomous Sensor," *Proceedings of the ASME Dynamic Systems and Control Division*, **50**, pp. 183–191.
[15.4]   Koushanfar, F., Potkonjak, M. and Sangiovanni-Vincentelli, A., 2003, "On-line Fault Detection of Sensor Measurements," *Proceedings of the IEEE Sensors Conference*, **2**, pp. 974–979.
[15.5]   Biel, L. and Wide, P., 2002, "An Intelligent Model Approach for Combination of Sensor Information," *Proceedings of the IEEE International Workshop on Haptic Virtual Environments and Their Applications*, pp. 43–48.
[15.6]   Huadong, W., Siegel, M., Stiefelhagen, R. and Jie, Y., 2002, "Sensor Fusion Using Dempster-Shafer Theory [for Context-aware HCI]," *Proceedings of the 19th Instrumentation and Measurement Technology Conference*, **1**, pp. 7–12.
[15.7]   Muldoon, S.E., Kowalczyk, M. and Shen, J., 2002, "Vehicle Fault Diagnostics Using a Sensor Fusion Approach," *Proceedings of the IEEE Sensors Conference*, **2**, pp. 1591–1596.
[15.8]   Oza, N.C., Polikar, R., Kittler, J. and Roli, F., Eds., 2005, *Proceedings of the 6th International Workshop on Multiple Classifier Systems*.
[15.9]   http://www.aviationweek.com/shownews/03paris/topstor3_16.htm.
[15.10]  Ramohalli, G., 1994, "Honeywell's Aircraft Monitoring and Diagnostic Systems for the Boeing 777," *Proceedings of the 17th Symposium on Aircraft Integrated Monitoring Systems*, pp. 69–71, 73–87.
[15.11]  Zuniga, F.A., Maclise, D.C., Romano, D.J., Jize, N.N., Wysocki, P.F. and Lawrence, D.P., 2005, "Integrated Systems Health Management for Exploration Systems," *Proceedings 1st Space Exploration Conference: Continuing the Voyage of Discovery*, AIAA 2005-2586.
[15.12]  Davidson, M. and Stephens, J., 2004, "Advanced Health Management System for the Space Shuttle Main Engine," *Proceedings 40th AIEE/ASME/SAE/ASEE Joint Propulsion Conference and Exhibit*, AIAA 2004-3912.
[15.13]  http://www.nasa.gov/pdf/1975main_shuttle.pdf.
[15.14]  Jue, F. and Kuck, F., 2002, "Space Shuttle Main Engine (SSME) Options for the Future Shuttle," *Proceedings 38th AIEE/ASME/SAE/ASEE Joint Propulsion Conference and Exhibit*, AIAA 2002-3758.
[15.15]  Schweikhard, K.A., Theisen, J., Mouyos, W. and Garbo, R., 2001, "Flight Demonstration of X-33 Vehicle Health Management System Components on the F/A-18 Systems Research Aircraft," *NASA/TM-2001-209037*.
[15.16]  Bajwa, A. and Sweet, A., 2002, "The Livingstone Model of a Main Propulsion System," *RIACS Technical Report 03.04*. Available at http://www.riacs.edu/trs/.

[15.17] Prosser, W. H., Allison, S. G., Woodard, S. E., Wincheski, R. A., Cooper, E. G., Price, D. C., Hedley, M., Prokopenko, M., Scott, D. A., Tessler, A. and Spangler, J. L., 2004, "Structural Health Management for Future Aerospace Vehicles," *Proceedings of the 2^nd Australian Workshop on Structural Health Management.*

[15.18] Hammett, R. C., 2001, "Networking Intelligent Components to Create Intelligent Spacecraft," *Proceedings of the IEEE Aerospace Conference*, **5**, pp. 2209–2215.

[15.19] http://grouper.ieee.org/groups/1451/0/.

[15.20] Lee, K., 2001, "Sensor Networking and Interface Standardization," *Proceedings of the 18^th IEEE Conference on Instrumentation and Measurement Technology*, pp. 147–152.

[15.21] Liu, S. C., Tomizuka, M. and Shoureshi, R., 2003, "Strategic Research for Sensors and Smart Structures Technology," *Proceedings of 1^st International Conference on Structural Health Monitoring and Intelligent Infrastructure.*

[15.22] Schmalzel, J., Figueroa, F., Morris, J., Mandayam, S. and Polikar, R., 2004, "An Architecture for Intelligent Systems Based on Smart Sensors," *Proceedings of the 21^st IEEE Conference on Instrumentation and Measurement Technology*, pp. 71–75.

[15.23] http://www.gensym.com.

[15.24] http://www.systranfederal.com.

# Index

Printing: Krips bv, Meppel
Binding: Stürtz, Würzburg